The Designer's Guide to the Cortex-M Processor Family

The Designer's Guide to the Cortex-M Processor Family

Second Edition

Trevor Martin

AMSTERDAM • BOSTON • HEIDELBERG • LONDON
NEW YORK • OXFORD • PARIS • SAN DIEGO
SAN FRANCISCO • SINGAPORE • SYDNEY • TOKYO
Newnes is an imprint of Elsevier

Newnes is an imprint of Elsevier
The Boulevard, Langford Lane, Kidlington, Oxford OX5 1GB, UK
50 Hampshire Street, 5th Floor, Cambridge, MA 02139, USA

Notices

Knowledge and best practice in this field are constantly changing. As new research and experience broaden our understanding,
changes in research methods, professional practices, or medical treatment may become necessary.

Practitioners and researchers must always rely on their own experience and knowledge in evaluating and using any
information, methods, compounds, or experiments described herein. In using such information or methods they should be
mindful of their own safety and the safety of others, including parties for whom they have a professional responsibility.

To the fullest extent of the law, neither the Publisher nor the authors, contributors, or editors, assume any liability for any
injury and/or damage to persons or property as a matter of products liability, negligence or otherwise, or from any use or
operation of any methods, products, instructions, or ideas contained in the material herein.

British Library Cataloguing-in-Publication Data
A catalogue record for this book is available from the British Library.

Library of Congress Cataloging-in-Publication Data
A catalog record for this book is available from the Library of Congress.

ISBN: 978-0-08-100629-0

For Information on all Newnes publications
visit our website at https://www.elsevier.com/

Publisher: Joe Hayton
Acquisition Editor: Tim Pitts
Editorial Project Manager: Charlotte Kent
Production Project Manager: Julie-Ann Stansfield
Designer: Mark Rogers

Typeset by MPS Limited, Chennai, India

To my wife Sarah and my parents Ann and Maurice

Contents

Foreword

When this book was first published in 2013, ARM was fueling a silent revolution in the embedded industry. At the time of the launch of the second edition, ARM is clearly the winning architecture with over 3000 ARM Cortex processor-based microcontrollers in the market and unparalleled growth.

ARM has brought profound change to the industry by fostering innovation and competition. Designing processors is hard, and in the past few silicon vendors were capable of playing in the microcontroller market. By introducing the licensable Cortex-M range of microcontrollers, ARM enabled an ecosystem of silicon companies, tools vendors, and software partners to enter the market—building an ever-stronger ecosystem in the process. The combination of ARM's wide range of Cortex-M processors at different performance-power points with the capabilities of its multiple partners has brought a degree of choice to microcontroller users that they could only dream of.

The trend that has most impacted embedded software developers in recent years has been complexity. The requirement for better connectivity, graphical interfaces, and higher processing power in the end application has been matched by increased functionality in the microcontrollers themselves, and of course, in the size of the software running on them. In order to help developers cope with this complexity Cortex-M processors include advanced debug and trace capabilities, which are exploited by professional development tools such as Keil MDK. However, this is not enough: in order to keep software development costs under control software reuse has become mandatory. This is why ARM partnered with its ecosystem of tools and RTOS vendors to create the CMSIS standard, which enables efficient porting of off-the-shelf middleware and custom software between multiple toolchains and microcontrollers.

Today's new challenges are safety and security. As more and more connected embedded devices become part of our lives it is imperative that they are rock-solid, and ARM has the systems in place to provide the necessary security and privacy. The software development flow needs to be right to comply with safety standards, and the software needs to be structured to take advantage of the hardware security features available in the latest

ARM cores. Once more, software methodology, development tools, and standards have a critical role to play.

I have known Trevor for the last 10 years, during which he has helped hundreds of customers get to market. Having access to his experience and expertise synthesized in the pages of this book is a unique opportunity to both new and experienced embedded developers. I hope you make the most of it.

Javier Orensanz

Preface

ARM first introduced the Cortex-M processor family in 2004. Since then the Cortex-M processor has gained wide acceptance as a highly scalable general purpose processor for small microcontrollers. At the time of writing, there are well over 3000 standard devices that feature the Cortex-M processor. These are available from many leading semiconductor vendors and the pace of development shows no sign of slowing. The Cortex-M processor family has now established itself as an industry standard. As such the knowledge of how to use it is becoming a requisite skill for professional developers. This book is intended as both an introduction to the Cortex-M processor and a guide to the techniques used to develop application software to run on it. The book is written as a tutorial and the chapters are intended to be worked through in order. Each chapter contains a number of examples that present the key principles outlined in this book using a minimal amount of code. Each example is designed to be built with the evaluation version of the Microcontroller Development Kit for ARM. These examples are designed to run in a simulator so you can use most of the hands on examples in this book without the need for any additional hardware.

Chapter 1 "Introduction to the Cortex-M Processor Family": Provides an introduction and feature overview of each processor in the Cortex-M family.

Chapter 2 "Developing Software for the Cortex-M Family": Introduces you to the basics of building a "C" project for a Cortex-M processor.

Chapter 3 "Cortex-M Architecture": Provides an architectural description of the Cortex-M3 and its differences to the other Cortex-M processors.

Chapter 4 "Cortex Microcontroller Software Interface Standard": Introduces the Cortex Microcontroller Software Interface Standard (CMSIS) programming specifications for Cortex-M processors.

Chapter 5 "Advanced Architecture Features": Extends Chapter 3 "Cortex-M Architecture" by introducing the more advanced features of the Cortex-M architecture.

Chapter 6 "Cortex-M7 Processor": In this chapter, we look at the Cortex-M7 processor which introduces a more complex memory model and bus structure.

Chapter 7 "Debugging with CoreSight": Provides a description of the CoreSight debug system and its real-time features.

Chapter 8 "Practical DSP for Cortex-M4 and Cortex-M7": Looks at the math and digital signal processing (DSP) support available on the Cortex-M4 and how to design real-time DSP applications.

Chapter 9 "Cortex Microcontroller Software Interface Standard-Real-Time Operating System": Introduces the use of an real-time operating system (RTOS) on a Cortex-M processor.

Chapter 10 "RTOS Techniques": This chapter examines some real-world techniques that can be used when developing an RTOS-based project.

Chapter 11 "Test Driven Development": In this chapter, we look at how to use a developer test framework called Unity with a Cortex-M-based microcontroller.

Chapter 12 "Software Components": The CMSIS standards introduced in Chapter 4 "Cortex Microcontroller Software Interface Standard" helps to standardize how "C" code is written for a Cortex-M microcontroller. This chapter looks at how to develop software components and how to package them so they can be easily distributed and reused.

Chapter 13 "ARMv8-M": The final chapter presents an introduction to the new ARMv8-M architecture. ARMv8-M introduces the next generation of Cortex-M processors. ARMv8-M also brings ARM TrustZone technology to microcontrollers for the first time. TrustZone provides a hardware-based security model for even the smallest devices.

This book is useful for students, beginners, and advanced and experienced developers alike. However, it is assumed that you have a basic knowledge of how to use microcontrollers and that you are familiar with the programming in the "C" language. In addition, it is helpful to have basic knowledge on how to use the μVision debugger and IDE.

Acknowledgments

I would like to thank Charlotte Kent and Tim Pitts of Elsevier for helping produce the book and Joseph Yui, Richard York and Ian Johnson of ARM for their cooperation. I would also like to thank Tanya Wrycraft for proof reading and helpful advice.

CHAPTER 1

Introduction to the Cortex-M Processor Family

Since the first edition of this book was published in 2013, the number of Silicon Vendors providing Cortex-M-based devices has almost doubled and the number of microcontroller variants is now well over 3000 and continues to rise. A decade ago I was familiar with the key features of all the mainstream Cortex-M microcontrollers in use. Today I struggle to keep up with the range of devices available, could be old age of course but you get the picture. In this book we are going to learn about the Cortex-M processor itself and also the software techniques required to design effective and efficient application code. This book is arranged as a tutorial and it is best to work through it chapter by chapter. Each chapter contains a number of hands on examples and you will learn a lot more by actually doing the examples.

Cortex Profiles

In 2004, ARM introduced its new Cortex family of processors. The Cortex processor family is subdivided into three different profiles. Each profile is optimized for different segments of embedded systems applications (Fig. 1.1).

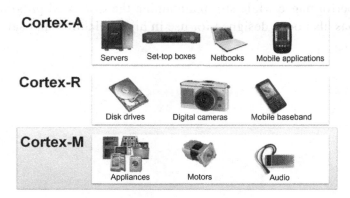

Figure 1.1
The Cortex processor family has three profiles Application, Real Time, and Microcontroller.

The Designer's Guide to the Cortex-M Processor Family.
DOI: http://dx.doi.org/10.1016/B978-0-08-100629-0.00001-3
© 2016 Elsevier Ltd. All rights reserved.

The Cortex-A profile has been designed as a high-end application processor. Cortex-A processors are capable of running feature-rich operating systems such as Win RT and Linux. The key applications for Cortex-A are consumer electronics such as smartphones, tablet computers, and set-top boxes. The second Cortex profile is Cortex-R. This is the real-time profile that delivers a high-performance processor which is the heart of an application-specific device. Very often a Cortex-R processor forms part of a "system-on-chip" design that is focused on a specific task such as hard disk drive (HDD) control, automotive engine management and medical devices. The final profile is Cortex-M or the microcontroller profile. Unlike earlier ARM CPUs, the Cortex-M processor family has been designed specifically for use within a small microcontroller. The Cortex-M processor currently comes in six variants: Cortex-M0, Cortex-M0 +, Cortex-M1, Cortex-M3, Cortex-M4, and Cortex-M7. The Cortex-M0 and Cortex-M0+ are the smallest processors in the family. They allow silicon manufacturers to design low-cost, low-power devices that can replace existing 8-bit microcontrollers while still offering 32-bit performance. The Cortex-M1 has much of the same features as the Cortex-M0 but has been designed as a "soft core" to run inside a Field Programmable Gate Array (FPGA) device. The Cortex-M3 is the mainstay of the Cortex-M family and was the first Cortex-M variant to be launched. It has enabled a new generation of high-performance 32-bit microcontrollers which can be manufactured at a very low cost. Today, there are many Cortex-M3-based microcontrollers available from a wide variety of silicon manufacturers. This represents a seismic shift where Cortex-M-based microcontrollers are starting to replace the traditional 8-/16-bit microcontrollers and even other 32-bit microcontrollers. The next highest performing member of the Cortex-M family is the Cortex-M4. This has all the features of the Cortex-M3 and adds support for digital signal processing (DSP). The Cortex-M4 also includes hardware floating point support for single precision calculations. The Corex-M7 is the Cortex-M processor with the highest level of performance while still maintaining the Cortex-M programmers model. The Cortex-M7 has also been designed for use in high reliability and safety critical systems (Fig. 1.2).

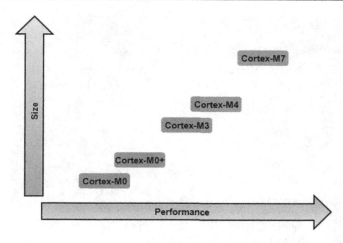

Figure 1.2
The Cortex-M profile has five different variants with a common programmers model.

In the late 1990s, various manufacturers produced microcontrollers based on the ARM7 and ARM9 CPUs. While these microcontrollers were a huge leap in performance and competed in price with existing 8-/16-bit architectures, they were not always easy to use. A developer would first have to learn how to use the ARM CPU and then have to understand how a specific manufacturer had integrated the ARM CPU into their microcontroller system. If you have moved to another ARM-based microcontroller you might have gone through another learning curve of the microcontroller system before you could confidently start development. Cortex-M changes all that; it is a complete Microcontroller (MCU) architecture, not just a CPU core. It provides a standardized bus interface, debug architecture, CPU core, interrupt structure, power control, and memory protection. More importantly, each Cortex-M processor is the same across all manufacturers, so once you have learned to use one Cortex-M-based processor you can reuse this knowledge with any other manufacturers of Cortex-M microcontrollers. Also within the Cortex-M family, once you have learned the basics of how to use a Cortex-M3, then you can use this experience to develop using any other Cortex-M processor. Through this book, we will use the Cortex-M3 as a reference device and then look at the differences between Cortex-M3 and Cortex-M0, Cortex-M0+, and Cortex-M4. The Cortex-M7 has a more advanced memory system which introduces features such as Tightly Coupled Memories and Caches, we will take a detailed look at the Cortex-M7 separately so by the end of this book you will have a practical knowledge of all the Cortex-M processors.

Cortex-M3

Today, the Cortex-M3 is the most widely used of all the Cortex-M processors. This is partly because it has been available not only for the longest period of time but also it meets the requirements for a general-purpose microcontroller. This typically means it has a good balance between high performance, low-power consumption, and low cost (Fig. 1.3).

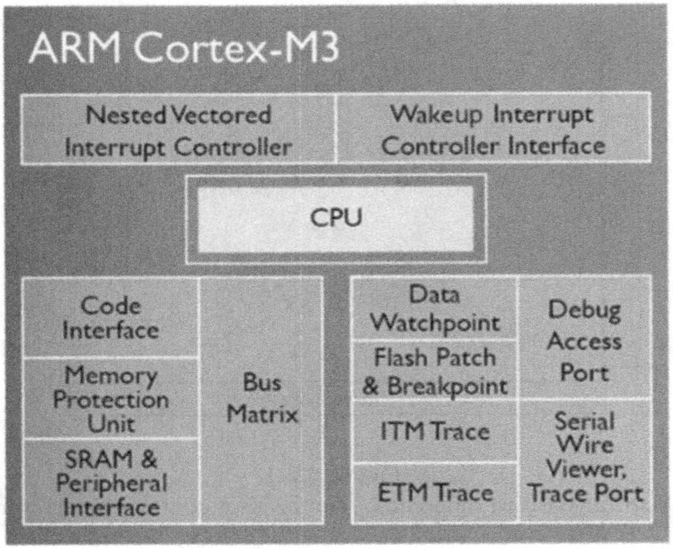

ARM Cortex-M3

Nested Vectored Interrupt Controller	Wakeup Interrupt Controller Interface

CPU

Code Interface		Data Watchpoint	Debug Access Port
Memory Protection Unit	Bus Matrix	Flash Patch & Breakpoint	
		ITM Trace	Serial Wire Viewer, Trace Port
SRAM & Peripheral Interface		ETM Trace	

Figure 1.3
The Cortex-M3 was the first Cortex-M device available. It is a complete processor for a general-purpose microcontroller.

The heart of the Cortex-M3 is a high-performance 32-bit CPU. Like the ARM7 this is a reduced instruction set computer where most instructions will execute in a single cycle (Fig. 1.4).

Fetch   Decode   Execute

Figure 1.4
The Cortex-M3 CPU has a three-stage pipeline with branch prediction.

This is partly made possible by a three-stage pipeline with separate fetch decode and execute units (Fig. 1.5).

Figure 1.5
The Cortex-M3 CPU can execute most instructions in a single cycle. This is achieved by the pipeline executing one instruction, decoding the next, and fetching a third.

So, while one instruction is being executed, a second is being decoded, and a third is being fetched. The same approach was used on the ARM7. This is great when the code is going in a straight line, however, when the program branches, the pipeline must be flushed and refilled with new instructions before execution can continue. This made branches on the ARM7 quite expensive in terms of processing power. However, the Cortex-M3 and Cortex-M4 include an instruction to fetch unit that can handle speculative branch target fetches which can reduce the bench penalty. The Cortex-M7 includes a full Branch Target Address cache unit which is even more efficient. This helps the Cortex-M3 and Cortex-M4 to have a sustained processing power of 1.25 DMIPS/MHz while the Cortex-M7 reaches 2.14 DMIPS/MHz.

In addition, the Cortex-M3 processor has a hardware integer math unit with hardware divide and single cycle multiply. The Cortex-M3 processor also includes a nested vector interrupt Controller (NVIC) that can service up to 240 interrupt sources. The NVIC provides fast deterministic interrupt handling and from an interrupt being raised to reaching the first line of "C" in the interrupt service routine takes just 12 cycles every time. The NVIC also contains a standard timer called the SysTick timer. This is a 24-bit countdown timer with an auto reload. This timer is present on all of the different Cortex-M processors. The SysTick timer is used to provide regular periodic interrupts. A typical use of this timer is to provide a timer tick for small footprint real-time operating systems (RTOS). We will have a look at such an RTOS in Chapter 9 "CMSIS-RTOS" Also next to the NVIC is the wakeup interrupt controller (WIC); this is a small area of the Cortex-M processor that is kept alive when the processor is in low-power mode. The WIC can use the interrupt signals from the microcontroller peripherals to wakeup the Cortex-M processor from a low-power mode. The WIC can be implemented in various ways and in some cases does not require a clock to function; also, it can be in a separate power region from the main Cortex-M processor. This allows 99% of the Cortex-M processor to be placed in a low-power mode with just minimal current being used by the WIC.

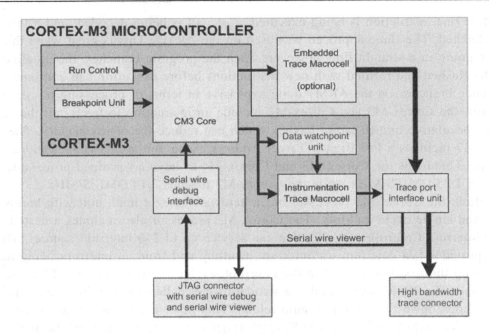

Figure 1.6
The Cortex-M debug architecture is independent of the CPU and contains upto three real-time trace units in addition to the run control unit.

The Cortex-M family also has a very advanced debug architecture called CoreSight (Fig. 1.6). The earlier ARM7/9 processors could be debugged through a "Joint Test Action Group" (JTAG) debug interface. This provided a means to download the application code into the on-chip flash memory and then exercise the code with basic run/stop debugging. While a JTAG debugger provided a low-cost way of debugging, it had two major problems. The first was a limited number of breakpoints, generally two with one being required for single stepping code and secondly, when the CPU was executing code the microcontroller became a black box with the debugger having no visibility to the CPU, memory, or peripherals until the microcontroller was halted. The CoreSight debug architecture within the Cortex-M processors is much more sophisticated than the old ARM7 or ARM9 processors. It allows up to eight hardware breakpoints to be placed in code or data regions. CoreSight also provides three separate trace units that support advanced debug features without intruding on the execution of the Cortex CPU. The Cortex-M3 and Cortex-M4 are always fitted with a data watchpoint and trace (DWT) unit and an instrumentation trace macrocell (ITM) unit. The debug interface allows a low-cost debugger to view the contents of memory and peripheral registers "on the fly" without halting the CPU, and the DWT can stream the contents of program variables in real time without using any processor resources. The second trace unit is called the instrumentation trace. This trace unit provides a debug communication method between the running code and the debugger user interface. During development, the standard IO channel can be redirected to a console window in the debugger. This allows you to instrument your code with printf() debug messages which can then be read in the debugger while the code is running. This can be useful for trapping complex runtime problems. The instrumentation trace is also very useful during software testing as it provides a way for a test harness to dump data to the PC without needing any specific hardware on the target. The instrumentation trace is actually more complex than a simple Universal asynchronous Receiver Transmitter (UART), as it provides 32 communication channels which can be used by different resources within the application code. For example, we can provide extended debug information about the performance of an RTOS by placing the code in the RTOS kernel that uses an instrumentation trace channel to communicate with the debugger. The final trace unit is called the embedded trace macrocell (ETM). This trace unit is an optional fit and is not present on all Cortex-M devices. Generally, a manufacturer will fit the ETM on their high-end microcontrollers to provide extended debug capabilities. The ETM provides instruction trace information that allows the debugger to build an assembler and High-Level Language trace listing of the code executed. The ETM also enables more advanced tools such as code coverage monitoring and timing performance analysis. These debug features are often a requirement for safety critical and high-integrity code development.

Advanced Architectural Features

The Cortex-M3 and Cortex-M4 can also be fitted with another unit to aid high-integrity code execution. The memory protection unit (MPU) allows developers to segment the Cortex-M memory map into regions with different access privileges. We will look at the operating modes of the Cortex-M processor in Chapter 5 "Advanced Architecture Features," but to put it in simple terms, the Cortex CPU can execute the code in a privileged mode or a more restrictive unprivileged mode. The MPU can define privileged and unprivileged regions over the 4 GB address space (ie, code, RAM, and peripheral). If the CPU is running in unprivileged mode and it tries to access a privileged region of memory, the MPU will raise an exception and execution will vector to the MPU fault service routine. The MPU provides hardware support for more advanced software designs. For example, you can configure the application code so that an RTOS and low-level device drivers have full privileged access to all the features of the microcontroller while the application code is restricted to its own region of code and data. Like the ETM, the MPU is an optional unit which may be fitted by the manufacturers during design of the microcontroller. The MPU is generally found on high-end devices which have large amounts of FLASH memory and

SRAM. Finally, the Cortex-M3 and Cortex-M4 are interfaced to the rest of the microcontroller through a Harvard bus architecture. This means that they have a port for fetching instructions and constants from code memory and a second port for accessing SRAM and peripherals. We will look at the bus interface more closely in Chapter 5 "Advanced Architecture Features," but in essence, the Harvard bus architecture increases the performance of the Cortex-M processor but does not introduce any additional complexity for the programmer.

The earlier ARM CPUs, ARM7 and ARM9, supported two instruction sets. This code could be compiled either as 32-bit ARM code or as 16-bit Thumb code. The ARM instruction set would allow code to be written for maximum performance, while Thumb code would achieve a greater code density (Fig. 1.7). During development, the programmer had to decide which function should be compiled with the ARM 32-bit instruction set and which should be built using the Thumb 16-bit instruction set. The linker would then interwork the two instruction sets together. While the Cortex-M processors are code compatible with the original Thumb instruction set, they are designed to execute an extended version of the Thumb instruction set called Thumb-2. Thumb-2 is a blend of 16- and 32-bit instructions that has been designed to be very C friendly and efficient. For even the smallest Cortex-M project, all of the code can be written in a high-level language, typically C, without any need to use an assembler.

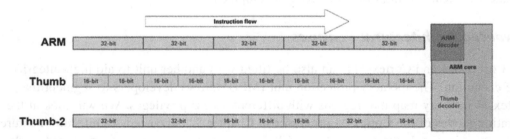

Figure 1.7
Earlier ARM CPU's had two instruction sets ARM (32 bit) and Thumb (16 bit). The Cortex-M processors have an instruction set called Thumb-2 which is a blend of 16-bit and 32-bit instructions.

With the Thumb-2 instruction set, the Cortex-M3 provides 1.25 DMIPS/MHz, while the ARM7 CPU using the ARM 32-bit instruction set achieves 0.95 DMIPS/MHz, and the Thumb 16-bit instruction set on ARM7 is 0.7 DMIPS/MHz (Fig. 1.8). The Cortex-M CPU has a number of hardware accelerators such as single cycle multiply, hardware division, and bit field manipulation instructions that help boost its performance compared to ARM7-based devices.

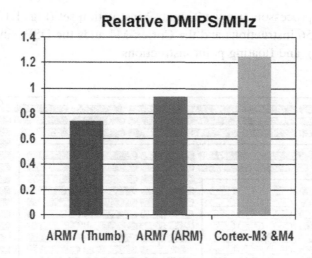

Figure 1.8
The Cortex-M3 and M4 Thumb-2 instruction set achieves higher performance levels than either the Thumb or ARM instruction set running on the ARM7.

The Thumb-2 instruction set is also able to achieve excellent code density that is comparable to the original 16-bit Thumb instruction set while delivering more processing performance than the ARM 32-bit instruction set (Fig. 1.9).

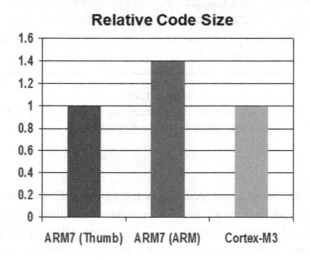

Figure 1.9
The Thumb-2 instruction set of the Cortex-M3 and -M4 achieves the same code density as the ARM7 Thumb (16 bit) instruction set.

All of the Cortex-M processors use the Thumb-2 instruction set (Fig. 1.10). The Cortex-M0 uses a subset of just 56 instructions and the Cortex-M4 adds the DSP, single instruction multiple data (SIMD), and floating point instructions.

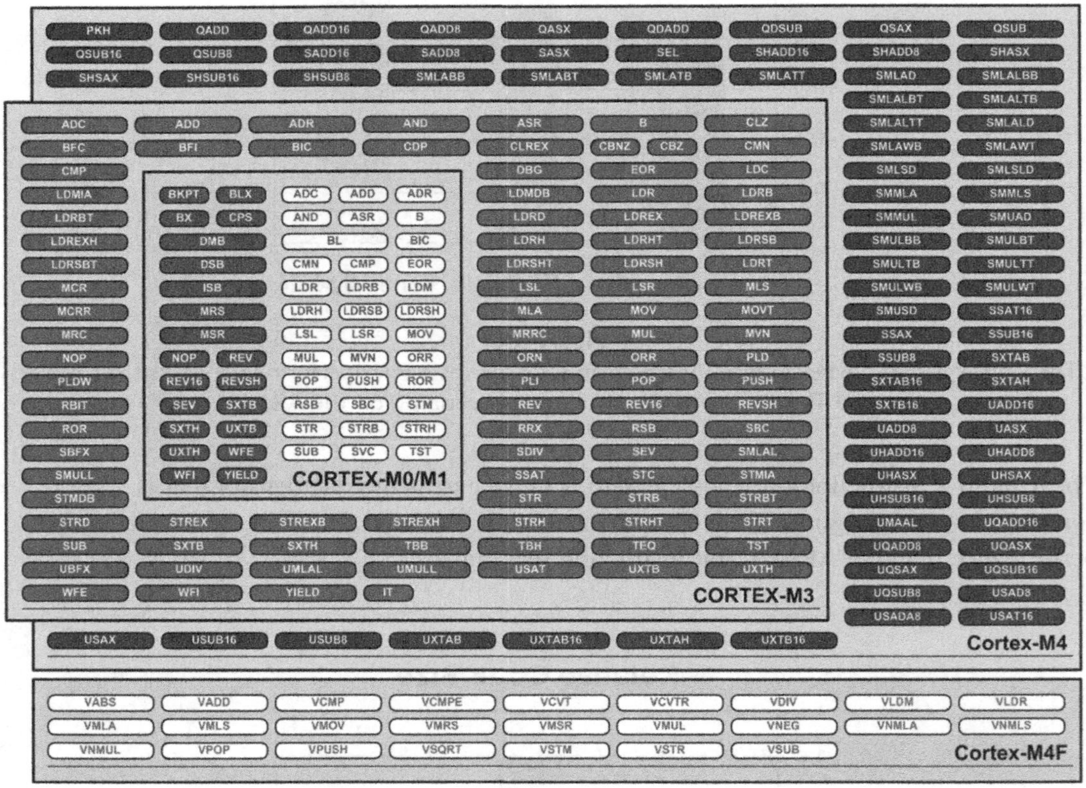

Figure 1.10
The Thumb-2 instruction set scales from 56 instructions on the Cortex-M0, -M0+ up to 169 instructions on the Cortex-M4.

Cortex-M0

The Cortex-M0 was introduced a few years after the Cortex-M3 was released and was in general use. The Cortex-M0 is a much smaller processor than the Cortex-M3 and can be as small as 12 K gates in minimum configuration. The Cortex-M0 is typically designed into microcontrollers that are intended to be very low-cost devices and/or intended for low-power operation. However, the important thing is that once you understand the Cortex-M3, you will have no problem using the Cortex-M0; the differences are mainly transparent to high-level languages (Fig. 1.11).

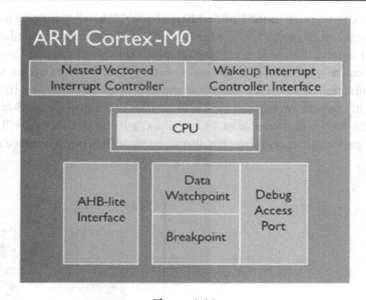

Figure 1.11
The Cortex-M0 is a reduced version of the Cortex-M3 while still keeping the same programmers model.

The Cortex-M0 processor has a CPU that can execute a subset of the Thumb-2 instruction set. Like the Cortex-M3, it has a three-stage pipeline but no branch speculation fetch, therefore branches and jumps within the code will cause the pipeline to flush and refill before execution can resume. The Cortex-M0 also has a Von Neumann bus architecture, so there is a single path for code and data. While this makes for a simple design, it can become a bottleneck and reduce performance. Compared to the Cortex-M3, the Cortex-M0 achieves 0.84 DMIPS/MHz, which while less than the Cortex-M3 is still about the same as an ARM7 which has three times the gate count. So, while the Cortex-M0 is at the bottom end of the Cortex-M family, it still packs a lot of processing power. The Cortex-M0 processor has the same NVIC as the Cortex-M3, but it is limited to a maximum of 32 interrupt lines from the microcontroller peripherals. The NVIC also contains the SysTick timer that is fully compatible with the Cortex-M3. Most RTOS that run on the Cortex-M3 and Cortex-M4 will also run on the Cortex-M0, though the vendor will need to do a dedicated port and recompile the RTOS code. As a developer, the biggest difference you will find between using the Cortex-M0 and the Cortex-M3 is its debug capabilities. While on the Cortex-M3 and Cortex-M4 there is extensive real-time debug support, the Cortex-M0 has a more modest debug architecture. On the Cortex-M0, the DWT unit does not support data trace and the ITM is not fitted, so we are left with basic run control (ie, run, halt, single stepping and breakpoints, and watchpoints) and on-the-fly memory/peripheral accesses. This is still an enhancement from the JTAG support provided on ARM7 and ARM9 CPUs.

While the Cortex-M0 is designed to be a high-performance microcontroller processor it has a relatively low gate count. This makes it ideal for both low-cost and low-power devices.

The typical power consumption of the Cortex-M0 is 16 µW/MHz when running and almost zero when in its low-power sleep mode. While other 8- and 16-bit architectures can also achieve similar low-power figures, they need to execute far more instructions than the Cortex-M0 to achieve the same end result (Fig. 1.12). This means extra cycles and extra cycles would mean more power consumption. If we pick a good example for Cortex-M0 such as a 16 × 16 multiply, then the Cortex-M0 can perform this calculation in 1 cycle. In comparison, an 8-bit typical architecture like the 8051 will need at least 48 cycles and a 16-bit architecture will need 8 cycles. This is not only a performance advantage but also an energy efficiency advantage as well (Table 1.1).

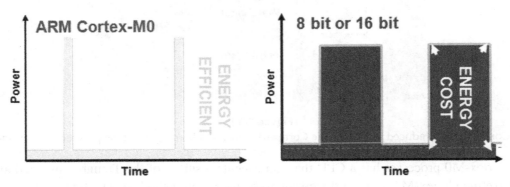

Figure 1.12

The Cortex-M0 is designed to support low-power standby modes. Compared to 8-bit or 16-bit MCU it can stay in sleep mode for much longer time because it needs to execute fewer instructions than an 8-/16-bit device to achieve the same result.

Table 1.1: Number of cycles taken for a 16 × 16 multiply against typical 8- and 16-bit architectures

8-Bit Example (8051)	16-Bit Example	ARM Cortex-M
MOV A,R7		
MOV B(0xF0),R5	MOV R1,&Mul0p1	MULS r0,r1,r0
MUL AB	MOV R2,&Mul0p2	
MOV R0,B(0xF0)	MOV SumLo,R3	
XCH A,R7	MOV SumHi,R4	
MOV B(0xF0),R4		
MUL AB	(Memory mapped multiply unit)	
ADD A,R0		
XCH A,R6		
MOV B(0xF0),R5		
MUL AB		
ADD A,R6		
MOV R6,A		
RET		
Time: 336 clock cycles	Time: 8 clock cycles	Time: 1 clock cycle
Code size: 18 bytes	Code size: 8 bytes	Code size: 2 bytes

Like the Cortex-M3, the Cortex-M0 also has the WIC feature. While the WIC is coupled to the Cortex-M0 processor, it can be placed in a different power domain within the microcontroller (Fig. 1.13).

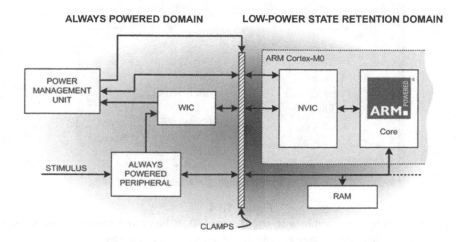

Figure 1.13
The Cortex-M processor is designed to enter low-power modes. The WIC can be placed in a separate power domain.

This allows the microcontroller manufacturer to use their expertise to design very low-power devices where the bulk of the Cortex-M0 processor is placed in a dedicated low-power domain which is isolated from the microcontroller peripheral power domain. These kinds of architected sleep states are critical for designs that are intended to run from batteries.

Cortex-M0+

The Cortex-M0+ processor is the second generation ultra-low power Cortex-M core. It has complete instruction set compatibility with the Cortex-M0 allowing you to use the same compiler and debug tools. As you might expect, the Cortex-M0+ has some important enhancements over the Cortex-M0 (Fig. 1.14).

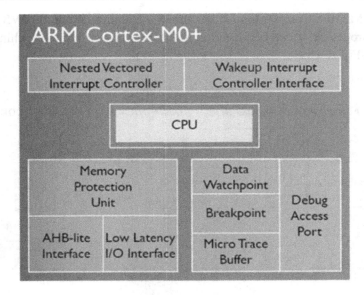

Figure 1.14
The Cortex-M0+ is fully compatible with the Cortex-M0. It has more advanced features, more processing power, and lower power consumption.

The defining feature of the Cortex-M0+ is its power consumption, which is just 9.8 μW/ MHz compared to 16 μW/MHz for the Cortex-M0 and 32 μW/MHz for the Cortex-M3. One of the Cortex-M0+ key architectural changes is a move to a two-stage pipeline. When the Cortex-M0 and Cortex-M0+ execute a conditional branch, the instructions in the pipeline are no longer valid. This means that the pipeline must be flushed every time there is a branch. Once the branch has been taken, the pipeline must be refilled to resume execution. While this impacts on performance it also means accessing the FLASH memory and each access costs energy as well as time. By moving to a two-stage pipeline, the number of FLASH memory accesses and hence the runtime energy consumption is also reduced (Fig. 1.15).

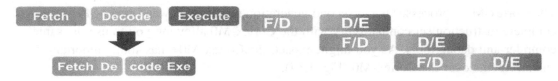

Figure 1.15
The Cortex-M0+ has a two-stage pipeline compared to the three-stage pipeline used in other Cortex-M processors.

Another important feature added to the Cortex-M0+ is a new peripheral I/O interface that supports single cycle access to peripheral registers. The single cycle I/O interface is a standard part of the Cortex-M0+ memory map and uses no special instructions or paged addressing. Registers located within the I/O interface can be accessed by normal "C" pointers from within your application code. The I/O interface allows faster access to peripheral registers with less energy use while being transparent to the application code. The single cycle I/O interface is separate from the advanced high-speed bus (AHB) lite external bus interface, so it is possible for the processor to fetch instructions via the AHB lite interface while making a data access to the peripheral registers located within the I/O interface.

The Cortex-M0+ is designed to support fetching instructions from 16-bit FLASH memories. Since most of the Cortex-M0+ instructions are 16-bit, this does not have a major impact on performance but does make the resulting microcontroller design simpler, smaller, and consequently cheaper. The Cortex-M0+ has some Cortex-M3 features missing on the original Cortex-M0. This includes the MPU, which we will look at in Chapter 5 "Advanced Architecture Features," and the ability to relocate the vector table to a different position in memory. These two features provide improved operating system (OS) support and support for more sophisticated software designs with multiple application tasks on a single device (Fig. 1.16).

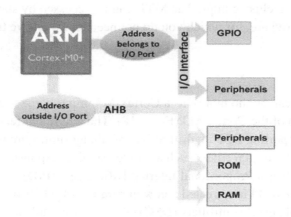

Figure 1.16
The I/O port allows single cycle access to GPIO and peripheral registers.

The Cortex-M0+ also has an improved debug architecture compared to the Cortex-M0. As we will see in Chapter 7 "Debugging with CoreSight" it supports the same real-time access to peripheral registers and SRAM as the Cortex-M3 and Cortex-M4. In addition, the Cortex-M0+ has a new debug feature called the "Micro Trace Buffer" (MTB) (Fig 1.17). The MTB

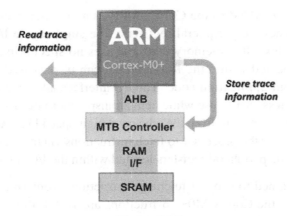

Figure 1.17
The Micro Trace Buffer can be configured to record executed instructions into a section of user SRAM. This can be read and displayed as an instruction trace in the PC debugger.

allows executed program instructions to be recorded into a region of SRAM setup by the programmer during development. When the code is halted this instruction trace can be downloaded and displayed in the debugger. This provides a snapshot of code execution immediately before the code was halted. While this is a limited trace buffer, it is extremely useful for tracking down elusive bugs. The MTB can be accessed by standard JTAG/Serial wire debug adaptor hardware, for which you do not need an expensive trace tool.

Cortex-M4

While the Cortex-M0 can be thought of as a Cortex-M3 minus some features, the Cortex-M4 is an enhanced version of the Cortex-M3 (Fig. 1.18). The additional features on the Cortex-M4 are focused on supporting DSP algorithms. Typical algorithms are transforms such as fast Fourier transform (FFT), digital filters such as finite impulse response (FIR) filters, and control algorithms such as a Proportional Internal Differential (PID) control loop. With its DSP features, the Cortex-M4 has created a new generation of ARM-based devices that can be characterized as digital signal controllers (DSC). These devices allow you to design devices that combine microcontroller type functions with real-time signal processing. In Chapter 8 "Practical DSP for the Cortex-M4 and Cortex-M7" we will look at the Cortex-M4 DSP extensions in more detail and also how to construct software that combines real-time signal processing with typical event-driven microcontroller code.

The Cortex-M4 has the same basic structure as the Cortex-M3 with the same CPU programmers modes, NVIC, CoreSight debug architecture, MPU, and bus interface. The enhancements over the Cortex-M3 are partly to the instruction set where the Cortex-M4 has

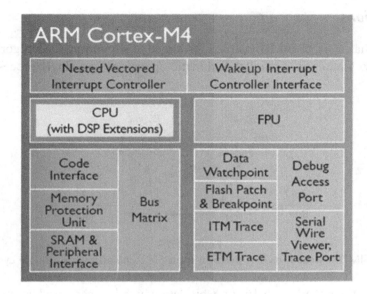

Figure 1.18
The Cortex-M4 is fully compatible with the Cortex-M3 but introduces a hardware floating point unit and additional DSP instructions.

additional DSP instructions in the form of SIMD instructions. The hardware multiply accumulate (MAC) has also been improved so that many of the 32 × 32 arithmetic instructions are single cycle (Table 1.2).

Table 1.2: Single cycle MAC instructions on the Cortex-M4

Operation	Instruction
16 × 16 = 32	SMULBB, SMULBT, SMULTB, SMULTT
16 × 16 + 32 = 32	SMLABB, SMLABT, SMLATB, SMLATT
16 × 16 + 64 = 64	SMLALBB, SMLALBT, SMLALTB, SMLALTT
16 × 32 = 32	SMULWB, SMULWT
(16 × 32) + 32 = 32	SMLAWB, SMLAWT
(16 × 16) ± (16 × 16) = 32	SMUAD, SMUADX, SMUSD, SMUSDX
(16 × 16) ± (16 × 16) + 32 = 32	SMLAD, SMLADX, SMLSD, SMLSDX
(16 × 16) ± (16 × 16) + 64 = 64	SMLALD, SMLALDX, SMLSLD, SMLSLDX
32 × 32 = 32	MUL
32 ± (32 × 32) = 32	MLA, MLS
32 × 32 = 64	SMULL, UMULL
(32 × 32) + 64 = 64	SMLAL, UMLAL
(32 × 32) + 32 + 32 = 64	UMAAL
32 ± (32 × 32) = 32(upper)	SMMLA, SMMLAR, SMMLS, SMMLSR
(32 × 32) = 32(upper)	SMMUL, SMMULR

DSP Instructions

The Cortex-M4 has a set of SIMD instructions aimed at supporting DSP algorithms. These instructions allow a number of parallel arithmetic operations in a single processor cycle (Fig. 1.19).

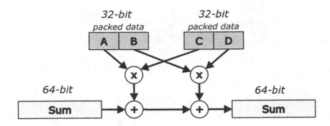

Figure 1.19
The SIMD instructions can perform multiple calculations in a single cycle.

The SIMD instructions work with 16- or 8-bit data which has been packed into 32-bit word quantities. So, for example, we can perform two 16-bit multiplies and sum the result into a 64-bit word. It is also possible to pack the 32-bit works with 8-bit data and perform a quad 8-bit addition or subtraction. As we will see in Chapter 8, the SIMD instructions can be used to vastly enhance the performance of DSP algorithms such as digital filters that are basically performing lots of multiply and sum calculations on a pipeline of data. The Cortex-M4 processor may also be fitted with a hardware FPU. This choice is made at the design stage by the microcontroller vendor, so like the ETM and MPU you will need to check the microcontroller datasheet to see if it is present. The FPU supports single precision floating point arithmetic calculations using the IEEE 754 standard (Table 1.3).

Table 1.3: Cortex-M4 floating point unit cycle times

Operation	Cycle Count
Add/Subtract	1
Divide	14
Multiply	1
Multiply accumulate (MAC)	3
Fused MAC	3
Square root	14

On small microcontrollers, floating point math has always been performed by software libraries provided by the compiler tool. Typically, such libraries can take hundreds of instructions to perform a floating point multiply. So, the addition of floating point hardware that can do the same calculation in a single cycle gives an unprecedented performance boost. The FPU can be thought of as a coprocessor that sits alongside the Cortex-M4 CPU. When a calculation is performed, the floating point values are transferred directly from the FPU registers to and from the SRAM memory store, without the need to use the CPU registers. While this may sound involved, the entire FPU transaction is managed by the compiler. When you build an application for the Cortex-M4, you can compile code to automatically use the FPU rather than software libraries. Then any floating point calculations in your "C" code will be carried out on the FPU.

With optimized code, the Cortex-M4 can run DSP algorithms far faster than standard microcontrollers and even some dedicated DSP devices (Fig. 1.20). Of course, the weasel word here is "optimized," this means having a good knowledge of the processor and the DSP algorithm you are implementing and then hand coding the algorithm making use of compiler intrinsics to get the best level of performance. Fortunately ARM provide a full open source DSP library which implements many commonly required DSP algorithms as easy to use library functions. We will look at using this library in Chapter 8 "Practical DSP for the Cortex-M4 and Cortex-M7"

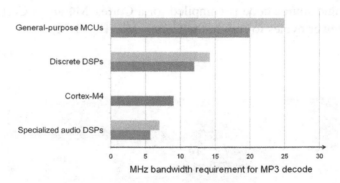

Figure 1.20
MP3 decode benchmark.

Cortex-M7

At the time of writing the latest Cortex-M processor to be released by ARM is the Cortex-M7. The Cortex-M7 is the highest performance Cortex-M7 processor currently available. This is actually a bit of an understatement. The benchmark figures for the Cortex-M7 vs the Cortex-M4 are shown below. Please note that these figures are shown per MHz of the CPU frequency. As well as significantly outperforming the Cortex-M4 the Cortex-M7 can run at much higher frequencies. In short it leaves the Cortex-M4 in the dust. However, the Cortex-M7 still maintains the Cortex-M programmer's model so if you have used an earlier Cortex-M processor moving to the Cortex-M7 is not a major challenge (Fig. 1.21).

Benchmark	ARM data	
	Cortex-M4	Cortex-M7
CoreMark/MHz	3.4	5
DMIPS/MHz	1.25	2.14

Figure 1.21
Cortex-M4 versus Cortex-M7 benchmark.

The Cortex-M7 achieves this boost in performance levels with some architectural enhancements and a more sophisticated memory system. The Cortex-M7 CPU has a superscalar architecture; this means it has two parallel three-stage pipelines which can dual issue instructions. The CPU is also capable of processing different groups of instructions in parallel. The CPU also has a "Branch Target Address Cache" which improves the performance of high-level statements such as conditional branches and more importantly loops. When the same source code is compiled for a Cortex-M4 and a Cortex-M7 it will take significantly fewer cycles to run on the Cortex-M7 (Fig. 1.22).

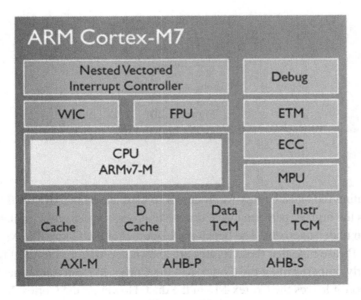

Figure 1.22
Cortex-M7 Processor.

The Cortex-M7 also introduces a more complex bus structure. The first ARM-based microcontrollers used an ARM7 processor which used a bus called the Advanced High-Performance bus (AHB). As the complexity of microcontrollers grew, this bus became a bottleneck when there were several bus masters (CPU and DMA units) within the microcontroller. With the introduction of the Cortex-M family the AHB was upgraded to the AHB matrix lite. This in effect is a set of parallel busses that allow multiple bus masters to access memory in parallel. Internal bus arbitration only occurs if two bus masters try to access the same group of peripherals or block of memory. The Cortex-M7 introduces the Advanced Extensible Interface (AXI-M). The AXI-M is a high-performance 64-bit interface that supports multiple outstanding memory transactions. It also opens up a lot of possibilities for silicon developers to design multicore systems and is a step towards network on chip (NoC) designs.

For a Developer the most significant difference between a Cortex-M4 and a Cortex-M7 is a more complex memory system. In order for the Cortex-M7 processor to achieve very high levels of performance it has a memory hierarchy. The CPU has two regions of memory called the Instruction and Data Tightly Coupled Memories (TCM). The I-TCM and D-TCM are blocks of zero wait state memory which can be upto 16K in size. This ensures that any critical routines or data can be accessed by the processor without any delays. The processor can also include two caches of upto 64K. These units provide upto 64K of Instruction and Data cache for system memory which is located on the AXI-M bus. When you are developing an application you need to understand and manage this memory system. Like the Cortex-M4 the Cortex-M7 has additional DSP capability in the form of SIMD instructions and can be fitted with a single or double precession Floating Point Unit. The Cortex-M7 can also be qualified for safety critical applications. The Cortex-M7 processor can be synthesized with additional safety feature such as "Error Correcting Codes" on its bus interfaces and a "Built In Self Test" unit. The Cortex-M7 is also the first Cortex-M processor to have a complete safety documentation pack to allow Silicon Vendors to produce a fully safety qualified device.

Conclusion

This book is really covering two topics. An introduction to the Cortex-M processor hardware and also in introduction to developing software for Cortex-M-based microcontrollers. With the introduction of the Cortex-M processor we now have a low-cost hardware platform that is capable of supporting more sophisticated software design and the last decade has seen the adoption of RTOSs and middleware libraries needed to support the more complex peripherals found on Cortex-M devices. So alongside understanding the low-level features of the Cortex-M processors we also need to use more sophisticated design techniques (Fig. 1.23).

ARM Cortex-M Area, Power, Performance					
	M0	M0+	M3	M4	M7
90 nm LP dynamic power (μW/MHz)	16	9.8	32	33	n/a
90 nm LP area mm²	0.04	0.035	0.12	0.17	n/a
40 nm G dynamic power (μW/MHz)	4	3	7	8	n/a
40 nm G area mm²	0.01	0.009	0.03	0.04	n/a
Dhrystone (official) DMIPS/MHz	0.84	0.94	1.25	1.25	2.14
Dhrystone (max options) DMIPS/MHz	1.21	1.31	1.89	1.95	3.23
CoreMark/MHz	2.33	2.42	3.32	3.40	5.04

Figure 1.23
Performance and power figures for the Cortex-M processor family.

CHAPTER 2

Developing Software for the Cortex-M Family

Introduction

One of the big advantages of using a Cortex-M processor is that it has wide and growing range development tool support. There are toolchains available at zero cost up to several thousand dollars depending on the depth of your pockets and the type of application you are developing. Today there are five main toolchains that are used for Cortex-M development (Table 2.1).

Table 2.1: Cortex-M processor toolchains

Development Tool
GNU GCC with free and commercial IDE
Greenhills
IAR Embedded Workbench for ARM
Keil Microcontroller Development Kit for ARM (MDK-ARM)
Tasking VX Toolset for ARM

Strictly speaking the GNU GCC is a compiler linker toolchain and does not include an integrated development environment or a debugger. A number of companies have created a toolchain around the GCC compiler by adding their own IDE and debugger to provide a complete development system. Some of these are listed in the appendix; there are quite a few, so this is not a complete list.

Keil Microcontroller Development Kit

In this tutorial we are going to use the Keil MDK-ARM (Microcontroller Development Kit for ARM) toolchain. The Keil MDK-ARM provides a complete development environment for all Cortex-M-based microcontrollers (Fig. 2.1).

The Designer's Guide to the Cortex-M Processor Family.
DOI: http://dx.doi.org/10.1016/B978-0-08-100629-0.00002-5
23

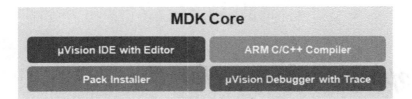

Figure 2.1
The MDK-ARM Core installation contains an IDE, compiler, and debugger. Device and middleware support is added through software packs.

The MDK-ARM includes its own development environment called μVision (Micro Vision) which acts as an editor project manager and debugger. One of the great strengths of the MDK-ARM is that it uses the ARM "C" compiler. This is a very widely used "C\C++" compiler which has been continuously developed by ARM since their first CPU's were created. The MDK-ARM also includes an integrated Real-Time Operating System (RTOS) called RTX. All of the Cortex-M processors are capable of running an operating system and we will look at using an RTOS in Chapter 9 "CMSIS-RTOS". As well as including an RTOS, the MDK-ARM includes a DSP library which can be used on the Cortex-M4 and also the Cortex-M3. We will look at this library in Chapter 8 "Practical DSP for Cortex-M4 and Cortex-M7."

Software Packs

The MDK-ARM is installed as a core toolchain. This consists of the μVision IDE, the compiler and debugger plus a utility called the pack installer. The core toolchain does not contain any support for specific Cortex-M microcontrollers. Support for a specific family of Cortex-M microcontrollers is installed through a software pack system. The pack installer allows you to select and install support for a family of microcontrollers. Once selected, a "Device Family Pack" will be downloaded from a Pack Repository website and installed into the toolchain. This software pack system can also be used to distribute software libraries and other software components. We will look at how to make a software component for code reuse in Chapter 12 "Software Components."

The Tutorial Exercises

There are a couple of key reasons for using the MDK-ARM as the development environment for this book. First, it includes the ARM "C" compiler which is the industry reference compiler for ARM processors. Second, it includes a software simulator which models each of the Cortex-M processors and the peripherals for a range of Cortex-M-based microcontrollers. This allows you to run most of the tutorial examples in this book without the need for a hardware debugger or evaluation board. The simulator is a very good way to learn how each Cortex-M processor works as you can get as much detailed debug information from the simulator as from a hardware debugger. The MDK-ARM also includes the first RTOS to support the CMSIS-RTOS specification. We will see more of this in Chapter 9 "CMSIS-RTOS," but it is basically a universal API for Cortex-M RTOS. While we can use the simulator to experiment with the different Cortex-M processors there comes a point when you will want to run your code on some real hardware. Now there are a number of very low-cost modules which include debugger hardware. The appendix provides URLs to websites where these boards can be purchased. In Chapter 6 "Cortex-M7 Processor," we will look at the Cortex-M7 and the practical examples will be run on the STM32F7 discovery board. This is a low-cost evaluation board and is widely available from most electronics catalog suppliers (Fig. 2.2).

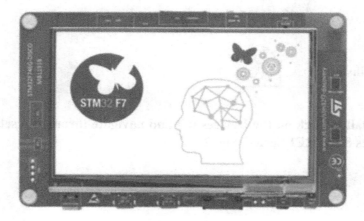

Figure 2.2
Low-cost Cortex-M modules include the ST32F7 Discovery board. We will use this board for the Cortex-M7 and hardware debug exercise.

The tutorial exercises described in this book are geared toward using the software simulator to demonstrate the different features of the varied Cortex-M processors. Towards the end of this chapter, we will look at how to connect and download a project to a real hardware board.

Installation

For the practical exercises in this book we need to install the MDK-ARM lite toolchain and the "Device Family Support" packs for the microcontrollers we will use and the tutorial example pack.

First Download the Keil MDK-ARM lite from www.keil.com.

The MKD-ARM lite is a free version of the full toolchain which allows you to build application code up to a 32K image size. It includes the fully working compiler, RTOS and simulator and works with compatible hardware debuggers. The direct link for the download is:

https://www.keil.com/arm/demo/eval/arm.htm.

If this changes over time, visit the main Keil web page and follow the link to the MDK-ARM tools.

Run the downloaded executable to install the MDK-ARM onto your PC.

Start the μVision IDE

Open the pack installer

In the pack installer, click on the Devices tab and navigate through to select the ST Microelectronics STM32F1 Series (Fig. 2.3).

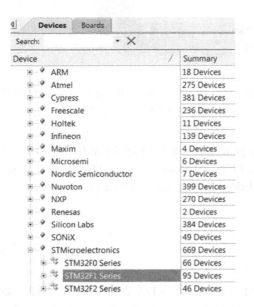
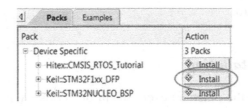

Figure 2.3
Use the pack installer to download support files for the STM32F1 family of microcontrollers.

In the Packs tab, select the Keil::STM32F1xx_DFP and press the Install button.

Repeat this process to install support for the following device families (Table 2.2).

Table 2.2: Required device family support packs

Silicon Vendor	Device	Device Pack
ST Microelectronics	STM32F7 series	Keil::STM32F7xx_DFP
NXP	LPC1700 series	Keil::LPC1700_DFP

In the Generic section of the Pack window select and install or update the following packs (Table 2.3).

Table 2.3: Required generic software packs

Generic Pack
ARM::CMSIS
ARM::CMIS-Driver_Validation
Keil::ARM_Compiler

Download the example pack for the practical exercises.

The tutorial examples can be found at http://booksite.elsevier.com/9780081006290.

Download the Tutorial Examples file and save it on your hard disk. Make sure it has a .pack extension; some browsers may rename the file by giving it a .zip extension.

On the Tutorial pages website there is also a Utilities pack. Download this pack to your hard drive and check it has a .pack extension.

Once you have downloaded the Tutorial Examples Pack and the Utilities pack, double click on each file in turn to install their contents into the MDK-ARM Toolchain.

To view the examples, open the pack installer.

Select the Boards tab and select the "Designers Guide Tutorial."

Now click on the Examples tab to see the tutorial examples.

To show them in order, click on the gray block at the head of the column.

Exercise 2.1 Building a First Program

Now that the toolchain and the examples are installed, we can look at setting up a project for a typical small Cortex-M-based microcontroller. Once the project is configured, we can get familiar with the µVision IDE, build the code, and take our first steps with the debugger.

The Blinky Project

In this example, we are going to build a simple project called Blinky. The code in this project is designed to read a voltage using the microcontroller's ADC. The value obtained from the ADC is then displayed as a bar graph on a small LCD display (Fig. 2.4).

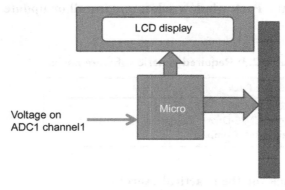

Figure 2.4
The Blinky project hardware consists of an analog voltage source, external LCD display, and a bank of LEDs.

The code also flashes a group of LEDs in sequence. There are eight LEDs attached to port pins on GPIO port B. The speed at which the LEDs flash is set by the ADC value.

Start the µVision IDE by clicking on the UV4 icon.

Open the pack installer.

Click on the Boards tab and select "The Designers Guide Tutorial Examples" entry.

Select the Examples tab and press the copy button for "Ex 1.1 First Project."

Set the install directory to a suitable place on your hard drive. In the following instructions, this location will be called <path>\First Project.

In this exercise, we will go through the steps necessary to recreate this project from scratch.

In μVision, close the current project by selecting project\close project from the main menu bar (Fig. 2.5).

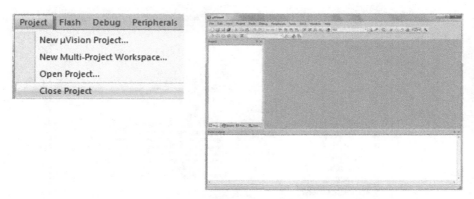

Figure 2.5
Close any open project.

Start a new project by selecting Project\new μVision project (Fig. 2.6).

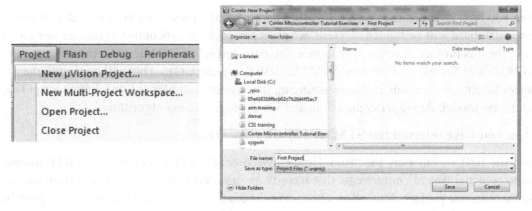

Figure 2.6
Create a new project and save it into the First Project directory.

This will open a menu asking for a project name and directory.

You can give the project any name you want but make sure you select the <path> \First Project Directory.

This directory contains the "C" source code files which we will use in our project.

Enter your project name and click Save.

Next, select the microcontroller to use in the project (Fig. 2.7).

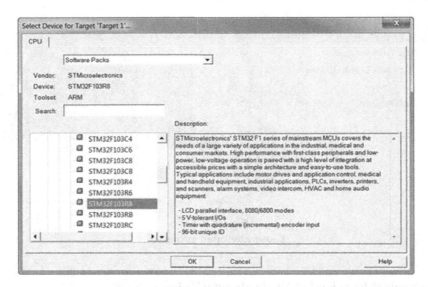

Figure 2.7
Select the STM32F103RB from the device database.

Once you have selected the project directory and saved the project name, a new dialog with a device database will be launched. Here, we must select the microcontroller that we are going to use for this project. Navigate the device database and select ST Microelectronics, the "STM32F103 Series" and then the STM32F103RB and click OK. This will configure the project for this device; this includes setting up the correct compiler options, linker script file, simulation model, debugger connection, and Flash programming algorithms.

When you have selected the STM32F103RB, click OK.

Now, the IDE will display the "Run-Time Environment" (RTE) Manager. The RTE allows you to select software components that have been installed through the pack system and add them to our project. This allows you to build complex software platforms very quickly. For help on any of the software components, click on the blue link in the description column (Fig. 2.8).

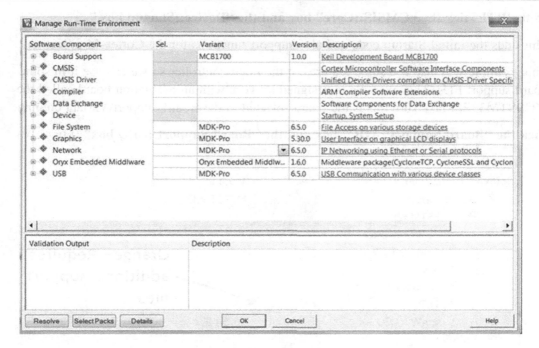

Figure 2.8
The Run-Time Environment manager allows you to add software components to your project to rapidly build a development "platform."

For now, we need to add the minimum support for our initial project (Fig. 2.9).

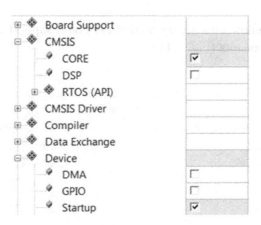

Figure 2.9
Select the "Core" and "Startup" components.

In the RTE, tick the "CMSIS::Core" box and the "Device::Startup box."

This adds the initial Startup code and also support functions for the Cortex-M processor.

This example will also use some features of the microcontroller so we can use it in some board support files. This example is designed to work with an evaluation board called the MCBSTM32E. Make sure this is the board selected in the Board Support\Variant column.

Tick the "Board Support::ADC" box and the "Board Support::LED box" (Fig. 2.10).

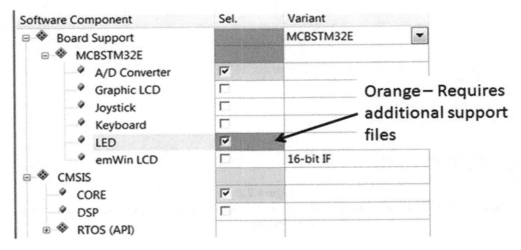

Figure 2.10
The board support components require subcomponents to work. Until this is resolved, the "sel." Column will be colored Orange (Dark Grey in the print versions).

The validation output window now shows us that the board support components require some additional subcomponents (Fig. 2.11).

Validation Output	Description
⚠ Keil.MCBSTM32E::Board Support:MCBS...	Additional software components required
require Device:GPIO	Select component from list
Keil::Device:GPIO	GPIO driver used by RTE Drivers for STM32F1 Series

Figure 2.11
The LED functions require the GPIO driver to be added to the project.

To add in the GPIO driver you can either open the Device section of the RTE and manually add in the GPIO driver or simply press the RTE resolve button and all of the component dependencies will be resolved automatically (Fig. 2.12).

Figure 2.12
Select the GPIO support file manually or press the resolve button to add the required components automatically.

Now that we have selected all of the components required by our project, press the OK button, and the support files will be added to the project window (Fig. 2.13).

Figure 2.13
Initial project with selected software components.

Double click on the startup_stm32F10x.md.s file to open it in the editor.

Click on the configuration Wizard tab at the bottom of the editor window (Fig. 2.14).

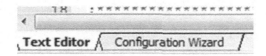

Figure 2.14
Configuration files can be viewed as Text or as configuration wizards.

This converts the plain text source file to a view that shows the configuration options within the file (Fig. 2.15).

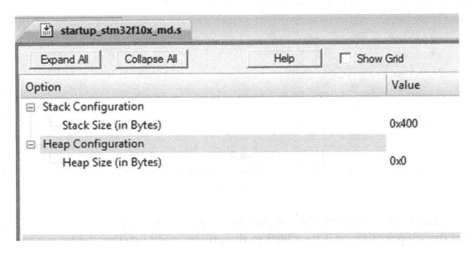

Figure 2.15
The configuration wizard allows you to view and modify #defines within a header or source file.

This view is created by XML tags in the source file comments. Changing the values in the configuration wizard modifies the underlying source code. In this case we can set the size of the stack space and heap space.

In the project view, click the Books tab at the bottom of the window (Fig. 2.16).

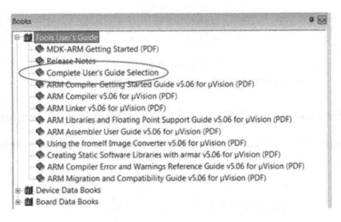

Figure 2.16
Toolchain help.

In the Books window, the "Complete Users Guide" opens the help system for the μVision and compiler manuals (Fig. 2.17).

Figure 2.17
Software component help.

In the RTE, the blue hyperlinks in the Description open the help files for a specific software component.

Switch back to the project view and add the project "C" source files.

Highlight the "Source Group" folder in the project window, right click and select "Add Existing files to Group Source Group 1" (Fig. 2.18).

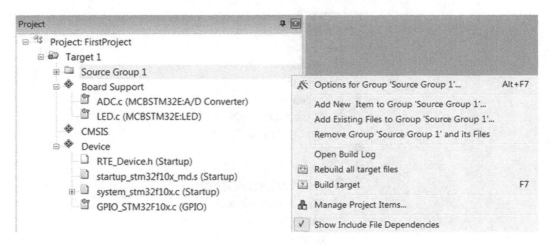

Figure 2.18
Add an existing file to the project source group.

This will open an "Add files to Group" dialog. In the dialog, add the Blinky.c project file (Fig. 2.19).

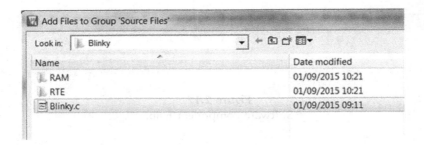

Figure 2.19
Add Blinky.c.

The project should now contain Blinky.c and the RTE components (Fig. 2.20).

Figure 2.20
The complete project.

Build the project by selecting project\build target (Fig. 2.21).

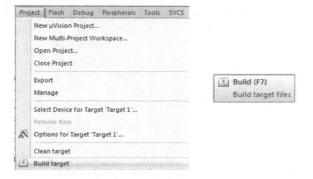

Figure 2.21
Build the project.

This will compile each of the ".c" modules in turn and then link them together to make a final application program. The output window shows the result of the build process and reports any errors or warnings (Fig. 2.22).

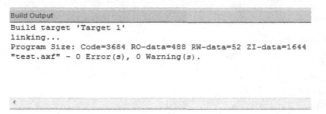

Figure 2.22
The final program size is reported in the Build Output window.

The program size is also reported (Table 2.4).

Table 2.4: Linker section memory use

Section	Description
Code	Size of the executable image
RO data	Size of the code constants in the Flash memory
RW data	Size of the initialized variable in SRAM
ZI data	Size on uninitialized variables in the SRAM

If errors or warnings are reported in the build window clicking on them will take you to the line of code in the editor window.

Open the Options for Target dialog (Fig. 2.23).

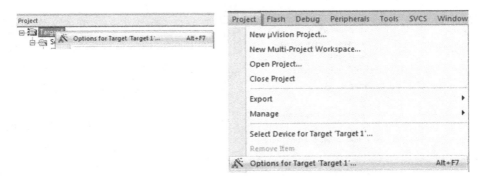

Figure 2.23
Open the project global options.

This can be done in the project menu by right clicking the project name and selecting "Options for Target" or by selecting the same option in the project menu from the main toolbar (Fig. 2.24).

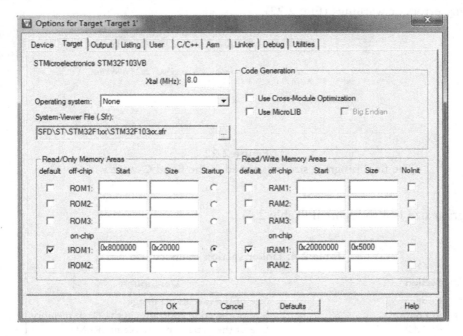

Figure 2.24
The target menu defines the project memory map.

The "Options for Target" dialog holds all of the global project settings.

Now select the Debug tab (Fig. 2.25).

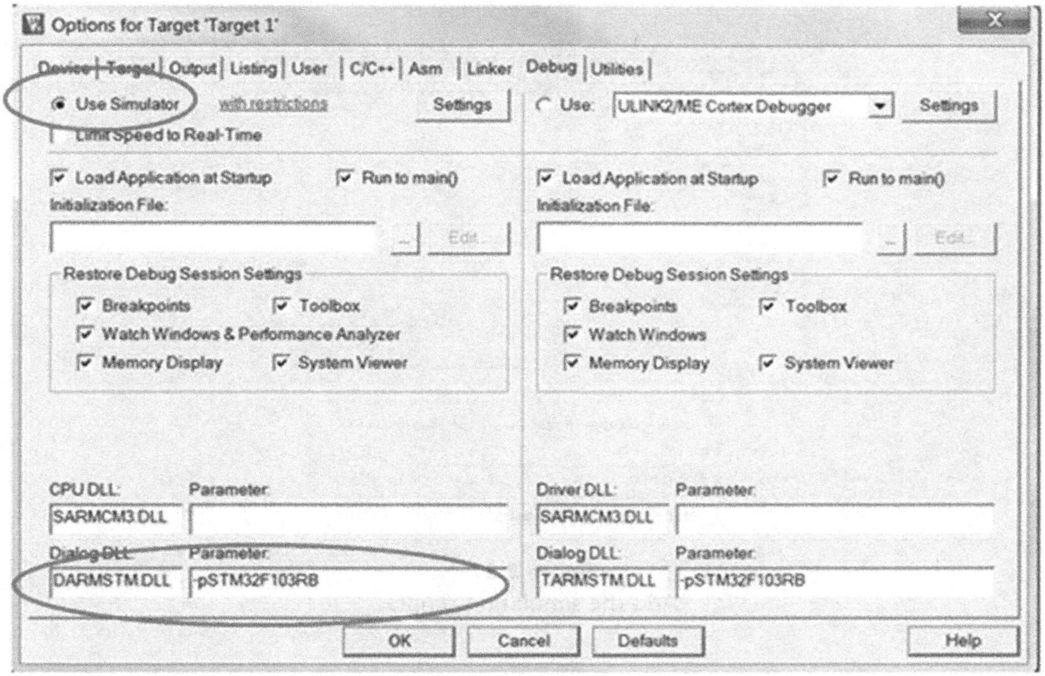

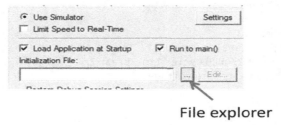

Figure 2.25
Select the simulator and the STM32F103RB simulation model.

The Debug menu is split into two halves. The simulator options are on the left and the hardware debugger is on the right.

Click the Use Simulator radio box.

Check that the Simulator Dialog DLL is set to DARMRM.DLL and the parameter is −pSTM32F103RB.

Now we can add a simulation script to provide some input values to the ADC (Fig. 2.26).

Figure 2.26
Open the file explorer to select the simulation script.

Press the File explorer button and add the Dbg_sim.ini file which is in the first project directory as the debugger initialization file (Fig. 2.27).

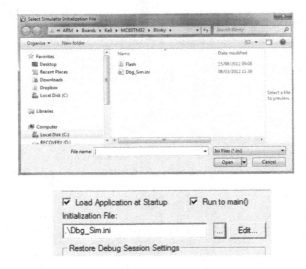

Figure 2.27
Add the simulation script.

The script file uses a "C"-like language to model the external hardware. All of the simulated microcontroller "pins" appear as virtual registers which can be read from and written to by the script. The debug script generates a simulated voltage for the ADC, the script for this is shown below. This generates a signal that ramps up and down and it is applied to the virtual register ADC1_IN1 which is channel 1 of ADC convertor 1. The twatch function reads the simulated clock of the processor and halts the script for a specified number of cycles.

```
Signal void Analog (float limit) {
  float volts;
  printf ("Analog (%f) entered.\n", limit);
  while (1) {          /* forever */
    volts = 0;
    while (volts <= limit) {
      ADC1_IN1 = volts;    /* analog input-2 */
      twatch (250000);     /* 250000 Cycles Time-Break */
      volts + = 0.1;       /* increase voltage */
    }
    volts = limit;
    while (volts >= 0.0) {
      ADC1_IN1 = volts;
```

```
    twatch (250000);    /* 250000 Cycles Time-Break */
    volts -= 0.1;       /* decrease voltage */
   }
  }
}
```

Click OK to close the options for Target dialog.

Now start the Debugger and run the code (Fig. 2.28).

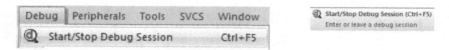

Figure 2.28
Start the Debugger.

This will connect μVision to the Simulation model and download the project image into the simulated memory of the microcontroller. Once the program image has been loaded, the microcontroller is reset and the code is run until it reaches main() ready to start debugging (Fig. 2.29).

Figure 2.29
The debug view.

The μVision debugger is divided into a number of windows that allow you to examine and control the execution of your code. The key windows are shown in Figs. 2.30−2.35.

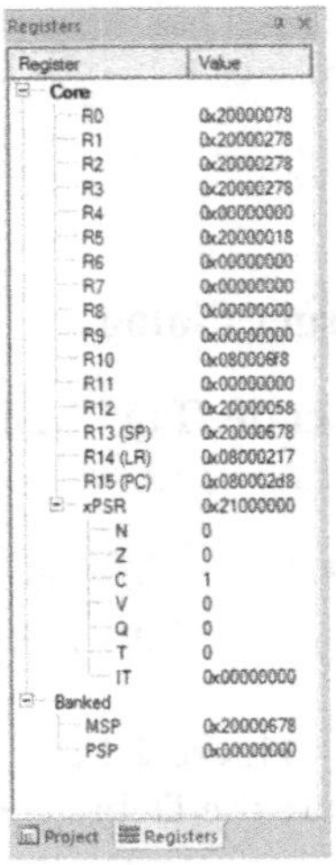

Figure 2.30
Register window.

The Register window displays the current contents of the CPU register file (R0-R15). The Program Status Register (xPSR) and also the Main Stack Pointer (MSP) and the Process Stack Pointer (PSP). We will look at all of these registers in the next chapter (Fig. 2.31).

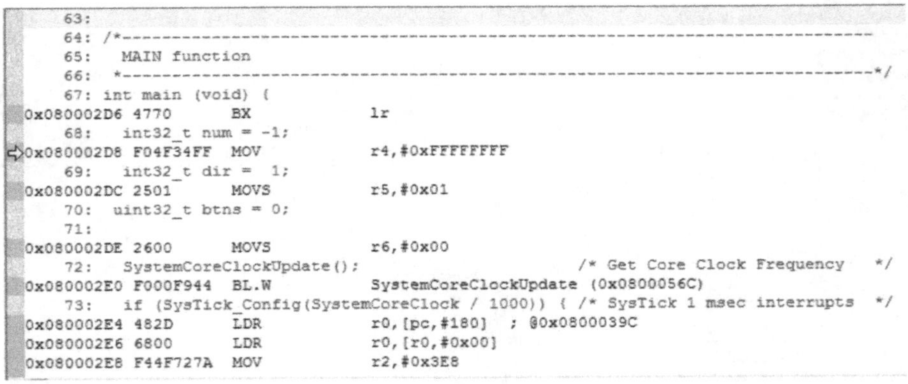

Figure 2.31
Disassembly window.

As its name implies, the Disassembly window will show you the low-level assembler listing interleaved with the high-level "C" code listing. One of the great attractions of the Cortex-M family is that all of your project code can be written in a high-level language such as "C\C + +." You never or very rarely need to write low-level assembly routines. However, it is useful to be able to "read" the low-level assembly code to see what the compiler is doing. The disassembly window shows the absolute address of the current instruction, next, it shows the OP Code, this is either a 16-bit instruction or 32-bit instruction. The raw OP

Code is then displayed as an assembler mnemonic. The current location of the program counter is shown by the yellow arrow in the left hand margin. The dark gray blocks indicate the location of executable lines of code (Fig. 2.32).

```
Blinky.c    startup_stm32f10x_hd.s    system_stm32f10x.c    core_cm3.h    ADC.c
 42    Main function
 43    *-------------------------------------------------------------------------*/
 44    
 45  int32_t adc = -1, adcVal=0;
 46
 47 int main (void) {
 48     int32_t max_num = LED_Num() - 1;
 49     int32_t num = 0;
 50     int32_t dir = 1;
 51
 52     char text[5] = "";
 53
 54     SysTick_Config(SystemCoreClock/100);      /* Generate interrupt each 10 ms */
 55     LED_Initialize();                          /* LED Initialization            */
 56     ADC_Initialize();                          /* A/D Converter Init            */
 57     while (1) {
 58       if (LEDOn) {
 59         LEDOn = 0;
 60         LED_On (num);                          /* Turn specified LED on         */
```

Figure 2.32
Editor window.

The source code window has a similar layout the disassembly window. This window just displays the high-level "C" source code. The current location of the program counter is shown by the yellow arrow in the left hand margin. The blue arrow shows the location of the cursor. Like the disassembly window, the dark gray blocks indicate the location of executable lines of code. The source window allows you to have a number of project modules open. Each source module can be reached by clicking the tab at the top of the window.

The command window allows you to enter debugger commands to directly configure and control the debugger features. These commands can also be stored in a text file and executed as a script when the debugger starts (Fig. 2.33).

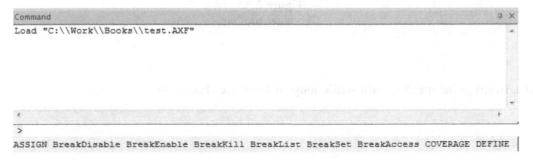

Figure 2.33
Debugger command line.

Next to the command window is a group of watch windows. These windows allow you to view local variables, global variables, and the raw memory (Fig. 2.34).

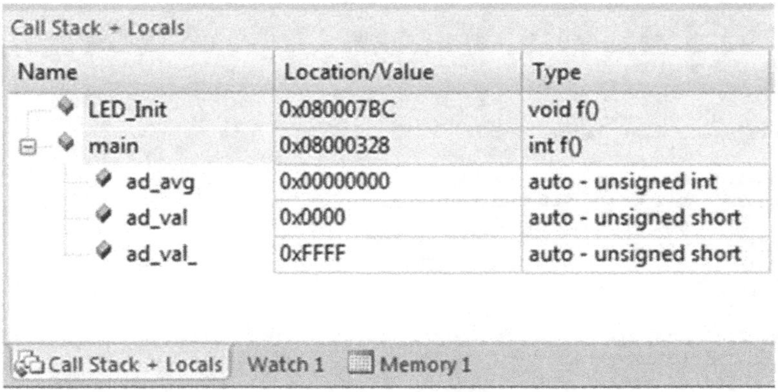

Call Stack + Locals
Name
Location/Value
Type
LED_Init
0x080007BC
void f()
main
0x08000328
int f()
ad_avg
0x00000000
auto - unsigned int
ad_val
0x0000
auto - unsigned short
ad_val_
0xFFFF
auto - unsigned short
Call Stack + Locals
Watch 1
Memory 1

Figure 2.34
Variable watch window.

You can control execution of the code through icons on the toolbar (Fig. 2.35). The code can be single stepped "C" or assembled a line at a time or run at full speed and halted. The same commands are available through the debug menu that also shows the function key shortcuts which you may prefer.

Reset
Halt
Locate program counter
Run
Single Step

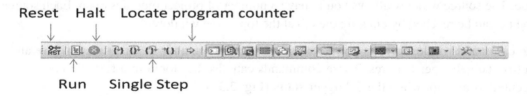

Figure 2.35
Debugger toolbar.

Set a breakpoint on the main while loop in Blinky.c (Fig. 2.36).

```
50
51     SysTick_Config(SystemCoreClock / 100);
52
53 □   while (1) {
54
55       /* AD converter input
56 □     if (AD_done) {
57         AD_done = 0;
58
```

Figure 2.36
A breakpoint is displayed as a red dot (Next to line 53).

You can set a breakpoint by moving the mouse cursor into a dark gray block next to the line number and left clicking. A breakpoint is marked by a red dot.

Start the code execution.

With the simulation script in place we will be able to execute all of the LCD code and reach the breakpoint (Fig. 2.37).

```
51     SysTick_Config(SystemCoreClock / 100);
52
53 □   while (1) {
54
55       /* AD converter input
56 □     if (AD_done) {
57         AD_done = 0;
```

Figure 2.37
Set a breakpoint on the main while loop. If you get here, the initialization code ran successfully.

Now spend a few minutes exploring the debugger run control (Fig. 2.38).

Step   Step out of function
 ↓   ↓

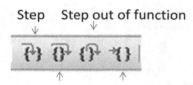

 ↑   ↑
Step over function   Run to Cursor (Blue arrow)

Figure 2.38
Toolbar single step options.

Use the single step commands, set a breakpoint and start the simulator running at full speed. If you lose what's going on, exit the debugger by selecting debug/start/stop the debugger and then restart again.

Add a variable to the Watch window.

Once you have finished familiarizing yourself with the run control commands within the debugger locate the main() function within Blinky.c. Just above main() is the declaration for a variable called ADC_DbgValue. Highlight this variable, right click, and select "Add ADC_DbgValue" to Watch 1 (Fig. 2.39).

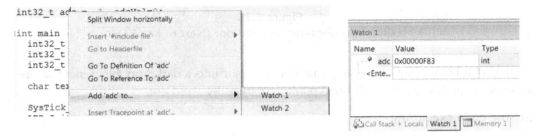

Figure 2.39
Add the ADC_DbgValue to Watch 1.

Now start the code running and you will be able to see the ADC_DbgValue variable updating in the watch window.

The simulation script is feeding a voltage to the simulated microcontroller which in turn provides converted results to the application code.

Now add the same variable to the Logic Trace window.

The μVision debugger also has a logic trace feature that allows you to visualize the historical values of a given global variable. In the watch window (or the source code window), highlight the AD_DbgValue variable name, right click, and select add AD_DbgValue to Logic Analyzer (Fig. 2.40).

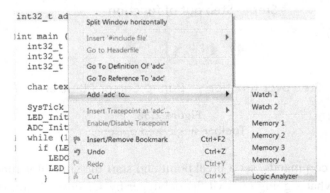

Figure 2.40
Adding a variable to the Logic Analyzer.

If the Logic Analyzer window does not open automatically select it from the toolbar (Fig. 2.41).

Figure 2.41
Opening the Logic Analyzer.

Now with the code running, press the Min\Max Auto button which will set the min and max values and also click the zoom out button to get to a reasonable time scale. Once this is done, you will be able to view a trace of the values stored in the AD_DbgVal variable. You can add any other global variables or peripheral registers to the logic analyzer window (Fig. 2.42).

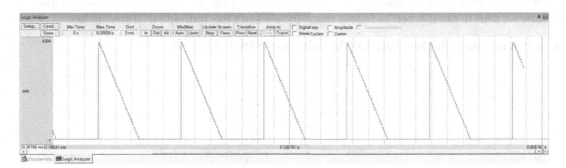

Figure 2.42
Logic Analyzer.

Now view the state of the user peripherals.

The simulator has a model of the whole microcontroller not just the Cortex-M processor, so it is possible to examine the state of the microcontroller peripherals directly.

Select peripherals\General purpose IO\GPIOB (Fig. 2.43).

Figure 2.43
Peripheral GPIO B window.

This will open a window that displays the current state of the microcontroller GPIO port B. As the simulation runs we can see the state of the port pins. If the pins are configured as inputs, we can manually set and clear them by clicking the individual "Pins" boxes.

You can do the same for the ADC by selecting ADC1 (Fig. 2.44).

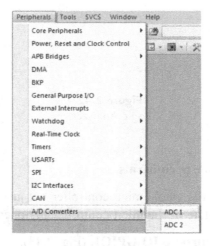

Figure 2.44
Opening the ADC peripheral view window.

When the code is running, it is possible to see the current configuration of the ADC and the conversion results. You can also manually set the input voltage by entering a fresh value in the Analog Inputs boxes (Fig. 2.45).

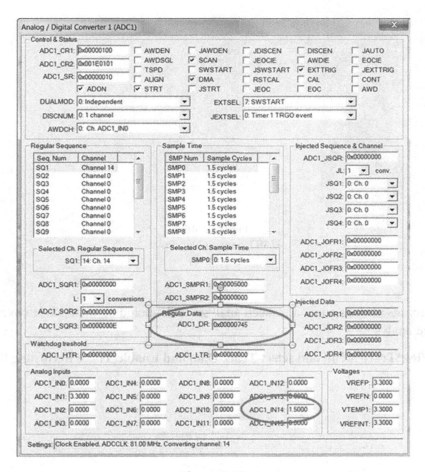

Figure 2.45
ADC peripheral window.

The simulator also includes a terminal that provides an I\O channel for the microcontrollers UARTS.

Select the view\serial windows\UART #1 window.

This opens a console type window that displays the output from a selected UART and also allows you to input values (Fig. 2.46).

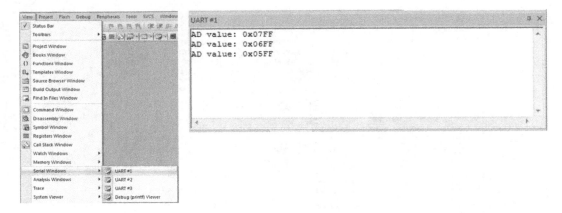

Figure 2.46
UART console window.

The simulator also boasts some advanced analysis tools including trace, code coverage, and performance analysis.

Open the View\Trace menu and select Trace Data and Enable Trace Recording (Fig. 2.47).

Figure 2.47
Enabling instruction the Trace.

This will open the instruction Trace window. The Trace records the history of each instruction executed (Fig. 2.48).

Nr.	Time	Address	Opcode	Instruction	Src Code
32,738	9.383 351 014 s	0x080006C4	4818	LDR r0,[pc,#96] ; @0x08000728	if (clock_1s) {
32,739	9.383 351 042 s	0x080006C6	7800	LDRB r0,[r0,#0x00]	
32,740	9.383 351 069 s	0x080006C8	B130	CBZ r0,0x080006D8	
32,741	9.383 351 111 s	0x080006D8	E7CF	B 0x0800067A	while (1) { /* Lo...
32,742	9.383 351 153 s	0x0800067A	4825	LDR r0,[pc,#148] ; @0x08000710	if (AD_done) { /* L..
32,743	9.383 351 181 s	0x0800067C	7800	LDRB r0,[r0,#0x00]	
32,744	9.383 351 208 s	0x0800067E	B170	CBZ r0,0x0800069E	
32,745	9.383 351 250 s	0x0800069E	EA940006	EORS r0,r4,r6	if (ad_val ^ ad_val_) { /...
32,746	9.383 351 264 s	0x080006A2	D00F	BEQ 0x080006C4	
32,747	9.383 351 306 s	0x080006C4	4818	LDR r0,[pc,#96] ; @0x08000728	if (clock_1s) {

Figure 2.48
Instruction trace.

Now open the View\Analysis\Code Coverage and View\Analysis\Performance Analyzer windows (Fig. 2.49).

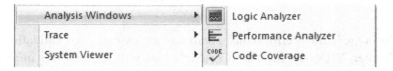

Figure 2.49
Selecting the Performance Analyzer.

The "Performance Analysis" window shows the number of calls to a function and its cumulative run time (Fig. 2.50).

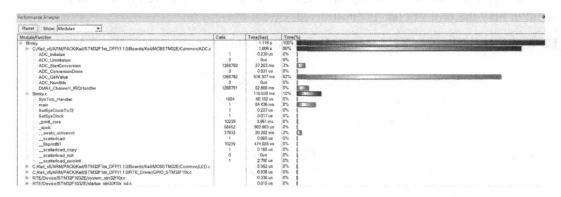

Figure 2.50
Performance Analyzer window.

The code coverage window provides a digest of the number of lines executed and partially executed in each function (Fig. 2.51).

Code Coverage

| Update | Clear | Module: | <All Modules> ▼ |

Modules/Functions	Execution percentage
⊟ RTE/Device/STM32F103ZE/system_stm32f10x.c	
SetSysClockTo72	94% of 92 instructions, 5 condjump(s) not fully executed
SetSysClock	100% of 3 instructions
SystemInit	100% of 33 instructions
SystemCoreClockUpdate	0% of 65 instructions
⊞ RTE/Device/STM32F103ZE/startup_stm32f10x_hd.s	
⊟ C:/Keil_v5/ARM/PACK/Keil/STM32F1xx_DFP/1.1.0/RTE_Driver/GPIO_STM32F10x.c	
GPIO_PortClock	11% of 128 instructions, 3 condjump(s) not fully executed
GPIO_PinConfigure	56% of 44 instructions, 3 condjump(s) not fully executed
GPIO_AFConfigure	0% of 30 instructions
⊟ C:/Keil_v5/ARM/PACK/Keil/STM32F1xx_DFP/1.1.0/Boards/Keil/MCBSTM32E/Common/LED.c	
LED_Initialize	88% of 34 instructions, 1 condjump(s) not fully executed
LED_Uninitialize	0% of 15 instructions
LED_On	76% of 17 instructions, 1 condjump(s) not fully executed
LED_Off	76% of 17 instructions, 1 condjump(s) not fully executed
LED_Out	0% of 17 instructions
LED_Num	100% of 2 instructions
⊟ C:/Keil_v5/ARM/PACK/Keil/STM32F1xx_DFP/1.1.0/Boards/Keil/MCBSTM32E/Common/ADC.c	
ADC_Initialize	100% of 100 instructions, 2 condjump(s) not fully executed
ADC_Uninitialize	0% of 1 instructions
ADC_StartConversion	100% of 11 instructions
ADC_ConversionDone	0% of 7 instructions
ADC_GetValue	100% of 11 instructions
ADC_NumBits	0% of 2 instructions
DMA1_Channel1_IRQHandler	100% of 11 instructions, 1 condjump(s) not fully executed

Figure 2.51
Code Coverage window.

Both "Code Coverage" and "Performance Analysis" are essential for validating and testing software. In Chapter 7 "Debugging with CoreSight" we will see how this information can be obtained from a real microcontroller.

Project Configuration

Now that you are familiar with the basic features of the debugger we can look in more detail at how the project code is constructed.

First quit the debugger by selecting Debug\Start\Stop debug session (Fig. 2.52).

Figure 2.52
Exit the debugger.

Open the "Options for Target" dialog.

All of the key global project settings can be found in the "Options for Target" dialog (Fig. 2.53).

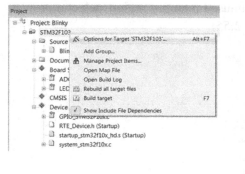

Figure 2.53
Options for Target dialog.

The Target tab defines the memory layout of the project. A basic template is defined when you create the project. On this microcontroller there is 128K of internal Flash memory and 20K of SRAM. If you need to define a more complex memory layout it is possible to create additional memory regions to subdivide the volatile and nonvolatile memories (Fig. 2.54).

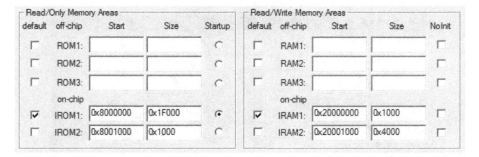

Figure 2.54
The Target menu defines the project memory map.

The more complex memory map above has split the internal FLASH into two blocks and defined the lower FLASH block as the default region for code and constant data. Nothing will be placed into the upper block unless you explicitly tell the linker to do this. Similarly, the SRAM has been split into two regions and the upper region is unused unless you explicitly tell the linker to use it. When the linker builds the project it looks for the "RESET" code label. The linker then places the reset code at the base of the code region designated as the Startup region. The initial Startup code will write all of the internal SRAM to zero unless you tick the NoInit box for a given SRAM region. Then the SRAM will be left with its startup garbage values. This may be useful if you want to allow for a soft reset where some system data is maintained.

If you want to place objects (code or data) into an unused memory region, select a project module, right click and open its local options (Fig. 2.55).

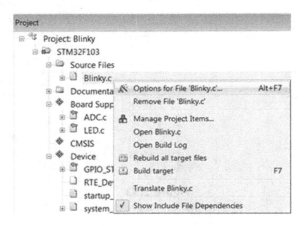

Figure 2.55
Opening the local project options.

In its local options, the memory assignment boxes allow you to force the different memory objects in a module into a specific code region (Fig. 2.56).

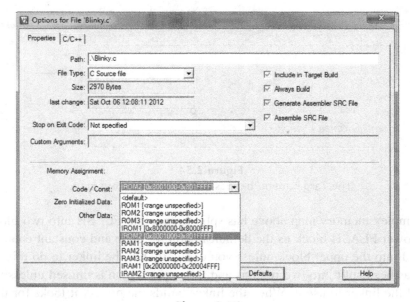

Figure 2.56
Selecting the memory regions for a module. This overrides the global settings.

Back in the main options for Target menu there is an option to set the External Crystal frequency used by the microcontroller (Fig. 2.57).

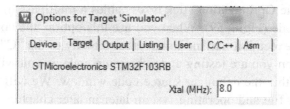

Figure 2.57
The Crystal (Xtal) frequency is only used by the simulator.

Often this will be a standard value that can be multiplied by internal phase locked loop oscillator of the microcontroller to reach the maximum clock speed supported by the microcontroller. This option is only used to provide the input frequency for the simulation model and nothing else.

The Keil MDK-ARM comes with two ANSI library sets (Fig. 2.58). The first is the standard library that comes with the ARM compiler. This is fully compliant with the current ANSI standard and as such has a large code footprint for microcontroller use. The second library set is the Keil MicroLIB, this library has been written to an earlier ANSI standard, the C99 standard. This version of the ANSI standard is more in tune with the needs of microcontroller users (Table 2.5).

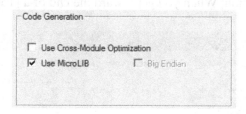

Figure 2.58
Selecting the MicroLIB library.

Table 2.5: Size comparison between the standard ARM ISO libraries and the Keil MicroLIB library

Processor	Object		Standard	MicroLIB	Saving (%)
Cortex-M0(+)	Thumb	Library otal	16,452	5996	64
		RO total	19,472	9016	54
Cortex-M3\M4	Thumb-2	Library total	15,018	5796	63
		RO total	18,616	8976	54

By selecting MicroLIB you will save at least 50% of the ANSI library code footprint verses the ARM compiler libraries. So try to use MicroLIB wherever possible. However, it does have some limitations most notably it does not support all of the functions in standardlib and double precision floating point calculations. In a lot of applications you can live with this.

The "Use Cross-Module Optimization" tick box enables a multipass linking process that fully optimizes your code (Fig. 2.59). When you use this option, the code generation is changed and the code execution may no longer map directly to the "C" source code. So do not use this option when you are testing and debugging code as you will not be able to accurately follow it within the debugger source code window. We will look at the other options, system viewer file and operating system later in later chapters.

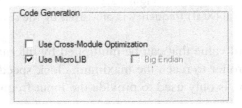

Figure 2.59
Cross-Module Optimization enables a multipass compile process for best code generation.

The output menu allows you to control the final image of your project (Fig. 2.60). Here, we can choose between generating a standalone program by selecting "Create Executable" or we can create a library which can be added to another project. The default is to create a standalone project with debug information. When you get toward the end of a project you will need to

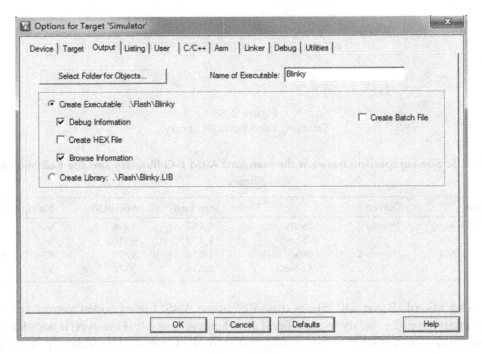

Figure 2.60
The output dialog.

select the "create hex file" option to generate a HEX32 file which can be used with a production programmer. If you want to build the project outside of μVision, select Create Batch File and this will produce a <Project name>.bat DOS batch file which can be run from another program to rebuild the project outside of the IDE. By default, the name of the final image is always the same as your project name. If you want to change this simply change the "Name of Executable" field. You can also select a directory to store all of the project, compiler, and linker generated files. This ensures that the original project directory only contains your original source code. This can make life easier when you are archiving projects.

The Listing tab allows you to enable compiler listings and linker map files (Fig. 2.61). By default, the linker map file is enabled. A quick way to open the map file is to select the project window and double click on the project root. The linker map file contains a lot of information which can seem incomprehensible at first, but there are a couple of important sections that you should learn to read and keep an eye on when developing a real project. The first is the "memory map of the image." This shows you a detailed memory layout of your project. Each memory segment is shown against its absolute location in memory. Here you can track which objects have been located to which memory region. You can also see

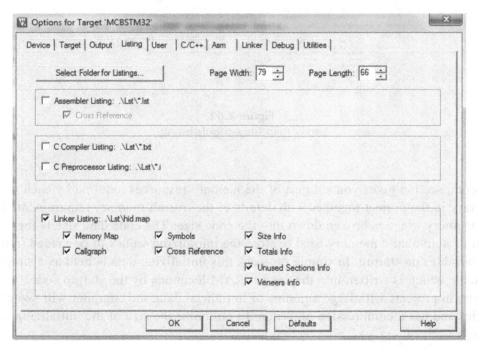

Figure 2.61
The Listing dialog.

the total amount of memory resources allocated, the location of the stack, and also if it is enabled the location of the heap memory (Fig. 2.62).

```
Image Entry point : 0x000000cd

Load Region LR_IROM1 (Base: 0x00000000, Size: 0x00000d98, Max: 0x00040000, ABSOLUTE)

    Execution Region ER_IROM1 (Base: 0x00000000, Size: 0x00000d78, Max: 0x00040000, ABSOLUTE)

    Base Addr    Size         Type    Attr    Idx    E Section Name        Object

    0x00000000   0x000000cc   Data    RO        3      RESET               startup_lpc17xx.o
    0x000000cc   0x00000000   Code    RO      267    * .ARM.Collect$$$00000000    mc_w.1(entry.o)
    0x000000cc   0x00000004   Code    RO      533      .ARM.Collect$$$00000001    mc_w.1(entry2.o)
    0x000000d0   0x00000004   Code    RO      536      .ARM.Collect$$$00000004    mc_w.1(entry5.o)
    0x000000d4   0x00000000   Code    RO      538      .ARM.Collect$$$00000008    mc_w.1(entry7b.o)
    0x000000d4   0x00000008   Code    RO      539      .ARM.Collect$$$00000009    mc_w.1(entry8.o)
    0x000000dc   0x00000004   Code    RO      534      .ARM.Collect$$$00002712    mc_w.1(entry2.o)
    0x000000e0   0x00000024   Code    RO        5      .text               startup_lpc17xx.o
    0x00000104   0x00000030   Code    RO      100      .text               retarget.o
    0x00000134   0x00000080   Code    RO      127      .text               serial.o
    0x000001b4   0x000000ac   Code    RO      154      .text               led.o
    0x00000260   0x00000074   Code    RO      180      .text               irq.o
    0x000002d4   0x0000001e   Code    RO      542      .text               mc_w.1(llsh1.o)
    0x000002f2   0x00000002   Code    RO      566      i.__scatterload_null    mc_w.1(handlers.o)
    0x000002f4   0x00000008   PAD
    0x000002fc   0x00000004   Code    RO        4      .ARM.__at_0x02FC    startup_lpc17xx.o
    0x00000300   0x00000310   Code    RO       15      .text               system_lpc17xx.o
    0x00000610   0x000000d8   Code    RO      207      .text               blinky.o
    0x000006e8   0x000000c4   Code    RO      239      .text               adc.o
    0x000007ac   0x00000062   Code    RO      270      .text               mc_w.1(uldiv.o)
    0x0000080e   0x00000020   Code    RO      544      .text               mc_w.1(llushr.o)
    0x0000082e   0x00000002   PAD
    0x00000830   0x00000024   Code    RO      557      .text               mc_w.1(init.o)
    0x00000854   0x00000020   Code    RO      479      i.__0printf$8       mc_w.1(printf8.o)
    0x00000874   0x00000028   Code    RO      481      i.__0sprintf$8      mc_w.1(printf8.o)
    0x0000089c   0x0000000e   Code    RO      565      i.__scatterload_copy    mc_w.1(handlers.o)
    0x000008aa   0x0000000e   Code    RO      567      i.__scatterload_zeroinit    mc_w.1(handlers.o)
    0x000008b8   0x00000420   Code    RO      486      i._printf_core      mc_w.1(printf8.o)
    0x00000cd8   0x00000026   Code    RO      487      i._printf_post_padding    mc_w.1(printf8.o)
    0x00000cfe   0x00000030   Code    RO      488      i._printf_pre_padding    mc_w.1(printf8.o)
    0x00000d2e   0x0000000a   Code    RO      490      i._sputc            mc_w.1(printf8.o)
    0x00000d38   0x00000020   Data    RO      155      .constdata          led.o
    0x00000d58   0x00000020   Data    RO      563      Region$$Table       anon$$obj.o

    Execution Region RW_IRAM1 (Base: 0x10000000, Size: 0x00000000, Max: 0x00004000, ABSOLUTE)

    **** No section assigned to this execution region ****

    Execution Region RW_IRAM2 (Base: 0x2007c000, Size: 0x00000230, Max: 0x00008000, ABSOLUTE)

    Base Addr    Size         Type    Attr    Idx    E Section Name        Object

    0x2007c000   0x00000004   Data    RW       16      .data               system_lpc17xx.o
    0x2007c004   0x00000008   Data    RW      101      .data               retarget.o
    0x2007c00c   0x00000009   Data    RW      181      .data               irq.o
    0x2007c019   0x00000001   PAD
    0x2007c01a   0x00000004   Data    RW      240      .data               adc.o
    0x2007c01e   0x00000002   PAD
    0x2007c020   0x0000000a   Zero    RW      208      .bss                blinky.o
    0x2007c02a   0x00000006   PAD
    0x2007c030   0x00000200   Zero    RW        1      STACK               startup_lpc17xx.o
```

Figure 2.62
Linker map file symbols listing.

The second section gives you a digest of the memory resources required by each module and library in the project together with details of the overall memory requirement. The image memory usage is broken down into the code size. The code data size is the amount of nonvolatile memory used to store the initializing values to be loaded into RAM variables on startup. In simple projects, this initializing data is held as a simple ROM table which is written into the correct RAM locations by the startup code. However, in projects with large amounts of initialized data, the compiler will switch strategies and use a compression algorithm to minimize the size of the initializing data.

On startup, this table is decompressed before the data is written to the variable locations in memory. The RO data entry lists the amount of nonvolatile memory used to store code literals. The SRAM usage is split into initialized RW data and uninitialized ZI data (Fig. 2.63).

```
Image component sizes

    Code (inc. data)   RO Data   RW Data   ZI Data   Debug   Object Name
         196      30         0         4         0   16831   adc.o
         216      50         0         0        10   16612   blinky.o
         116      20         0        13         0     641   irq.o
         172      16        32         0         0    1317   led.o
          48       0         0         8         0    2658   retarget.o
         128      18         0         0         0   17327   serial.o
          40      12       204         0       512     880   startup_lpc17xx.o
         784      76         0         4         0    5105   system_lpc17xx.o
    ----------------------------------------------------------------------
        1700     222       268        32       528   61371   Object Totals
           0       0        32         0         0       0   (incl. Generated)
           0       0         0         3         6       0   (incl. Padding)
    ----------------------------------------------------------------------

    Code (inc. data)   RO Data   RW Data   ZI Data   Debug   Library Member Name
           0       0         0         0         0       0   entry.o
           8       4         0         0         0       0   entry2.o
           4       0         0         0         0       0   entry5.o
           0       0         0         0         0       0   entry7b.o
           8       4         0         0         0       0   entry8.o
          30       0         0         0         0       0   handlers.o
          36       8         0         0         0      68   init.o
          30       0         0         0         0      68   llshl.o
          32       0         0         0         0      68   llushr.o
        1224      62         0         0         0     504   printf8.o
          98       0         0         0         0      92   uldiv.o
    ----------------------------------------------------------------------
        1480      78         0         0         0     800   Library Totals
          10       0         0         0         0       0   (incl. Padding)
    ----------------------------------------------------------------------

    Code (inc. data)   RO Data   RW Data   ZI Data   Debug   Library Name
        1470      78         0         0         0     800   mc_w.l
    ----------------------------------------------------------------------
        1480      78         0         0         0     800   Library Totals
    ----------------------------------------------------------------------
    ======================================================================

    Code (inc. data)   RO Data   RW Data   ZI Data   Debug
        3180     300       268        32       528   61027   Grand Totals
        3180     300       268        32       528   61027   ELF Image Totals
        3180     300       268        32         0       0   ROM Totals
    ======================================================================

    Total RO  Size (Code + RO Data)                 3448 (   3.37kB)
    Total RW  Size (RW Data + ZI Data)               560 (   0.55kB)
    Total ROM Size (Code + RO Data + Rw Data)       3480 (   3.40kB)
```

Figure 2.63
Linker map file sections listing.

The next tab is the User tab. This allows you to add external utilities to the build process. The menu allows you to run a utility program to pre- or postprocess files in the project. A utility program can also be run before each module is compiled. Optionally, you can also start the debugger once the build process has finished (Fig. 2.64).

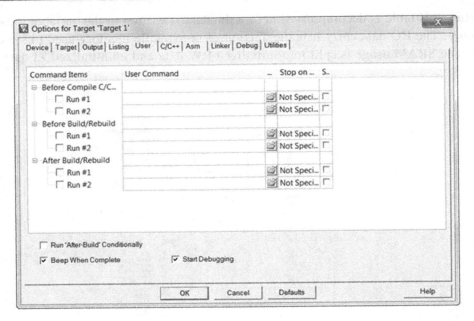

Figure 2.64
The User dialog.

The code generated by the compiler is controlled by the C\C++ tab. This controls the code generation for the whole project. However, the same menu is available in the local options for each source module. This allows you to have global build options and then different build options for selected modules. In the local options menu, the option tick boxes are a bit unusual in that they have three states (Fig. 2.65).

Figure 2.65
Tick boxes can have three states.

They can be unchecked, checked with a solid black tick or checked with a grey tick. Here, the grey tick means "inherit the global options" and this is the default state.

The most important option in this menu is the optimization control (Fig. 2.66). During development and debugging you should leave the optimization level at zero. Then, the generated code maps directly to the high-level "C" source code and it is easy to debug. As you increase the optimization level, the compiler will use more and more aggressive techniques to optimize the code. At the high optimization level, the generated code no longer maps closely to the original source code which then makes using the debugger very difficult. For example, when you single step the code, its execution will no longer follow the expected path through the source code. Setting a break point can also be hit and miss as the generated code may not exist on the same line as the source code. By default, the compiler will generate the smallest image. If you need to get the maximum performance, you can select the "Optimize for Time" option. Then, the compiler strategy will be changed to generate the fastest executable code (Fig. 2.67).

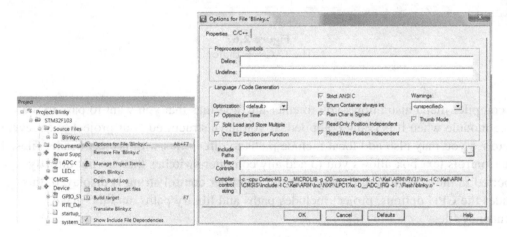

Figure 2.66
Local module options.

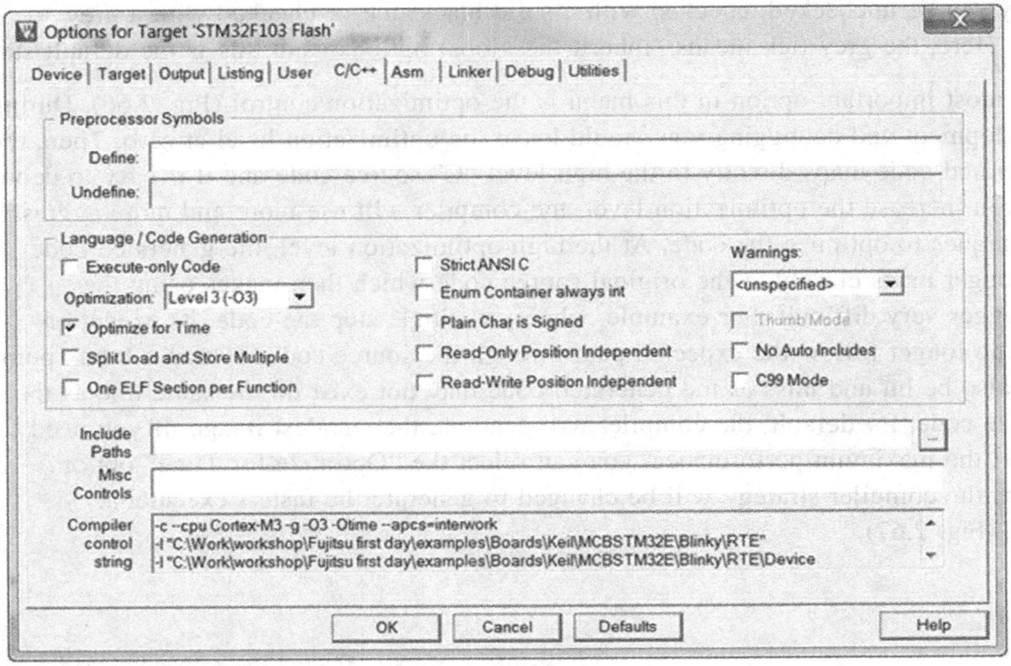

Figure 2.67
The C\C++ dialog.

The compiler menu also allows you to enter any #defines that you want to pass to the source module when it is being compiled. If you have structured your project over several directories you may also add local include paths to directories with project header files. The Misc Controls text box allows you to add any compiler switches that are not directly supported in the main menu. Finally, the full compiler control string is displayed, this includes the CPU options, project includes paths and library paths, and the make dependency files.

There is also an assembler Options window which includes many of the same options as the "C\C++" menu (Fig. 2.68). However, most Cortex-M project is written completely in "C \C++," so with luck you will never have to use this menu!.

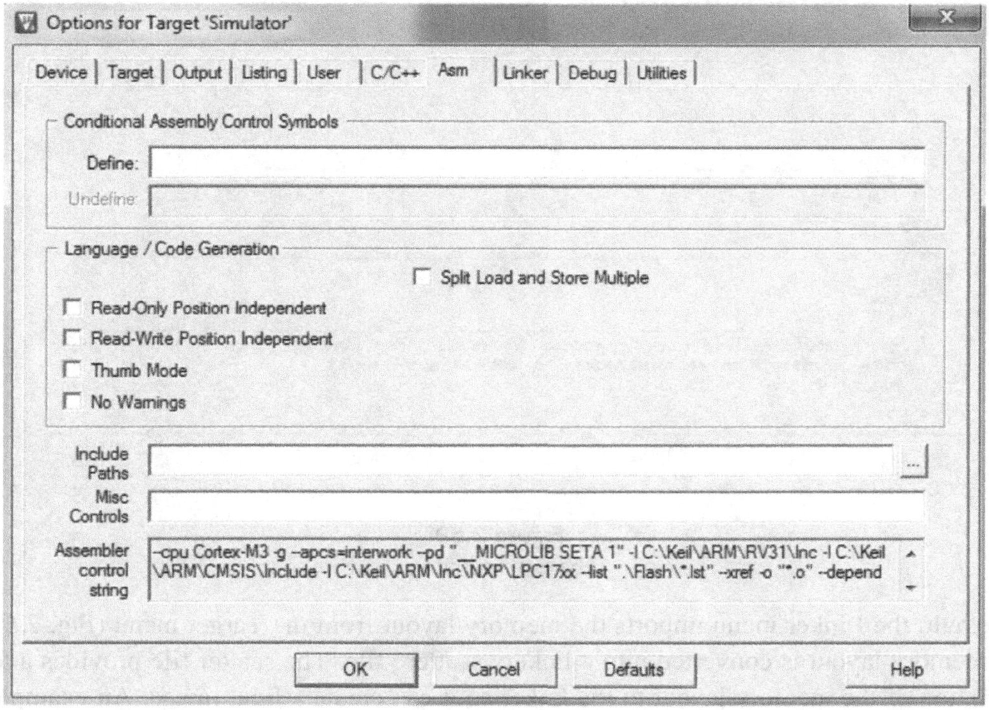

Figure 2.68
The Assembler dialog.

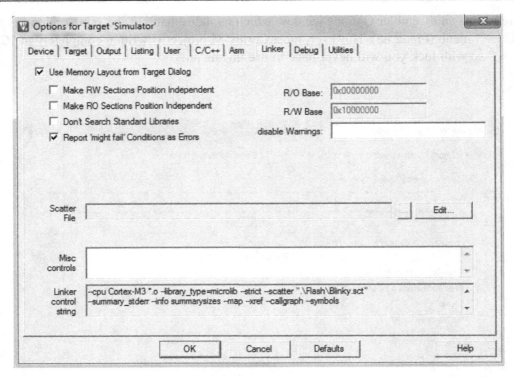

Figure 2.69
The Linker dialog.

By default, the Linker menu imports the memory layout from the Target menu (Fig. 2.69). This memory layout is converted into a linker "scatter" file. The scatter file provides a text description of the memory layout to the linker so it can create a final image. An example of a scatter file is shown below.

```
*****************************************************************
; *** Scatter-Loading Description File generated by uVision ***
; *****************************************************************
LR_IROM1 0x00000000 0x00040000 {   ; load region size_region
  ER_IROM1 0x00000000 0x00040000 { ; load address = execution address
   *.o (RESET, +First)
   *(InRoot$$Sections)
   .ANY (+RO)
   }
  RW_IRAM1 0x10000000 0x00004000 {  ; RW data
   .ANY (+RW + ZI)
   }
  RW_IRAM2 0x2007C000 0x00008000 {
   .ANY (+RW + ZI)
   }
 }
```

The scatter file defines the ROM and RAM regions and the program segments that need to be placed in each segment. The example scatter file first defines a ROM region of 256K. All of this memory is allocated in one bank. The scatter file also tells the linker to place the reset segment containing the vector table at the start of this section. Next, the scatter file tells the linker to place all the remaining nonvolatile segments in this region. The scatter file also defines two banks of SRAM of 16K and 32K. The linker is then allowed to use both pages of SRAM for initialized and uninitialized variables. This is a simple memory layout that maps directly onto the microcontrollers memory. If you need to use a more sophisticated memory layout, you can add extra memory regions in the Target menu and this will be reflected in the scatter file. If, however, you need a complex memory map which cannot be defined through the Target menu then you will need to write your own scatter file. The trick here is to get as close as you can with the Target menu and then hand edit the scatter file.

If you are using your own scatter file, you must then uncheck the "use Memory Layout from Target Dialog" and then manually select the new scatter file using the Scatter File text box (Fig. 2.70).

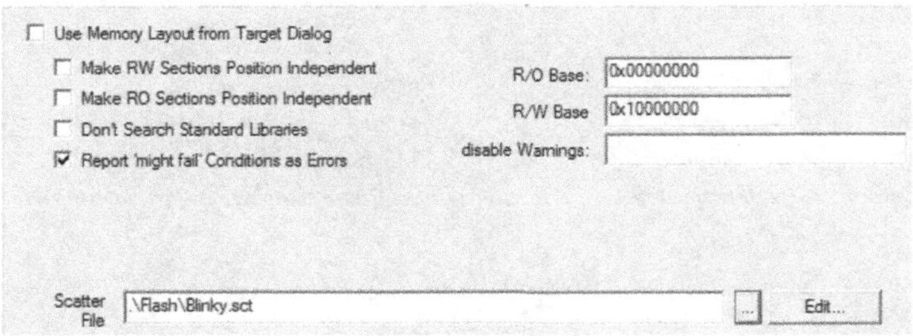

Figure 2.70
Using a custom scatter file.

Exercise 2.2 Hardware Debug

In this section, we will look at configuring the debugger to use a hardware debug interface rather than the internal simulator. In this exercise, we will use the STM32F7 Discovery board but the instructions below can be applied to any other hardware debug setup. If you are using a different board you must install the device family pack for the microcontroller used on the board. There will typically be a Blinky project in the device pack examples.

Open the Pack Installer.

Select the Boards tab and "The Designers Guide Tutorial Examples."

Select the Example tab and Copy "Ex 2.2 Hardware Debug."

The project shown here is for the STM32F7 Discovery board.

Connect the Discovery board to the PC using the USB mini connector.

Build the project.

The project is created and builds in exactly the same way as the simulator version.

Open the Options for Target dialog and the Debug tab (Fig. 2.71).

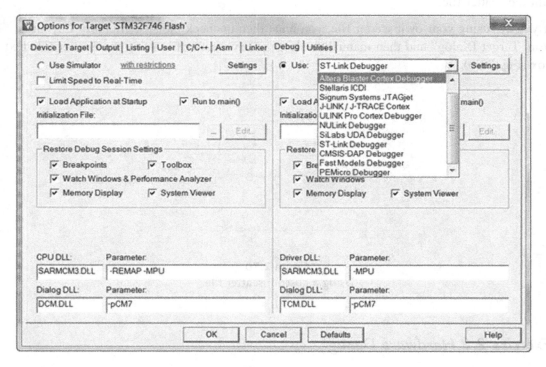

Figure 2.71
Selecting the hardware debug interface.

In the Debug menu, the "Use" option has been switched to select a hardware debugger rather than the simulator. We can also select the debug hardware interface. For the Discovery board, the ST-Link debugger is used.

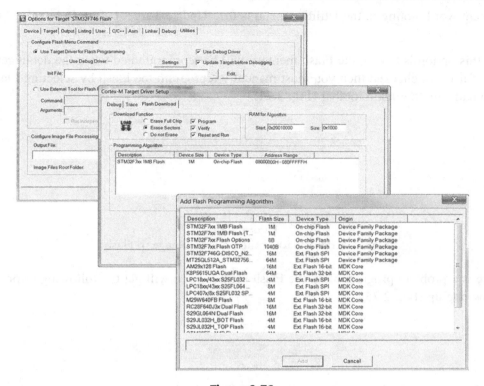

Figure 2.72
Setting the Flash programming algorithm from the Utilities Settings dialog.

Now open the utilities menu (Fig. 2.72).

When the project is created, the Utilities menu will have the "Use Debug Driver" box ticked. This will force the Flash programmer to use the same hardware interface selected in the Debug menu. However, the Utilities menu allows you to select a tool to program the microcontroller Flash memory. This can be another debug interface or an external tool such as a Silicon Vendor Bootloader tool.

The most common Flash programming problems are listed below (Fig. 2.73).

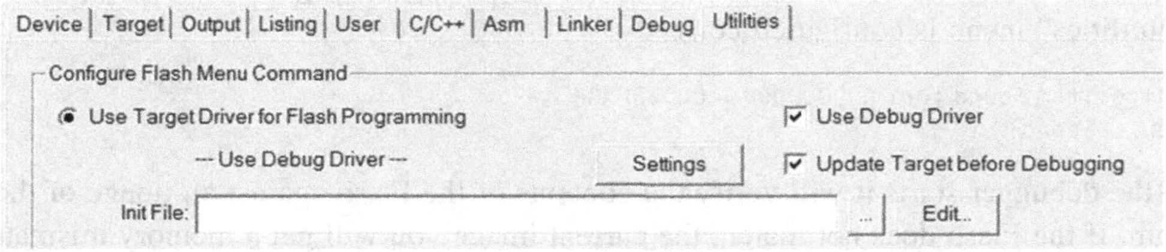

Figure 2.73
Update Target must be selected to automatically program the Flash when the debugger is started.

One point worth noting in the Utilities menu is the "Update Target before Debugging" tick box.

When this option is ticked, the Flash memory will be reprogrammed when the debugger starts. If it is not checked then you must manually reprogram the Flash by selecting Flash \Download from the main toolbar (Fig. 2.74).

Figure 2.74
Manually downloading the program image to Flash.

If there is a problem programming, the Flash memory you will get the following error window pop up (Fig. 2.75).

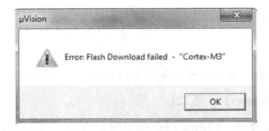

Figure 2.75
Flash programming error.

The build output window will report any further diagnostic messages. The most common error is a missing Flash algorithm. If you see the following message check the "options for target\utilities" menu is configured correctly.

```
No Algorithm found for: 00000000H — 000032A3H
Erase skipped!
```

When the debugger starts it will verify the contents of the Flash against an image of the program. If the Flash does not match, the current image you will get a memory mismatch error and the debugger will not start. This means that the Flash image is out of date and the current version needs to be downloaded into the Flash memory (Fig. 2.76).

Figure 2.76
Flash validation error.

Select cancel to close both of these dialogs without making any changes.

Start the debugger.

When the debugger starts, it is now connected to the hardware and will download the code into the Flash memory of the microcontroller and allow the debugger interface to control the real microcontroller in place of the simulation model.

Experiment with the debugger interface now that it is connected to the real hardware.

You will notice that some of the features available in the simulator are not present when using the hardware module. These are the Instruction Trace, Code Coverage, and Performance Analysis windows. These features are available with hardware debug but you need a more advanced hardware interface to get them. We will have a deeper look at the CoreSight debug system in Chapter 7 "Debugging with CoreSight."

Conclusion

By the end of this chapter, you should be able to setup a basic Cortex-M project, build the code and be able to debug it in the simulator or on a suitable hardware module. In the next chapter, we will start to look at the Cortex-M family of processors in more detail and then go on to look at some practical issues involved in developing software to run on them.

CHAPTER 3

Cortex-M Architecture

Introduction

In this chapter, we will have a closer look at the Cortex-M processor architecture. The bulk of this chapter will concentrate on the Cortex-M3 processor. Once we have a firm understanding of the Cortex-M3, we will look at the key differences in the Cortex-M0, Cortex-M0+ and Cortex-M4. There are some significant additions in the Cortex-M7 processor and we will look at these in Chapter 6 "Cortex-M7 processor." Through the chapter there are a number of exercises. These will give you a deeper understanding of each topic and can be used as a reference when developing your own code.

Cortex-M Instruction Set

As we have seen in Chapter 1 "Introduction to Cortex-M processor family," the Cortex-M processors are RISC-based processors and as such have a small instructions set. The Cortex-M0 has just 56 instructions, Cortex-M3 has 74, and Cortex-M4 has 137 with an option of additional 32 for the floating point unit. The ARM CPUs ARM7 and ARM9 which were originally used in microcontrollers have two instruction sets the ARM (32 bit) instruction set and the Thumb (16 bit) instruction set. The ARM instruction set was designed to get maximum performance from the CPU while the Thumb instruction set gave a good code density to allow programs to fit into the limited memory resources of a small microcontroller. The developer had to decide which function was compiled with the ARM instruction set and which was compiled with the Thumb instruction set. Then the two groups of functions could be "interworked" together to build the final program. The Cortex-M instruction set is based on the earlier 16-bit Thumb instruction set found in the ARM processors but extends the Thumb instruction to create a combined instruction set with a blend of 16- and 32-bit instructions (Fig. 3.1).

The Designer's Guide to the Cortex-M Processor Family.
DOI: http://dx.doi.org/10.1016/B978-0-08-100629-0.00003-7
© 2016 Elsevier Ltd. All rights reserved.

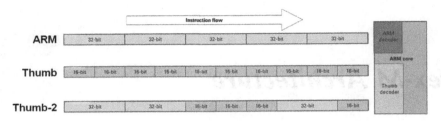

Figure 3.1
The ARM7 and ARM9 CPU had separate 32-bit and 16-bit instruction sets.
The Cortex-M processor has a single instruction set that is a blend of 16- and 32-bit
instructions.

The Cortex-M instruction set, called Thumb-2, is designed to be a good target for
compiler tools. The good news is that your whole Cortex-M project can be coded
in a high-level language such as "C/C++" without the need for any hand coded
assembler. It is useful to be able to "read" Thumb-2 assembly code via a debugger
Disassembly window to check what the compiler is up to, but you will rarely if ever
need to write an assembly routine. There are some useful Thumb-2 instructions that
are not reachable using the "C" language but most compiler toolchains provide
intrinsic instructions which can be used to access these instructions from within
your "C" code.

Programmer's Model and CPU Registers

The Cortex-M processors inherit the ARM RISC, load, and store method of operation.
This means that to do any kind of data processing, instruction such as ADD or SUBTRACT
the data must first be loaded into the CPU registers; the data processing instruction is
executed and the result is then stored back in the main memory. This means that
code executing on a Cortex-M processor revolves around the central CPU registers
(Fig. 3.2).

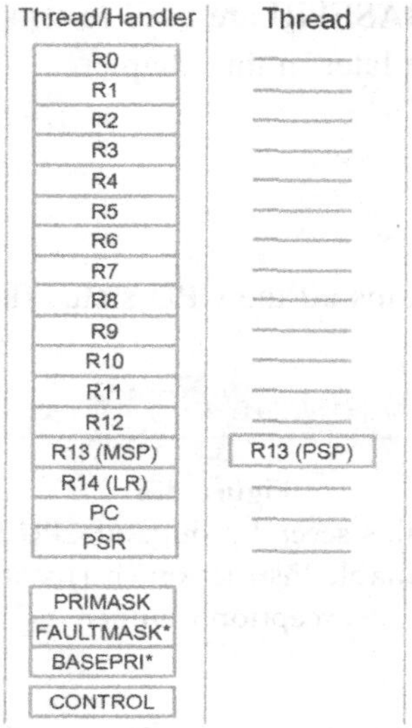

```
          Thread/Handler    Thread
             R0           ─────────
             R1           ─────────
             R2           ─────────
             R3           ─────────
             R4           ─────────
             R5           ─────────
             R6           ─────────
             R7           ─────────
             R8           ─────────
             R9           ─────────
             R10          ─────────
             R11          ─────────
             R12          ─────────
          R13 (MSP)     R13 (PSP)
          R14 (LR)       ─────────
             PC           ─────────
             PSR          ─────────

           PRIMASK
          FAULTMASK*
           BASEPRI*

          CONTROL
```

Figure 3.2
The Cortex-M CPU registers consist of 16 data registers, a program status register, and four special function registers. R13—R15 are used as the stack pointer, link register, and program counter. R13 is a banked register which allows the Cortex-M CPU to operate with dual stacks.

On all Cortex-M processors, the CPU register file consists of 16 data registers followed by the Program Status Register (PSR) and a group of configuration registers. All of the data registers (R0—R15) are 32 bits wide and can be accessed by all of the Thumb-2 load and store instructions. The remaining CPU registers may only be accessed by two dedicated instructions, Move Register to Special Register (MRS) and Move Special Registers to Register (MSR). The Registers R0—R12 are general user registers and are used by the compiler as it sees fit. The registers R13—R15 have special functions. R13 is used by the compiler as the stack pointer, this is actually a banked register with two R13 registers. When the Cortex-M processor comes out of reset, the second R13 register is not enabled and the processor runs in a "simple" mode with one stack pointer referred to as the Main Stack Pointer (MSP). It is possible to enable the second R13 register by writing to the Control register then the processor will be configured to run with two stacks. We will look at this in more detail in Chapter 5 "Advanced Architecture Features" but for now we will use the Cortex-M processor in its default mode. After the stack pointer, we have R14 the link register. When a procedure is called the return address is automatically stored in R14, the link register. Since the Thumb-2 instruction set does not contain a RETURN instruction when the processor reaches the end of a procedure it uses the branch instruction on R14 to return. Finally, R15 is the program counter, you can operate on this register just like all the others but you will not need to do this during normal program execution. The CPU registers

PRIMASK, FAULTMASK, and BASEPRI are used to temporarily disable interrupt handling and we will look at these later in this chapter.

Program Status Register

The PSR as its name implies contains all the CPU status flags (Fig. 3.3).

Figure 3.3
The Program Status Register contains several groups of CPU flags. These include the condition codes (NZCVQ), Interrupt Continuable Instruction (ICI) status bits, If Then flag, and current exception number.

The PSR has a number of alias fields that are masked versions of the full register. The three alias registers are the Application PSR, Interrupt PSR, and the Execution PSR. Each of these alias registers contains a subset of the full register flags and can be used as a shortcut if you need to access part of the PSR. The PSR is generally referred to as the xPSR to indicate the full register rather than any of the alias subsets (Fig. 3.4).

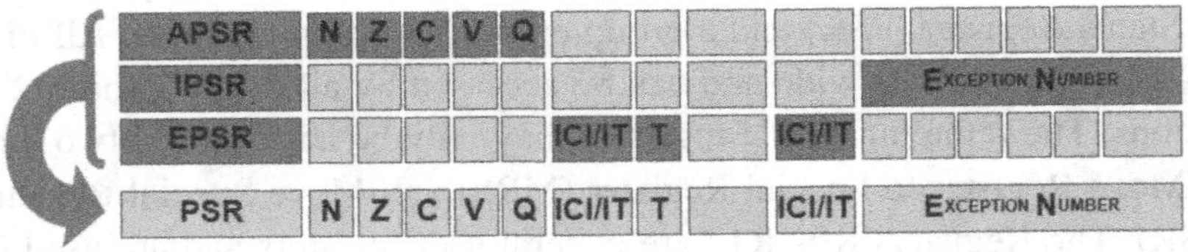

Figure 3.4
The Program Status Register has three alias registers which provide access to specific subregions or the Program Status Register. Hence, the generic name for the program status register is xPSR.

In a normal application program, your code will not make explicit access to the xPSR or any of its alias registers. Any use of the xPSR will be made by compiler generated code. As a programmer you need to have an awareness of the xPSR and the flags contained in it.

The most significant four bits of the xPSR are the condition code bits, **N**egative, **Z**ero, **C**arry, and o**V**erflow. These will be set and cleared depending on the results of a data processing instruction. The result of Thumb-2 data processing instructions can set or clear these flags. However, updating these flags is optional.

```
SUB  R8 ,R6, #240 Perform a subtraction and do not update the condition code flags
SUBS R8, R6, #240 Perform a subtraction and update the condition code flags
```

This allows the compiler to perform an instruction that updates the condition code flags, then perform some additional instructions that do not modify the flags, and then perform a conditional branch on the state of the xPSR condition codes. Following the four condition code flags is a further instruction flag the Q bit.

Q Bit and Saturated Maths Instructions

The Q bit is the saturation flag, the Cortex-M3/M4 and Cortex-M7 processors have special set of instructions called the saturated maths instructions. If a normal variable reaches its maximum value and you increment it further, it will roll round to zero. Similarly, if a variable reaches its minimum value and is then decremented it will roll round to the maximum value (Fig. 3.5).

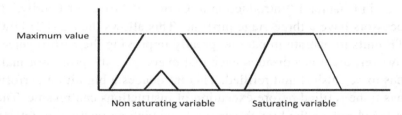

Figure 3.5
A normal variable will rollover to zero when it hits its maximum value. This is very dangerous in a control algorithm. The Cortex-M CPU supports saturated maths instructions which stick at their maximum and minimum values.

While this is a problem for most applications, it is especially serious for applications such as motor control and safety critical applications. The Cortex-M3/M4 and M7 saturated maths instructions prevent this kind of "roll round." When you use the saturated maths instructions, if the variable reaches its maximum or minimum value it will stick (saturate) at that value. Once the variable has saturated the Q bit will be set. The Q bit is a "sticky" bit and must be cleared by the application code. The standard maths instructions are not used by the "C" compiler by default. If you want to make use of the saturated maths instructions, you have to access them by using compiler intrinsics or CMSIS-Core functions.

```
uint32_t __SSAT(uint32_t value, uint32_t sat)
uint32_t __USAT(uint32_t value, uint32_t sat)
```

Interrupts and Multicycle Instructions

The next field in the PSR is the "Interrupt Continuable Instruction" (ICI) and "If Then" (IT) instruction flags. Most of the Cortex-M processor instructions are executed in a single cycle. However, some instructions such as Load Store Multiple, Multiply, and Divide take multiple cycles. If an interrupt occurs while these instructions are executing they have to be suspended while the interrupt is served. Once the interrupt has been served, we have to resume the multicycle instruction. The ICI field is managed by the Cortex-M processor so you do not need to do anything special in your application code. It does mean that when an exception is raised a multi cycle instruction can will be suspended, the ICI field will be updated and reaching the start of your interrupt routine will always take the same amount of cycles regardless of the instruction currently being executed by the CPU.

Conditional Execution—If Then Blocks

As we have seen in Chapter 1 "Introduction to Cortex-M Processor Family," most of the Cortex-M processors have a three-stage pipeline. This allows the FETCH DECODE and EXECUTE units to operate in parallel greatly improving the performance of the processor. However, there is a disadvantage that every time the processor makes a jump, the pipeline has to be flushed and refilled. This introduces a big hit on performance as the pipeline has to be refilled before execution of instructions can resume. The Cortex-M3 and -M4 reduce the branch penalty by having an instruction fetch unit that can carry out speculative branch target fetches from the possible branch address so the execution of the branch targets can start earlier. However, for small conditional branches the Cortex-M processor has another trick up its sleeve. For a small conditional branch, for example

```
If(Y==0x12C){
  I++;
}else{
I-;}
```

which compiles to less than 4 instructions, the Cortex-M processor can compile the code as an IF THEN condition block. The instructions inside the IF THEN block are extended with a condition code Table 3.1. This condition code is compared to the state of the Condition Code Flags in the PSR. If the condition matches the state of the flags, then the instruction will be executed if it does not and the instruction will still enter the pipeline but will be executed as a "No Operation" (NOP) instruction. This technique eliminates the branch and hence avoides the need to flush and refill the pipeline. Even though we are inserting NOP instructions, we still get better performance levels than by using a more standard compare and branch approach.

Table 3.1: Instruction condition codes

Condition Code	xPSR Flags Tested	Meaning
EQ	Z = 1	Equal
NE	Z = 0	Not equal
CS or HS	C = 1	Higher or same (Unsigned)
CC or LO	C = 0	Lower (Unsigned)
MI	N = 1	Negative
PL	N = 0	Positive or zero
VS	V = 1	Overflow
VC	V = 0	No overflow
HI	C = 1 and Z = 0	Higher (Unsigned)
LS	C = 0 or Z = 1	Lower or same (Unsigned)
GE	N = V	Greater than or equal (Signed)
LT	N! = V	Less than (Signed)
GT	Z = 0 and N = V	Greater than (Signed)
LE	Z = 1 and N! = V	Less than or equal (Signed)
AL	None	Always execute

To trigger an IF THEN block, we use the data processing instructions to update the PSR condition codes. By default, most instructions do not update the condition codes unless they have an S suffix added to the assembler opcode. This give the compiler a great deal of flexibility in applying the IF THEN condition.

```
ADDS R1,R2,R3      //perform and add and set the xPSR flags
ADD R2,R4,R5       //Do some other instructions but do not modify the xPSR
ADD R5,R6,R7
IT VS              //IF THEN block conditional on the first ADD instruction
SUBVS R3,R2,R4
```

So, our "C" IF THEN ELSE statement can be compiled to four instructions.

```
CMP     r6,#0x12C
  ITE     EQ
STREQ   r4,[r0,#0x08]
  STRNE   r5,[r0,#0x04]
```

The CMP compare instruction is used to perform the test and will set or clear the Zero Z flag in the PSR. The IF THEN block is created by the IF THEN (IT) instruction. The IT instruction is always followed by one conditionally executable instruction and optionally up to four conditionally executable instructions. The format of the IT instruction is as follows:

```
IT x y z cond
```

The x, y, and z parameters enable the second, third, and fourth instructions to be part of the conditional block. There can be further THEN or ELSE instructions. The cond parameter is the condition applied to the first instruction. So,

```
ITTTE NE    A four instruction If Then block with three THEN instructions which execute
            when the Z = 1 followed by an ELSE instruction which executes when Z=1
ITE GE      A two instruction IF THEN block with one THEN instruction which executes when
            N = V and one ELSE instruction which executes when N != V
```

The use of conditional executable IF THEN blocks is left up to the compiler. Generally, at low levels of optimization IF THEN blocks are not used, this gives a good debug view. However, at high levels of optimization, the compiler will make use of IF THEN blocks. So, normally there will not be any strange side effects introduced by the conditional execution technique but there are a few rules to bear in mind.

First, conditional code blocks cannot be nested, generally the compiler will take care of this rule. Secondly, you cannot use a GOTO statement to jump into a conditional code block. If you do make this mistake, the compiler will warn you and not generate such illegal code.

Thirdly, the only time you will really notice execution of an IF THEN condition block is during debugging. If you single step the debugger through the conditional statement, the conditional code will appear to execute even if the condition is false. If you are not aware of the Cortex-M condition code blocks, this can be a great cause of confusion!

The next bit in the PSR is the T or Thumb bit. This is a legacy bit from the earlier ARM CPU's and is set to one, if you clear this bit, it will cause a fault exception. In previous CPUs, the T bit was used to indicate that the Thumb 16-bit instruction set was running. It is included in the Cortex-M PSR to maintain compatibility with earlier ARM CPUs and allow legacy 16-bit Thumb code to be executed on the Cortex-M processors. The Final field in the PSR is the Exception number field. The Cortex Nested Vector Interrupt Unit can support up to 256 exception sources. When an exception is being processed, the exception number is stored here. As we will see later this field is not used by the application code when handling an interrupt, though it can be a useful reference when debugging.

Exercise 3.1 Saturated Maths and Conditional Execution

In this exercise, we will use a simple program to examine the CPU registers and make use of the saturated maths instructions. We will also rebuild the project to use conditional execution.

Open the Pack Installer.

Select the Boards tab and "The Designers Guide Tutorial Examples."

Select the Example tab and Copy "Ex 3.1 Saturation and Conditional Execution."

```
int a,range=300;  char c;
int main (void){
while (1){
for(a = 0;a<range;a++){
    c = a;
}}
```

This program increments an integer variable "a" from 0 to 300 and copies it to a char variable "c."

Build the program and start the debugger.

Add the two variables to the Logic Analyzer and run the program (Fig. 3.6).

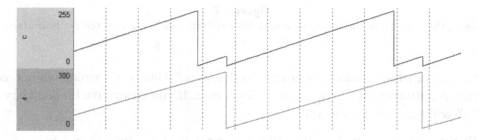

Figure 3.6
As we copy an integer into a char, the char value will rollover once its maximum value is reached.

In the Logic Analyzer window, we can see that while the integer variable performs as expected the char saturates when it reaches 255 and "rolls over" to zero and begins incrementing again.

Stop the debugger and modify the code as shown below to use saturated maths.

```
#define Q_FLAG 0x08000000
int a,range = 300;
char c;
unsigned int APSR;
int main (void){
while (1){
for(a=0;a<range;a++){
    c = __SSAT (a, 9);
    }
}}
```

This code replaces the equate with a saturated intrinsic function that saturates on the ninth bit of the integer value. This allows values 0–255 to be written to the byte value, any other values will saturate at the maximum allowable value of 255.

Build the project and start the debugger.

Run the code and view the contents of the variables in the Logic Analyzer (Fig. 3.7).

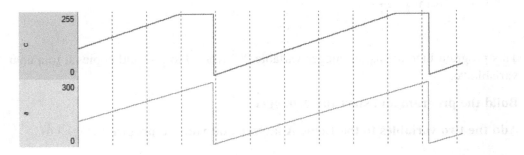

Figure 3.7
Using the saturation instructions, a variable cal "saturate" at a selected bit boundary.

Now the char variable saturates rather than "rollingover." This is still wrong but not as potentially catastrophically wrong as the rollover case. If you change the bit boundary value, "c" will saturate at lower values.

In the Registers window, click on the xPSR register to view the flags (Fig. 3.8).

⊟ xPSR	0x69000000
N	0
Z	1
C	1
V	0
Q	1
T	1
IT	Disabled
ISR	0

Figure 3.8
When a variable saturates, the Q bit in the xPSR will be set.

In addition to the normal NVCZ condition code flags the saturation Q bit is set.

Stop the debugger and modify the code as shown below.

```
#define Q_FLAG 0x08000000
int a,range = 300;  char c;  unsigned int APSR;
register unsigned int apsr __asm("apsr");
int main (void){
while (1){
for(a=0;a<range;a++){
    c = __SSAT(a, 9);
    }
APSR = __get_APSR ();
if(APSR&Q_FLAG){
    range--;
}
apsr = apsr&~Q_FLAG;
}}
```

Once we have written to the char variable, it is possible to read the APSR (application alias of the xPSR) and check if the Q bit is set. If the variable has saturated, we can take some corrective action and then clear the Q bit for the next iteration.

Build the project and start the debugger.

Run the code and observe the variables in the Watch window.

Now when the data is over range, the char variable will saturate and gradually the code will adjust the range variable until the output data fits into char variable.

Set breakpoints on lines 19 and 22 to enclose the Q bit test (Fig. 3.9).

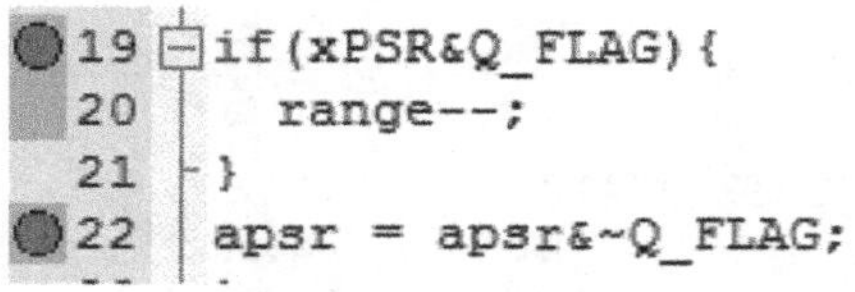

```
19  if(xPSR&Q_FLAG){
20      range--;
21  }
22  apsr = apsr&~Q_FLAG;
```

Figure 3.9
Breakpoints on lines 19 and 22.

Reset the program and then run the code until the first breakpoint is reached.

Open the Disassembly window and examine the code generated for the Q bit test.

```
    if(xPSR&Q_FLAG){
MOV    r0,r1
LDR    r0,[r0,#0x00]
TST    r0,#0x8000000
BEQ    0x080003E0
  20:     range--;
```

```
}else{
  LDR    r0,[pc,#44]  ; @0x08000400
LDR    r0,[r0,#0x00]
SUB    r0,r0,#0x01
LDR    r1,[pc,#36]  ; @0x08000400
STR    r0,[r1,#0x00]
B      0x080003E8
  locked = 1;
}
MOV    r0,#0x01
LDR    r1,[pc,#32]  ; @0x08000408
STR    r0,[r1,#0x00]
```

Also make a note of the value in the state counter, this is the number of cycles used since reset to reach this point (Fig. 3.10).

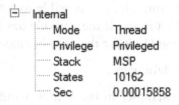

Figure 3.10
Value of the states counter at the first breakpoint.

Now run the code until it hits the next breakpoint and again make a note of the state counter (Fig. 3.11).

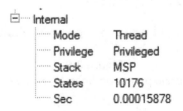

Figure 3.11
Value of the states counter at the second breakpoint.

Stop the debugger and open the "Options for Target/C" tab (Fig. 3.12).

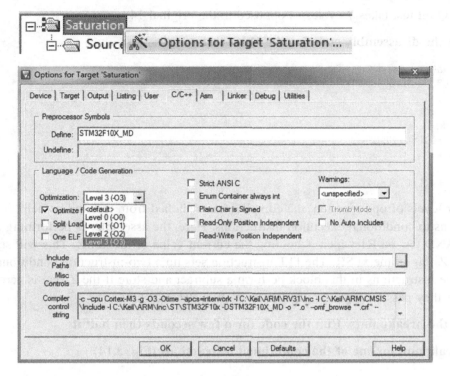

Figure 3.12
Change the optimization level and also "select optimize for speed."

Change the optimization level from Level 0 to Level 3 and check the "Optimize for speed" box.

Close the Options for Target and rebuild the project.

Now repeat the cycle count measurement by running to the two breakpoints (Fig. 3.13).

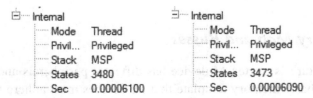

Figure 3.13
Cycle counts for first and second breakpoint.

Now the Q bit test takes 7 cycles as opposed to the original 14.

Examine the disassembly code for the Q bit test.

```
if(xPSR&Q_FLAG){
0x08000336 F0116F00 TST    r1,#0x8000000
range--;
}else{
ITTE   NE
STRNE  r1,[r0,#0x08]
locked = 1;
}
STREQ  r4,[r0,#0x04]
```

At higher levels of optimization, the compiler has switched from test and branch instructions to conditional execution instructions. Here, the assembler is performing a bitwise AND test on R1 which is holding the current value of the xPSR. This will set or clear the Z flag in the xPSR. The ITT instruction sets up a two-instruction conditional block. The instructions in this block perform a subtract and store if the Z flag is zero; otherwise they pass through the pipeline as NOP instructions.

Remove the breakpoints. Run the code for a few seconds then halt it.

Set a breakpoint on one of the conditional instructions (Fig. 3.14).

```
0x0800033A BF1A        ITTE        NE
0x0800033C 1E69        SUBNE       r1,r5,#1
0x0800033E 6081        STRNE       r1,[r0,#0x08]
```

Figure 3.14
IF THEN instruction followed by conditional instructions.

Start the code running again.

The code will hit the breakpoint even though the breakpoint is within an IF statement that should no longer be executed. This is simply because the conditional instructions are always executed.

Cortex-M Memory Map and Busses

While each manufacturer's Cortex-M device has different peripherals and memory sizes, ARM has defined a basic memory template that all devices must adhere to. This provides a standard layout so all the vendor provided memory and peripherals are located in the same blocks of memory (Fig. 3.15).

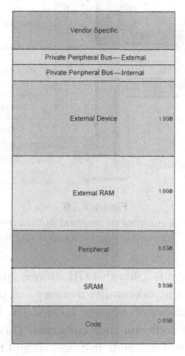

Figure 3.15
The Cortex-M memory map has a standard template which splits the 4 GB address range into specific memory regions. This memory template is common to all Cortex-M devices.

The Cortex-M memory template defines eight regions which cover the 4 GB address space of the Cortex-M processor. The first three regions are each 0.5 GB in size and are dedicated to the executable code space, internal SRAM, and internal peripherals. The next two regions are dedicated to external memory and memory mapped devices, both regions are 1 GB in size. The final three regions make up the Cortex-M processor memory space and contain the configuration registers for the Cortex-M processor and any vendor-specific registers (Fig. 3.16).

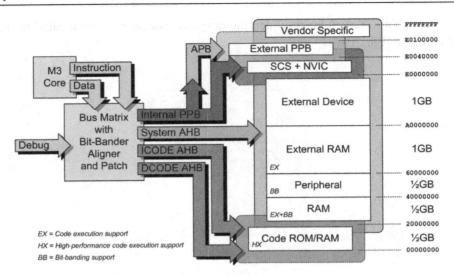

Figure 3.16
While the Cortex-M processor has a number of internal busses these are essentially invisible to the software developer. The memory appears as a flat 4 GB address space.

While the Cortex-M memory map is a linear 4 GB address space with no paged regions or complex addressing modes, the microcontroller memory and peripherals are connected to the Cortex-M processor by a number of different busses. The first 0.5 GB of the address space is reserved for executable code and code constants. This region has two dedicated busses, the ICODE bus is used to fetch code instructions and the DCODE bus is used to fetch code constants. The remaining user memory spaces (internal SRAM and peripherals plus the external SRAM and peripherals) are accessed by a separate system bus. The Cortex-M processor memory space has an additional private peripheral bus. While this may look complicated as far as your application code is concerned you have one seamless memory space. The Cortex-M processor has separate internal busses to optimize its access to different memory regions.

As mentioned earlier, most of the instructions in the Thumb-2 instruction set are executed in a single cycle and the Cortex-M3 can run up to 200 MHz and in fact some system on chip (SoC) designs manage to get the processor running even faster. However, current Flash memory used to store the program has an access time of around 50 MHz. So there is a basic problem of pulling instructions out of the FLASH memory fast enough to feed the Cortex-M processor. When you are selecting a Cortex-M microcontroller it is important to study the data sheet to see how the Silicon Vendor has solved this problem. Typically, the FLASH memory will be arranged as 64- or 128-bit wide memory, so one read from the FLASH memory can load multiple instructions. These instructions are then held in a "memory accelerator" unit which then feeds the instructions to the Cortex-M processor as required. The memory accelerator is a form of simple cache unit which is designed by the Silicon Vendor. Normally, this unit is disabled after reset, so you will need to enable it or the Cortex-M processor will be running directly from the FLASH memory. The overall performance of the Cortex-M processor will depend on how successfully this unit has been implemented by the designers of the microcontroller.

Write Buffer

The Cortex-M3 and Cortex-M4 contain a single entry data write buffer. This allows the CPU to make an entry into the write buffer and continue onto the next instruction while the write buffer completes the write to the real SRAM. This avoids stalling the processor while it waits for the write to complete. If the write buffer is full, the CPU is forced to wait until it has finished its current write. While this is normally a transparent process to the application code, there are some cases where it is necessary to wait until the write has finished before continuing program execution. For example, if we are enabling an external bus on the microcontroller it is necessary to wait until the write buffer has finished writing to the peripheral register and the bus is enabled before trying to access memory located on the external bus. The Cortex-M processor provides some memory barrier instructions to deal with these situations.

Memory Barrier Instructions

The memory barrier instructions halt execution of the application code until a memory write stage of an instruction has finished executing. They are used to ensure a critical section of code has completed before continuing execution of the application code (Table 3.2).

Table 3.2: Memory Barrier Instructions

Instruction	Description
DMB	Ensures all memory accesses are finished before a fresh memory access is made
DSB	Ensures all memory accesses are finished before the next instruction is executed
ISB	Ensures that all previous instructions are completed before the next instruction is executed. This also flushes the CPU pipeline

System Control Block

In addition to the CPU registers, the Cortex-M processors have a group of memory mapped configuration and status registers located near the top of the memory map starting at 0xE000 E008.

We will look at the key features supported by these registers in the rest of this book but a summary is given below (Table 3.3).

Table 3.3: The Cortex processor has memory mapped configuration and status registers located in the system control block

Register	Size in Words	Description
Auxiliary Control	1	Allows you to customize how some processor features are executed
CPU ID	1	Hardwired ID and revision numbers from ARM and the silicon manufacturer
Interrupt control and state	1	Provides Pend bits for the SysTick and NMI interrupts and extended interrupt Pending\active information
Vector table offset	1	Programmable address offset to move the Vector table to a new location in Flash or SRAM memory
Application interrupt and reset control	1	Allows you to configure the PRIGROUP and generate CPU and microcontroller resets
System control	1	Controls configuration of the processor sleep modes
Configuration and control	1	Configures CPU operating mode and some fault exceptions
System handler priority	3	These registers hold the 8-bit priority fields for the configurable processor exceptions
System handler control and state	1	Shows the cause of a bus, memory management, or usage fault
Configurable fault status	1	Shows the cause of a bus, memory management, or usage fault
Hard fault status	1	Shows what event caused a hard fault
Memory manager fault address	1	Holds the address of the memory location that generated the memory fault
Bus fault address	1	Holds the address of the memory location that generated the memory fault

The Cortex-M instruction set has addressing instructions that allow you to load and store 8-, 16-, and 32-bit quantities. Unlike the ARM7 and ARM9, the 16- and 32-bit quantities do not need to be aligned on a word or half-word boundaries (Fig. 3.17). This give the compiler and linker maximum flexibility to fully pack the SRAM memory. However, there is a penalty to be paid for this flexibility because unaligned transfers take longer to carry out. The Cortex-M instruction set contains load and store multiple instructions which can transfer multiple registers to and from memory in one instruction. This takes multiple processor cycles but uses only one 2-byte or 4-byte instruction. This allows for very efficient stack manipulation, block memory copy etc. The load and store multiple instructions only work for word aligned data. So, if you use unaligned data, the compiler is forced to use multiple individual load and store instructions to achieve the same thing. While you are making full use of the valuable internal SRAM, you are potentially increasing the application instruction size. Unaligned data is for user data only, you must ensure that the stacks are word aligned. The MSP initial value is determined by the linker, but the second stack pointer the Process Stack Pointer (PSP) is enabled and initialized by the user, so it is up to you to get it right. We will look at using the PSP in Chapter 5 "Advanced Architecture Features".

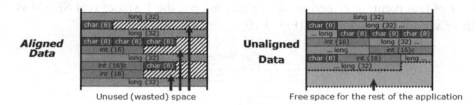

Figure 3.17
Unlike the earlier ARM7 and ARM9 CPUs, the Cortex processor can make unaligned memory accesses. This allows the compiler and linker to make best use of the device SRAM.

Bit Manipulation

In a small embedded system, it is often necessary to set and clear individual bits within the SRAM and peripheral registers. By using the standard addressing instructions, we can set and clear individual bits by using the "C" language bitwise AND and OR commands, while this works ok, the Cortex-M processors provide a more efficient bit manipulation method.

The Cortex-M processor provides a method called "bit banding" which allows individual SRAM and peripheral register bits to be set and cleared in a very efficient manner (Fig. 3.18).

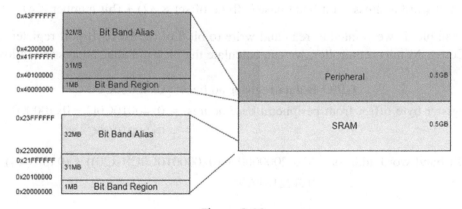

Figure 3.18
Bit banding is a technique that allows the first 1 MB of the SRAM region and the first 1 MB of the peripheral region to be bit addressed.

The first 1 MB or the SRAM region and the first 1 MB of the peripheral region are defined as bit band regions. This means that every memory location in these regions is bit addressable. So, in practice for today's microcontrollers, all of their internal SRAM and peripheral registers are bit addressable. Bit banding works by creating an alias word address for each bit of real memory or peripheral register bit. So, the 1 MB of real SRAM is aliased to 32 MB of virtual word addresses (Fig. 3.19).

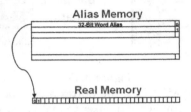

Figure 3.19
Each bit in the real memory is mapped to a word address in the alias memory.

This means that each bit of real SRAM or peripheral register is mapped to a word address in the bit band alias region and by writing 1s and 0s to the alias word address we can set and clear the real memory bit location. Similarly, if we write a word to the real memory location, we can read the bit band alias address to check the current state of a bit in that word. To use bit banding, you simply need to calculate the word address in the bit band region that maps to the bit location in the real memory that you want to modify. Then create a pointer to the word address in the bit band region. Once this is done, you can control the real bit memory location by reading and writing to the alias region via the pointer. The calculation for the word address in the bit band alias region is as follows:

Bit band word address = bit band base + (byte offset × 32) + (bit number × 4)

So, for example, is we wanted to read and write to bit 8 of the GPIO B port register on a typical Cortex-M microcontroller, we can calculate the bit band alias address as follows:

GPIO B data register address = 0x40010C0C
Register byte offset from peripheral base address = 0x40010C0C − 0x40000000
= 0x00010C0C

Bit band word address = (0x42000000 + (0x000010C0C*0x20)) + (0x8 *0x4)
= 0x422181A0

We can define a pointer to this address.

```
#define GPIO_PORTB_BIT8 = (*((volatile unsigned long *)0x422181A0))
```

Now by reading and writing to this word address, we can directly control the individual port bit.

```
GPIO_PORTB_BIT8 = 1  //set the port pin
```

This will compile to the following assembler instructions:

```
Opcode    Assembler
F04F0001  MOV    r0,#0x01
4927      LDR    r1,[pc,#156] ; @0x080002A8
6008      STR    r0,[r1,#0x00]
```

This sequence uses one 32-bit instruction and two 16-bit instructions or a total of 8 bytes. If we compare this to setting the port pin by using a logical OR to write directly to the port register.

```
GPIOB->ODR | = 0x00000100;   //LED on
```

We then get the following code sequence:

```
Opcode    Assembler
481E      LDR    r0,[pc,#120] ; @0x080002AC
6800      LDR    r0,[r0,#0x00]
F4407080  ORR    r0,r0,#0x100
491C      LDR    r1,[pc,#112] ; @0x080002AC
6008      STR    r0,[r1,#0x00]
```

This uses four 16-bit instructions and one 32-bit instruction or 12 bytes.

The use of bit banding gives us a win-win situation, smaller code size, and faster operation. So, as a simple rule, if you are going to repetitively access a single bit location you should use bit banding to generate the most efficient code.

You may find some compiler tools or Silicon Vendor software libraries provide macro functions to support bit banding. You should generally avoid using such macros as they may not yield the most efficient code and may be toolchain dependent.

Exercise 3.2 Bit Banding

In this exercise, we will look at defining a bit band variable to toggle a port pin and compare its use to bitwise AND and OR instructions.

Open the Pack Installer.

Select the Boards tab and "The Designers Guide Tutorial Examples."

Select the Example tab and copy "Ex 3.2 Bit Banding."

In this exercise, we want to toggle an individual port pin. We will use the Port B bit 8 pin as we have already done the calculation for the alias word address.

So, now in the "C" code we can define a pointer to the bit band address.

```
#define PortB_Bit8 (*((volatile unsigned long *)0x422181A0))
```

And in the application code, we can set and clear the port pin by writing to this pointer.

```
PortB_Bit8 = 1;
PortB_Bit8 = 0;
```

Build the project and start the debugger.

Enable the timing analysis with the debug\execution profiling\Show Time menu (Fig. 3.20).

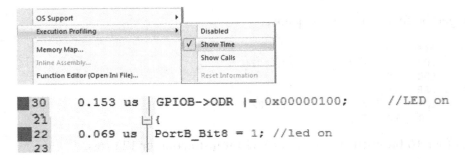

Figure 3.20
Enable the timing profile to show runtime per line of code.

This opens an additional column in the debugger which will display execution time for each line. Compare the execution time for the bit band instruction compared to the AND and OR instructions.

Open the Disassemble window and examine the code generated for each method of setting the port pin.

The bit banding instructions are the best way to set and clear individual bits. You should use them in any part of your code that repetitively manipulates a bit.

Dedicated Bit Manipulation Instructions

In addition to the bit band support, the Thumb-2 instruction set has some dedicated bit orientated instructions. Some of these instructions are not directly accessible from the "C" language and are supported by compiler "intrinsic" calls (Table 3.4).

**Table 3.4: In addition to bit banding, the Cortex-M3
processor has some dedicated bit manipulation
instructions**

BFC	Bit field clear
BFI	Bit field insert
SBFX	Signed bit field extract
SXTB	Sign extend a byte
SXTH	Sign extend a halfword
UBFX	Unsigned bit field extract
UXTB	Zero extend a byte
UXTH	Zero extend a halfword

SysTick Timer

All of the Cortex-M processors contain a standard timer. This is called the SysTick timer and is a 24-bit countdown timer with autoreload (Fig. 3.21). Once started, the SysTick timer will countdown from its initial value, once it reaches zero it will raise an interrupt and a new count value will be loaded from the reload register. The main purpose of this timer is to generate a periodic interrupt for a real-time operating system (RTOS) or other event driven software. If you are not running an RTOS, you can also use it as a simple timer peripheral.

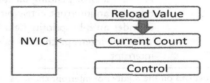

Figure 3.21
The SysTick timer is a 24-bit countdown timer with autoreload. It is generally used to provide a periodic interrupt for an RTOS scheduler.

The default clock source for the SysTick timer is the Cortex-M CPU clock. It may be possible to switch to another clock source but this will vary depending on the actual microcontroller you are using. While the SysTick timer is common to all the Cortex-M processors, its registers occupy the same memory locations within the Cortex-M3/-M4 and -M7. In the Cortex-M0 and Cortex-M0+, the SysTick registers are located in the system control block and have different symbolic names to avoid confusion. The SysTick timer interrupt line and all of the microcontroller peripheral lines are connected to the Nested Vector Interrupt Controller (NVIC).

Nested Vector Interrupt Controller

Aside from the Cortex-M CPU, the next major unit within the Cortex-M processor is the NVIC. The NVIC is the same to use between all Cortex-M processors, once you have set up an interrupt on a Cortex-M3, the process is the same for Cortex-M0 to Cortex-M7. The NVIC is designed for fast and efficient interrupt handling, on a Cortex-M3 you will reach the first line of "C" code in your interrupt routine after 12 cycles for zero wait state memory system. This interrupt latency is fully deterministic so from any point in the background (noninterrupt) code you will enter the interrupt with the same latency. As we have seen multicycle instructions can be halted with no overhead and then resumed once the interrupt has finished. On the Cortex-M3/M4 and Cortex-M7 the NVIC supports up to 240 interrupt sources, the Cortex-M0 can support up to 32. The NVIC supports up to 256 interrupt priority levels on Cortex-M3 and -M4 and four priority levels on Cortex-M0 (Table 3.5).

Table 3.5: The Nested Vector Interrupt Controller consists of seven register groups that allows you to enable, set priority levels, and monitor the user interrupt peripheral channels

Register	Maximum Size in Words[a]	Description
Set enable	8	Provides an interrupt enable bit for each interrupt source
Clear enable	8	Provides an interrupt disable bit for each interrupt source
Set pending	8	Provides a set pending bit for each interrupt source
Clear pending	8	Provides a clear pending bit for each interrupt source
Active	8	Provides an interrupt active bit for each interrupt source
Priority	60	Provides an 8-bit priority field for each interrupt source
Software trigger	1	Write the interrupt channel number to generate a software interrupt

[a]The actual number of words used will depend on the number of interrupt channels implemented by the microcontroller manufacturer

Operating Modes

While the Cortex-M CPU is executing background (noninterrupt code), the CPU is in an operating mode called thread mode. When an interrupt is raised, the NVIC will cause the processor to jump to the appropriate interrupt service routine (ISR). When this happens, the CPU changes to a new operating mode called handler mode. In simple applications without OS, you can use the default configuration of the Cortex-M processor out of reset and there is no major functional difference in these operating modes and they can be ignored. The Cortex-M processors can be configured with a more complex operating model that introduces operating differences between thread and handler mode and we will look at this in Chapter 5 "Advanced Architecture Features."

Interrupt Handling—Entry

When a microcontroller peripheral raises an interrupt line, the NVIC will cause two things to happen in parallel. First, the exception vector is fetched over the ICODE bus. This is the address of the entry point into the ISR. This address is pushed into R15, the program counter forcing the CPU to jump to the start of the interrupt routine. In parallel with this, the CPU will automatically push key registers onto the stack. This stack frame consists of the following registers, xPSR, PC, LR, R12, R3, R2, R1, and R0. This stack frame preserves the state of the processor and provides R0–R3 for use by the ISR. If the ISR needs to use more CPU registers, it must PUSH them onto the stack and POP them on exit (Fig. 3.22).

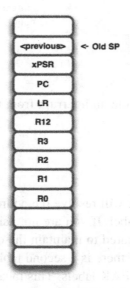

Figure 3.22
When an interrupt or exception occurs, the CPU will automatically push a stack frame. This consists of the xPSR, PC, LR, R12 and registers R0–R3. At the end of the interrupt or exception, the stack frame is automatically unstacked.

The interrupt entry process takes 12 cycles on the Cortex-M3/M4/M7 and 16 cycles on the Cortex-M0. All of these actions are handled by microcode in the CPU and do not require any dedicated entry instructions like LJMP or PUSH to be part of the application code.

The exception vectors are stored in an interrupt vector table. The interrupt vector table is located at the start of the address space. The first 4 bytes are used to hold the initial value of the main stack pointer. The starting value of the main stack pointer is calculated by the compiler and linker and will be stored in the first 4 bytes of the image. After reset, this value will be automatically loaded into R13 as part of the CPU reset. The interrupt vector table then has address locations every 4 bytes growing upwards through the address space. The vector table holds the address of the ISR for each of the possible interrupt sources within the microcontroller. The vector table for each microcontroller comes predefined as part of the startup code as we will see in the next chapter. A label for each ISR is stored at each interrupt vector location. To create your ISR you simply need to declare a void "C" function using the same name as the interrupt vector label.

```
            AREA   RESET, DATA, READONLY
            EXPORT __Vectors
__Vectors   DCD    __initial_sp           ; Top of Stack
            DCD    Reset_Handler          ; Reset Handler
          ; External Interrupts
            DCD    WWDG_IRQHandler        ; Window Watchdog
            DCD    PVD_IRQHandler         ; PVD through EXTI Line detect
            DCD    TAMPER_IRQHandler      ; Tamper
            DCD    RTC_IRQHandler         ; RTC
            DCD    FLASH_IRQHandler       ; Flash
            DCD    RCC_IRQHandler         ; RCC
```

So, to create the "C" routine to handle an interrupt from the real-time clock, we create a "C" function named as follows:

```
void RTC_IRQHandler(void) {
  ........
 }
```

When the project is built, the linker will resolve the address of the "C" routine and locate it in the vector table in place of the label. If you are not using this particular interrupt in your project, the label still has to be declared to maintain the order of the vector table. Following the interrupt vector table there is a second table which declares all of the ISR addresses. These are declared as WEAK labels. This means that this declaration can be overwritten if the label is declared elsewhere in the project. In this case, they act as a "backstop" to prevent any linker errors if the interrupt routine is not formally declared in the project source code.

```
EXPORT WWDG_IRQHandler      WEAK]
EXPORT PVD_IRQHandler       [WEAK]
EXPORT TAMPER_IRQHandler    [WEAK]
```

Interrupt Handling—Exit

Once the ISR has finished its task, it will force a return from the interrupt to the point in the background code from where it left off. However, the Thumb-2 instruction set does not have a return or return from interrupt instruction. The ISR will use the same return method as a noninterrupt routine namely a branch on the R14 the link register. During normal operation, the link register will contain the correct return address. However, when we entered the interrupt, the current contents of R14 was pushed onto the stack and in its place the CPU entered a special code. When the CPU tries to branch on this code instead of doing a normal branch, it is forced to restore the stack frame and resume normal processing (Table 3.6).

Table 3.6: At the start of an exception or interrupt R14 (Link Register) is pushed onto the stack. The CPU then places a control word in R14. At the end of the interrupt, the code will branch on R14. The control word is not a valid return address and will cause the CPU to retrieve a stack frame and return to the correct operating mode

Interrupt Return Value	Meaning
0xFFFFFFF9	Return to Thread mode and use the Main Stack Pointer
0xFFFFFFFD	Return to Thread mode and use the Process Stack Pointer
0xFFFFFFF1	Return to Handler mode

Interrupt Handling—Exit Important!

The interrupt lines that connect the user peripheral interrupt sources to the NVIC interrupt channels can be level sensitive or edge sensitive. In many microcontrollers the default is level sensitive, once an interrupt has been raised it will be asserted on the NVIC until it is cleared. This means that if you exit an ISR with the interrupt still asserted on the NVIC a new interrupt will be raised. To cancel the interrupt you must clear the interrupt status flags in the user peripheral before exiting the ISR. If the peripheral generates another interrupt while its interrupt line is asserted a further interrupt will not be raised. If you clear the interrupt status flags at the beginning of the interrupt routine then any further interrupts from the peripheral will be served. To further complicate things some peripherals will automatically clear some of their status flags. For example an ADC conversion complete flag may be automatically cleared when the ADC results register is read. Keep this in mind when you are reading the Microcontroller user manual.

Exercise 3.3 SysTick Interrupt

This project demonstrates setting up a first interrupt using the SysTick timer.

Open the Pack Installer.

Select the Boards tab and "The Designers Guide Tutorial Examples."

Select the Example tab and copy "Ex 3.2 SysTick Interrupt" (Fig. 3.23).

Figure 3.23
SysTick interrupt project layout.

This application consists of the minimum amount of code necessary to get the Cortex-M processor running and to generate a SysTick interrupt.

Open the main.c file

```
#include "stm32f10x.h"
#define SYSTICK_COUNT_ENABLE       1
#define SYSTICK_INTERRUPT_ENABLE   2
int main (void)
{
GPIOB->CRH = 0x33333333;
SysTick->VAL = 0x9000;
SysTick->LOAD  = 0x9000;
SysTick->CTRL = SYSTICK_INTERRUPT_ENABLE  | SYSTICK_COUNT_ENABLE;
while(1);
}
```

The main function configures a bank of port pins as outputs. Next, we load the SysTick timer and reload register and then enable the timer and its interrupt line to the NVIC. Once this is done the background code sits in a while() loop doing nothing.

When the timer counts down to zero, it will generate an interrupt which will run the SysTick ISR.

```
void SysTick_Handler (void)
{
static unsigned char count = 0,ledZero = 0x0F;
  if(count++>0x60)
  {
    ledZero = ledZero ^ 0xFF;
    LED_SetOut(ledZero);
    count = 0;
  }
}
```

The interrupt routine is then used to periodically toggle the GPIO lines.

Open the Device::STM32F10x.s file and locate the vector table.

```
SysTick_Handler PROC
EXPORT SysTick_Handler        [WEAK]
B     .
ENDP
```

The vector table provides standard labels for each interrupt source created as "weak" declarations. To create a "C" ISR, we simply need to use the label name as the name for a void function. The "C" function will then override the assembled stub and be called when the interrupt is raised.

Build the project and start the Debugger.

Without running the code, open the Register window and examine the state of the registers (Fig. 3.24).

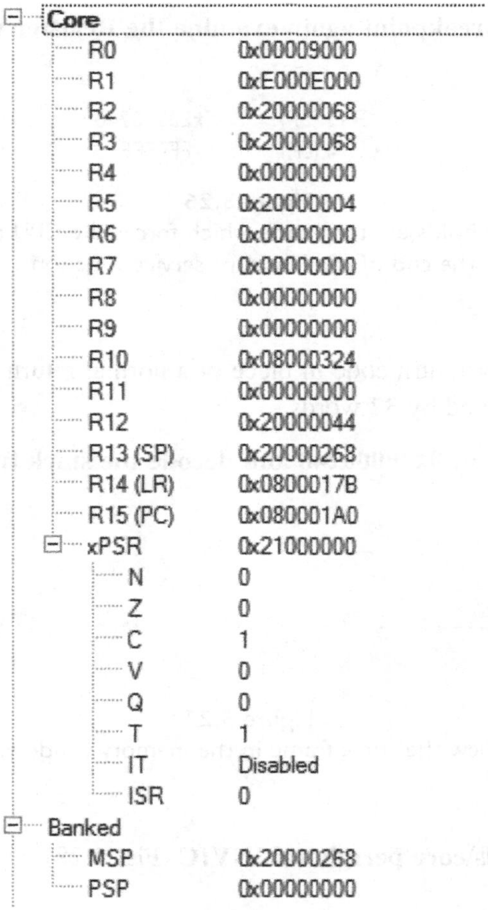

Core	
R0	0x00009000
R1	0xE000E000
R2	0x20000068
R3	0x20000068
R4	0x00000000
R5	0x20000004
R6	0x00000000
R7	0x00000000
R8	0x00000000
R9	0x00000000
R10	0x08000324
R11	0x00000000
R12	0x20000044
R13 (SP)	0x20000268
R14 (LR)	0x0800017B
R15 (PC)	0x080001A0
xPSR	0x21000000
N	0
Z	0
C	1
V	0
Q	0
T	1
IT	Disabled
ISR	0
Banked	
MSP	0x20000268
PSP	0x00000000

Figure 3.24
CPU register values at the start of the code.

In particular, note the value of the stack pointer (R13) the link register (R14) and the PSR.

Set a breakpoint in the interrupt routine and start the code running (Fig. 3.25).

```
26   void SysTick_Handler ( void)
27 ⊟ {
28     static unsigned char count = 0;
29       if(count++>0x60)
30 ⊟ {
31     GPIOB->ODR ^=0xFFFFFFFF;
32     count = 0;
33   }
34   }
```

Figure 3.25
Breakpoint set on entry to the interrupt.

When the code hits the breakpoint again examine the Register window (Fig. 3.26).

```
R13 (SP)    0x20000248
R14 (LR)    0xFFFFFFF9
```

Figure 3.26
The link register (R14) now holds a return code which forces the CPU to return from interrupt at the end of the interrupt service function.

Now, R14 has the interrupt return code in place of a normal return address and the stack pointer has been decremented by 32 words.

Open a Memory window at 0x20000248 and decode the stack frame (Fig. 3.27).

Address: 0x20000248
0x20000248: 00000003 E000E000 20000068 20000068 20000044 0800017B 080001A6 21000000 00000000

Figure 3.27
View the stack frame in the memory window.

Now open the peripherals\core peripherals\NVIC (Fig. 3.28).

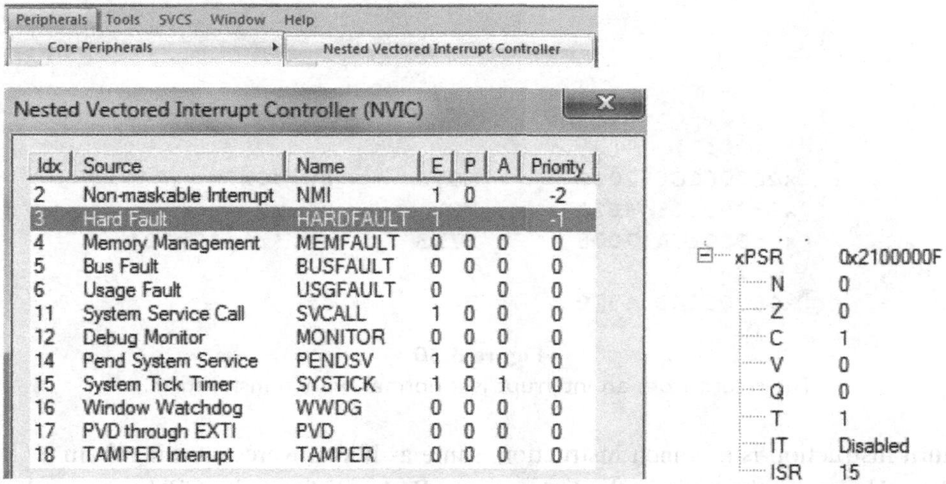

Figure 3.28
The NVIC peripheral window and the xPSR register view both show the SysTick timer as the active interrupt.

The NVIC peripheral window shows the state of each interrupt line. Line 15 is the SysTick timer and it is enabled (E) and active (A) (P = Pending). The idx column indicates the NVIC channel number and ties up with the ISR channel number in the PSR.

Now set a breakpoint on the closing brace of the interrupt function and run the code (Fig. 3.29).

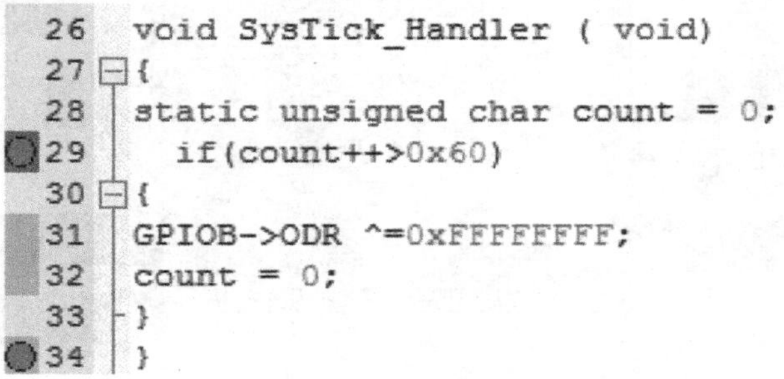

```
26    void SysTick_Handler ( void)
27    {
28    static unsigned char count = 0;
29       if(count++>0x60)
30    {
31    GPIOB->ODR ^=0xFFFFFFFF;
32    count = 0;
33    }
34    }
```

Figure 3.29
Breakpoints on the ISR entry and exit points.

Now open the disassembly window and view the return instruction (Fig. 3.30).

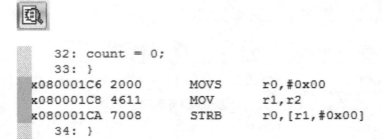

```
    32: count = 0;
    33: }
x080001C6 2000        MOVS      r0,#0x00
x080001C8 4611        MOV       r1,r2
x080001CA 7008        STRB      r0,[r1,#0x00]
    34: }
x080001CC 4770        BX        lr
```

Figure 3.30
The return from an interrupt is a normal branch instruction.

The return instruction is a branch instruction, same as if you were returning from a subroutine. However, the code in the link register R14 will force the CPU to unstack and return from the interrupt. Single step this instruction (F11) and observe the return to the background code and the stacked values return to the CPU registers.

Cortex-M Processor Exceptions

In addition to the peripheral interrupt lines, the Cortex-M processor has some internal exceptions and these occupy the first 15 locations of the vector table (Fig. 3.31).

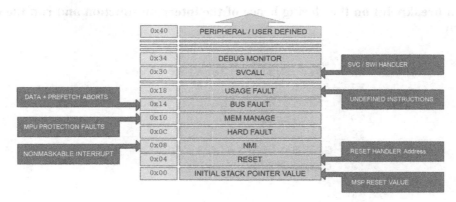

Figure 3.31
The first 4 bytes of memory hold the initial stack value. The vector table starts from 0x00000004.
The first 10 vectors are for the Cortex processor while the remainder are for user peripherals.

The first location in the vector table is the Reset handler. When the Cortex-M processor is reset, the address stored here will be loaded into the Cortex-M program counter forcing a jump to the start of your application code. The next location in the vector table is for a nonmaskable interrupt. How this is implemented will depend on the specific microcontroller you are using. It may, for example, be connected to an external pin on the microcontroller or to a peripheral such as a watchdog within the microcontroller. The next four exceptions are for handling faults that may occur during execution of the application code. All of these exceptions are present on the Cortex-M3/ M4 and Cortex-M7 but only the hard fault handler is implemented on the Cortex-M0. The type of faults that can be detected by the processor are usage fault, bus fault, memory manager fault and hard fault.

Usage Fault

A usage fault occurs when the application code has incorrectly used the Cortex-M processor. The typical cause is when the processor has been given an invalid opcode to execute. Most ARM compilers can generate code for a range of ARM processor cores. So, it is possible to incorrectly configure the compiler and to produce code that will not run on a Cortex-M processor. Other causes of a usage fault are shown below (Table 3.7).

Table 3.7: Possible causes of the usage fault exception

Undefined instruction
Invalid interrupt return address
Unaligned memory access using load and store multiple instructions
Divide by zero[a]
Unaligned memory access[a]

[a]This feature must be enabled in the system control block configurable fault usage register.

Bus Fault

A bus fault is raised when an error is detected on the AHB bus matrix (more about the bus matrix in Chapter 5 "Advanced Architecture Features"). The potential reasons for this fault are as follows (Table 3.8).

Table 3.8: Possible causes of the bus fault exception

Invalid memory region
Wrong size of transfer, that is, byte write to a word only peripheral register
Wrong processor privilege level (We will look at privilege levels in Chapter 5 "Advanced Architecture Features")

Memory Manager Fault

The memory protection unit (MPU) is an optional Cortex-M processor peripheral that can be added when the microcontroller is designed. It is available on all variants except the Cortex-M0. The MPU is used to control access to different regions of the Cortex-M address space depending on the operating mode of the processor. This will be looked at in more detail in Chapter 5 "Advanced Architecture Features." The MPU will raise an exception in the following cases (Table 3.9).

Table 3.9: Possible causes of the memory manager fault exception

Accessing an MPU region with the wrong privilege level
Writing to a read only region
Accessing a memory location outside of the defined MPU regions
Program execution from memory region that is defined as nonexecutable

Hard Fault

A hard fault can be raised in two ways. First, if a bus error occurs when the vector table is being read. Secondly, the hard fault exception is also reached through fault escalation. This means that if the usage, memory manager or bus fault exceptions are disabled, or if the exception service does not have a sufficient priority level, then the fault will escalate to a hard fault.

Enabling Fault Exceptions

The hard fault handler is always enabled and can only be disabled by setting the CPU FAULTMASK register. The other fault exceptions must be enabled in the System Control Block, "System Handler Control and State" register (SCB->SHCSR). The SCB->SHCSR register also contains Pend and Active bits for each fault exception.

We will look at the fault exceptions and tracking faults in Chapter 5 "Advanced Architecture Features."

Priority and Preemption

The NVIC contains a group of priority registers with an 8-bit field for each interrupt source. In its default configuration, the top 7 bits of the priority register allow you to define the preemption level. The lower the preemption level, the more important the interrupt. So, if an interrupt is being served and a second interrupt is raised with a lower preemption level, then the state of the current interrupt will be saved and the processor will serve the new interrupt. When it is finished, the processor will resume serving the original interrupt provided a higher priority interrupt is not pending (Fig. 3.32).

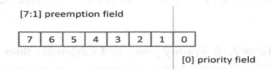

Figure 3.32
Each peripheral priority register consists of a configurable preemption field and a subpriority field.

The least significant bit (LSB) is the subpriority bit. If two interrupts are raised with the same preemption level, the interrupt with the lowest subpriority level will be served first. This means we have 128 preemption levels each with two subpriority levels (Fig. 3.33).

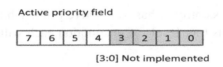

Figure 3.33
Each priority register is 8-bits wide. However, the silicon manufacturer may not implement all of the priority bits. The implemented bits always extend from the MSB toward the LSB.

When the microcontroller is designed, the manufacturer can define the number of active bits in the priority register. For Cortex-M3/M4 and Cortex-M7, this can be a minimum of three and up to a maximum of eight. For Cortex-M0 and M0 + it is always 2 bits. Reducing the number of active priority bits reduces the NVIC gate count and hence its power consumption. If the manufacturer does not implement the full 8 bits of the priority register, the LSBs will be disabled, this makes it safer to port code between microcontrollers with different numbers of active priority bits. You will need to check the manufacturer's data sheet to see how many bits of the priority register are active.

Groups and Subgroup

After a processor reset, the first 7 bits of the priority registers define the preemption level and the LSB defines the subpriority level. This split between preemption group and priority subgroup can be modified by writing to the "NVIC priority group" field in the "Application Interrupt and Reset Control" register. This register allows us to change the size of the preemption group field and priority subgroup. On reset, this register defaults to priority group zero (Table 3.10).

Table 3.10: Priority group and subgroup values

Priority Group	Preempt Group Bits	Subpriority Group Bits
0	7−1	0
1	7−2	1−0
2	7−3	2−0
3	7−4	3−0
4	7−5	4−0
5	7−6	5−0
6	7	6−0
7	None	7−0

So, for example, if our microcontroller has four active priority bits we could select priority group 5, which would give us four levels of preemption each with four levels of subpriority (Fig. 3.34).

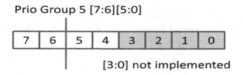

Figure 3.34

A priority register with four active bits and priority group 5. This yields four preempt levels and four priority levels.

The highest preemption level for a user exception is zero, however, some of the Cortex-M processor exceptions have negative priority levels so they will always preempt a user peripheral interrupt (Fig. 3.35).

	Exception	Name	Priority	Descriptions
Fault Mode and Startup Handlers	1	Reset	-3 (Highest)	Reset
	2	NMI	-2	Nonmaskable Interrupt
	3	Hard Fault	-1	Default fault if other hander not implemented
	4	Memory manage fault	Programmable	MPU violation or access to illegal locations
	5	Bus Fault	Programmable	Fault if AHB interface receives error
	6	Usage Fault	Programmable	Exceptions due to program errors
System Handlers	11	SVCall	Programmable	System SerVice call
	12	Debug Monitor	Programmable	Break points, watch points, external debug
	14	PendSV	Programmable	Pendable SerVice request for System Device
	15	Systick	Programmable	System Tick Timer
Custom Handlers	16	Interrupt #0	Programmable	External Interrupt #0
	...	...	...	...
	...	...	...	...
	...	...	...	...
	255	Interrupt #239	Programmable	External Interrupt #239

Figure 3.35
Cortex-M processor exceptions and possible priority levels.

Runtime Priority Control

There are three CPU registers which may be used to dynamically disable interrupt sources within the NVIC. These are the PRIMASK, FAULTMASK, and BASEPRI registers (Table 3.11).

Table 3.11: The CPU PRIMASK, FAULTMASK, and BASEPRI registers are used to dynamically disable interrupts and exceptions

CPU Mask Register	Description
PRIMASK	Disables all exceptions except hard fault and NMI
FAULTMASK	Disables all exceptions except NMI
BASEPRI	Disables all exceptions at the selected Preemption level and lower preempt level

These registers are not memory mapped, they are CPU registers and may only be accessed with the MRS and MSR instructions. When programming in "C" they may be accessed by dedicated compiler intrinsic instructions, we will look at these intrinsics more closely in Chapter 4 "Cortex Microcontroller Software Interface Standard."

Exception Model

When the NVIC serves a single interrupt, there is a delay of 12 cycles until we reach the ISR and a further 10 cycles at the end of the ISR until the Cortex-M processor resumes execution of the background code. This gives us fast deterministic handling of interrupts in a system which may have only one or two active interrupt sources (Fig. 3.36).

12 Cycles

Figure 3.36
When an exception is raised, a stack frame is pushed in parallel with the ISR address being fetched from the vector table. On the Cortex-M3 and Cortex-M4 this is always 12 cycles. On the Cortex-M0 it takes 16 cycles. The Cortex-M0+ takes 15 cycles.

In more complex systems, there may be many active interrupt sources all demanding to be served as efficiently as possible. The NVIC has been designed with a number of optimizations to ensure fast interrupt handling in such a heavily loaded system. All of the interrupt handling optimizations described below are an integral part of the NVIC and as such are performed automatically by the NVIC and do not require any configuration by the application code.

NVIC Tail Chaining

In a very interrupt driven design, we can often find that while the CPU is serving a high-priority interrupt a lower priority interrupt is also pending. In the earlier ARM CPUs and many other processors, it was necessary to return from the interrupt by POPing the CPU context from the stack back into the CPU registers and then performing a fresh stack PUSH before running the pending ISR. This is quite wasteful in terms of CPU cycles as it performs two redundant stack operations (Fig. 3.37).

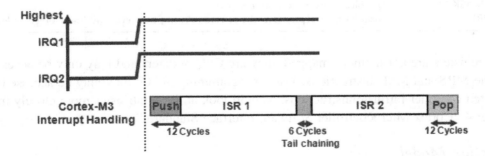

Figure 3.37
If an ISR is running and a lower priority interrupt is raised, it will automatically "tail chained" to run 6 cycles after the initial interrupt has terminated.

When this situation occurs on a Cortex-M processor, the NVIC uses a technique called tail chaining to eliminate the unnecessary stack operations. When the Cortex-M processor reaches the end of the active ISR and there is a pending interrupt then the NVIC simply forces the processor to vector to the pending ISR. This takes a fixed six cycles to fetch the start address of the pending interrupt routine and then execution of the next ISR can begin. Any further pending interrupts are dealt within the same way. When there are no further interrupts pending, the stack frame will be POPed back to the processor registers and the CPU will resume execution of the background code. As you can see from Fig. 3.37, tail chaining can significantly improve the latency between interrupt routines.

NVIC Late Arriving

Another situation that can occur is a "late arriving" high-priority interrupt. In this situation, a low-priority interrupt is raised followed almost immediately by a high-priority interrupt. Most microcontrollers will handle this by preempting the initial interrupt. This is undesirable because it will cause two stack frames to be pushed and delay the high-priority interrupt (Fig. 3.38).

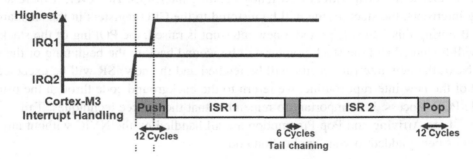

Figure 3.38
If the Cortex-M processor is entering and ISR and a higher priority interrupt is raised, the NVIC will automatically switch to serve the high-priority interrupt. This will only happen if the initial interrupt is in its first 12 cycles.

If this situation occurs on a Cortex-M processor and the high-priority interrupt arrives within the initial 12 cycle PUSH of the low-priority stack frame, then the NVIC will switch to serving the high-priority interrupt and the low-priority interrupt will be tail chained to execute once the high-priority interrupt is finished. For the "late arriving" switch to happen the high-priority interrupt must occur in the initial 12-cycle period of the low-priority interrupt. If it occurs any later than this, then it will preempt the low-priority interrupt which requires the normal stack PUSH and POP.

NVIC POP Preemption

The final optimization technique used by the NVIC is a called POP preemption (Fig. 3.39). This is kind of a reversal of the late arriving technique discussed above.

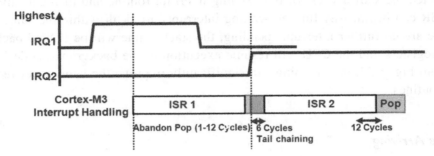

Figure 3.39

If an interrupt is raised while is in its exiting 12 cycles, the processor will "rewind" the stack and serve the new interrupt with a minimum delay of 6 cycles.

When a typical microcontroller reached the end of an ISR, it always has to restore the stack frame regardless of any pending interrupts. As we have seen above, the NVIC will use tail chaining to efficiently deal with any currently pending interrupts. However, if there are no pending interrupts, the stack frame will be restored to the CPU registers in the standard 10 cycles. If during this 12-cycle period a new interrupt is raised, the POPing of the stack frame will be halted and the stack pointer will be wound back to the beginning of the stack frame. Next, the new interrupt vector will be fetched and the new ISR will be executed. At the end of the new interrupt routine, we return to the background code through the usual 10-cycle POP process. It is important to remember that these three techniques, Tail Chaining, Late Arriving and Pop Preemption are all handled by the NVIC without any instructions being added to your application code.

Exercise 3.3 Working with Multiple Interrupts

This exercise extends our original SysTick exception exercise to enable a second ADC interrupt. We can use these two interrupts to examine the behavior of the NVIC when it has multiple interrupt sources.

Open the Pack Installer.

Select the Boards tab and "The Designers Guide Tutorial Examples."

Select the Example tab and Copy "Ex 3.3 Multiple Interrupts."

Open main.c and locate the main() function.

```
unsigned char BACKGROUND = 0;unsigned char ADC = 0;unsigned char SYSTICK = 0;
int main (void){
int i;
GPIOB->CRH = 0x33333333;          //Configure the Port B LED pins
SysTick->VAL  = 0x9000;           //Start value for the sys Tick counter
SysTick->LOAD = 0x9000;           //Reload value
SysTick->CTRL = SYSTICK_INTERRUPT_ENABLE | SYSTICK_COUNT_ENABLE;
init_ADC();                       //setup the ADC peripheral
ADC1->CR1  |= (1UL << 5);         // enable for EOC Interrupt
NVIC->ISER[0] = (1UL << 18);      // enable ADC Interrupt
ADC1->CR2  |= (1UL << 0);         // ADC enable
while(1){
  BACKGROUND = 1;
}}
```

We initialize the SysTick timer same as before. In addition, the ADC peripheral is also configured. To enable a peripheral interrupt, it is necessary to enable the interrupt source in the peripheral and also enable its interrupt channel in the NVIC by setting the correct bit in the NVIC ISER registers. In this example, we are writing directly to the NVIC- > ISER[0] register. In the next chapter, we will see a better more "standard" way to do this.

We have also added three variables: BACKGROUND, ADC, and SYSTICK. These will be set to logic one when the matching region of code is executing and zero at other times. This allows us to track execution of each region of code using the debugger Logic Analyzer.

```
void ADC_IRQHandler (void){
int i;
BACKGROUND   = 0;
SYSTICK      = 0;
for (i = 0;i<0x1000;i++){
  ADC  = 1;
}
ADC1->SR &= ~(1 << 1);          /* clear EOC interrupt          */
ADC = 0;
}
```

The ADC interrupt handler sets the execution region variables then sits in a delay loop. Before exiting it also writes to the ADC status register to clear the end of conversion flag. This deasserts the ADC interrupt request to the NVIC.

```
  void SysTick_Handler (void){
int i;
BACKGROUND = 0;
ADC = 0;
ADC1->CR2  |= (1UL << 22);
for (i = 0;i<0x1000;i++){
  SYSTICK = 1;
}
SYSTICK = 0;
}
```

The SysTick interrupt handler is similar to the ADC handler. It sets the region execution variables and sits in a delay loop before exiting. It also writes to the ADC control register to trigger a single ADC conversion.

Build the project and start the simulator.

Add each of the execution variables to the Logic Analyzer and start the code running (Fig. 3.40).

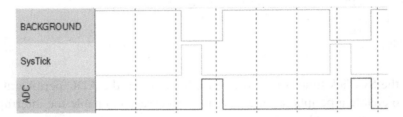

Figure 3.40
The SysTick interrupt is executed (logic high) then the ADC interrupt is tail chained and will run when the SysTick ends.

The SysTick interrupt is raised which starts the ADC conversion. The ADC finishes conversion and raises its interrupt before the SysTick interrupt completes so it enters a Pending state. When the SysTick interrupt completes, the ADC interrupt is tail chained and begins execution without returning to the background code.

Exit the debugger and comment out the line of code that clears the ADC end of conversion flag.

```
//ADC1->SR &= ~(1 << 1);
```

Build the code and restart the debugger and observe the execution of the interrupts in the Logic Analyzer window (Fig. 3.41).

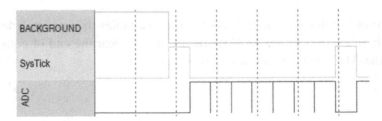

Figure 3.41
The ADC tail chains the SysTick and runs multiple times because the ADC status flag has not been cleared.

After the first ADC interrupt has been raised, the interrupt status flag has not been cleared and the ADC interrupt line to the NVIC stays asserted. This causes continuous ADC interrupts to be raised by the NVIC blocking the activity of the background code. The SysTick interrupt has the same priority as the ADC so it will be tail chained to run after the current ADC interrupt has finished. Neglecting to clear interrupt status flags is the most common mistake made when first starting to work with interrupts and the Cortex-M processors.

Exit the debugger and uncomment the end of conversion code.

```
ADC1->SR &= ~(1 ≪ 1);
```

Add the following lines to the background initializing code.

```
NVIC->IP[18] = (2 ≪ 4);
SCB->SHP[11] = (3 ≪ 4);
```

This programs the user peripheral NVIC interrupt priority registers to set the ADC priority level and the "System Handler Priority" registers to set the SysTick priority level. These are both byte arrays which cover the 8-bit priority field for each exception source. However, on this microcontroller the manufacturer has implemented four priority bits out of the possible eight. The priority bits are located in the upper nibble of each byte. On reset, the PRIGROUP is set to zero which creates a 7-bit preemption field and 1-bit priority field (Fig. 3.42).

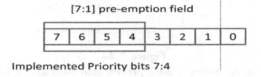

Figure 3.42
After reset a microcontroller with four implemented priority bits will have 16 levels of preemption.

On our device all of the available priority bits are located in the preemption field giving us 16 levels of priority preemption.

Build the code, restart the debugger, and observe the execution of the interrupts in the Logic Analyzer window (Fig. 3.43).

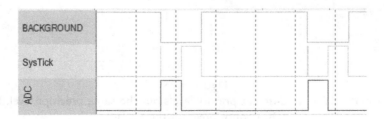

Figure 3.43
The ADC is at a higher priority than the SysTick so it pre-empts the SysTick interrupt.

The ADC now has the highest per emption value so as soon as its interrupt is raised it will preempt the SysTick interrupt. When it completes, the SysTick interrupt will resume and complete before returning to the background code.

Exit the debugger and uncomment the following lines in the background initialization code.

The AIRC register cannot be written to freely. It is protected by a key field which must be programmed with the value 0x5FA before a write is successful.

```
temp  =  SCB->AIRCR;
temp &= ~(SCB_AIRCR_VECTKEY_Msk | SCB_AIRCR_PRIGROUP_Msk);
temp  =  (temp|((uint32_t)0x5FA << 16) | (0x05 << 8));
SCB->AIRCR = temp;
```

This programs the PRIGROUP field in the AIRC register to a value of 5, which means a 2-bit preemption field and a 6-bit priority field. This maps onto the available 4-bit priority field giving four levels of preemption each with four levels of priority (Fig. 3.44).

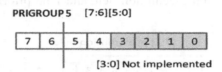

Figure 3.44
The PRI Group is set to define a 2-bit preemption field and a 2-bit priority field.

Build the code and restart the debugger and observe the execution of the interrupts in the Logic Analyzer window (Fig. 3.45).

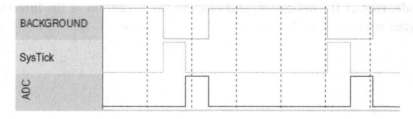

Figure 3.45
The SysTick and ADC have different priority levels but the same preempt level. Now, the ADC cannot preempt the SysTick.

The ADC interrupt is no longer preempting the SysTick timer despite them having different values in their priority registers. This is because they now have different values in the priority field but the same preempt value.

Exit the debugger and change the interrupt priorities as shown below.

```
NVIC->IP[18] = (2<<6 | 2<<4);
SCB->SHP[11] = (1<<6 | 3<<4);
```

Set the base priority register to block the ADC preempt group.

```
__set_BASEPRI (2<<6);
```

Build the code, restart the debugger, and observe the execution of the interrupts in the Logic Analyzer window (Fig. 3.46).

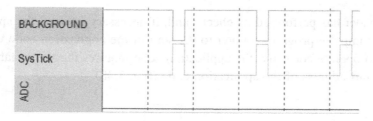

Figure 3.46
The setting the Base priority register disables the ADC interrupt.

Setting the BASEPRI register has disabled the ADC interrupt and any other interrupts that are on the same level preempt group or lower.

Bootloader Support

While the interrupt vector table is located at the start of memory when the Cortex-M processor is reset it is possible to relocate the vector table to a different location in memory. As software embedded in small microcontrollers becomes more sophisticated there is an increasing need to develop systems with a permanent bootloader program that can check the integrity of the main application code before it runs and check for a program update which can be delivered by various serial interfaces (eg, Ethernet, USB, UART) or an SD/multimedia card (Fig. 3.47).

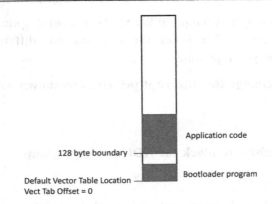

Figure 3.47
A bootloader program can be placed into the first sector in the Flash memory. It will check if there is an update to the application code before starting the main application program.

Once the bootloader has performed its checks and, if necessary updated the application program it will jump the program counter to the start of the application code which will start running. To operate correctly, the application code requires the vector table to be mapped to the start address of the application code (Fig. 3.48).

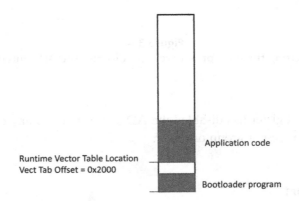

Figure 3.48
When the application code starts to run, it must relocate the vector table to the start of the application code by programming the NVIC vector Table Offset register.

The Vector table can be relocated by writing to a register in the system control block called the "Vector Table Offset" register. This register allows you to relocate the vector table to any 128-byte boundary in the Cortex processor memory map.

Exercise 3.4 Bootloader

This exercise demonstrates how a bootloader and application program can both be resident on the same Cortex-M microcontroller and how to debug such a system.

Open the Pack Installer.

Select the Boards tab and "The Designers Guide Tutorial Examples."

Select the Example tab and Copy "Ex 3.4 Bootloader."

This is a more advanced feature of the μVision IDE that allows you to view two or more projects at the same time (Fig. 3.49).

Figure 3.49
The bootloader and Blinky project in a multiproject workspace.

The workspace consists of two projects, the bootloader project which is built to run on the Cortex processor reset vector as normal, and the Blinky project which is our application. First, we need to build the Blinky project to run from an application address which not in the same Flash sector as the bootloader. In this example, the application address is chosen to be 0x2000.

Expand the Blinky project, right click on the workspace folder, and set it as the active project (Fig. 3.50).

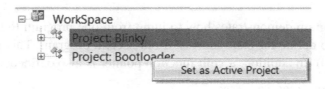

Figure 3.50
Select a project and right click to make it the active project.

Now click on the Blinky project folder and open the options for Target\Target tab (Fig. 3.51).

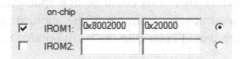

Figure 3.51
The Blinky project must have its code offset to the 0x2000 start address.

The normal start address for this chip is 0x8000000 and we have increased this to 0x8002000.

Open the system_stm32F10x.c file and locate line 128.

```
#define VECT_TAB_OFFSET 0x02000
```

This contains a #define for the Vector table offset register. Normally, this is zero but if we set this to 0x2000 the vector table will be re mapped to match our application code when it starts running.

Build the Blinky project.

Expand the bootloader project and set it as the active project (Fig. 3.52).

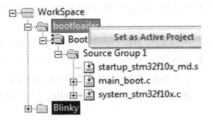

Figure 3.52
Select the bootloader as the active project.

Open main_boot.c

The bootloader program demonstrates how to jump from one program to the start of another. We need to define the start address of our second program. This must be a multiple of 128 bytes (0x200). In addition, a void function pointer is also defined:

```
#define APPLICATION_ADDRESS 0x2000
typedef void (*pFunction)(void);
pFunction Jump_To_Application;
uint32_t JumpAddress;
```

When the bootloader code enters main, it would perform custom checks on the application code Flash, such as a checksum and could also test other critical aspects of the hardware. The bootloader would then check to see if there is a new application ready to be programmed into the application area. This could be in response to a command from an upgrade utility via a serial interface, for example. If the application program checks fail or a new update is available, we would enter into the main bootloader code.

```
int main(void) {
uint32_t bootFlags;
  /* check the integrity of the application code */
```

```
/* Check if an update is available */
/* if either case is true set a bit in the bootflags register */
bootFlags = 0;
if (bootFlags != 0) {
//enter the Flash update code here
}
```

If the application code and the hardware is ok, then the bootloader will hand over to the application code. The reset vector of the application code is now located at the application address +4 and this can be loaded into the function pointer, which can then be executed resulting in a jump to the start of the application code. Before we jump to the application code, it is also necessary to load the stack pointer with the start address expected by the application code.

```
else {
    JumpAddress = *(__IO uint32_t*) (APPLICATION_ADDRESS + 4);
    Jump_To_Application = (pFunction) JumpAddress;
    // read the first four bytes of the application code and program this value into the
    stack pointer:
    //This sets the stack ready for the application code
    __set_MSP(*(__IO uint32_t*) APPLICATION_ADDRESS);
    Jump_To_Application();
}}
```

Build the Project.

Open the bootloader Options for Target\Debug tab and open the loadApp.ini file (Fig. 3.53).

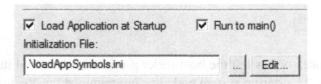

Figure 3.53
The debug script is used to load the application project symbols. This allows you to debug the bootloader and application code simultaneously.

```
Load  "..\\Blinky\\Flash\\Blinky.AXF" incremental
```

This script file can be used with the simulator or the hardware debugger. It is used to load the Blinky application code as well as the bootloader code. This allows us to debug seamlessly between the two separate programs.

Start the Debugger.

Single step the code through the bootloader checking that the correct stack pointer address is loaded into the MSP and that the Blinky start address is loaded into the function pointer.

Use the Memory window to view the Blinky application which starts at 0x800200 and check that the stack pointer value is loaded into R13 and the reset address is loaded into the function pointer (Fig. 3.54).

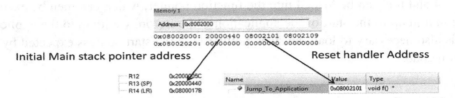

Figure 3.54
The first 8 bytes of the application image hold the initial stack pointer address and the reset handler address.

Open The blinky.c file in the Blinky project and set a breakpoint on main (Fig. 3.55).

```
35 ⊟int main (void) {
●36    uint32_t ad_avg = 0;
 37    uint16_t ad_val = 0, ad_val_ = 0xFFFF;
```

Figure 3.55
Set a breakpoint on main() in the Blinky project.

Run the Code

Now, the Cortex processor has left the bootloader program and entered the Blinky application program. The startup system code has programmed the Vector table offset register so now the hardware vector table matches the Blinky software.

Open the Peripherals\Core Peripherals\Nested Vector Interrupt Table (Fig. 3.56).

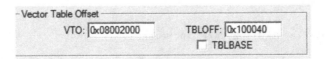

Figure 3.56
By programming the VTO, the hardware interrupt table has been moved to match the software vector table.

Now, the vector table is at 0x8002000 to match the Blinky code.

Open the IRQ.C file and set a breakpoint on the SysTick interrupt handler (Fig. 3.57).

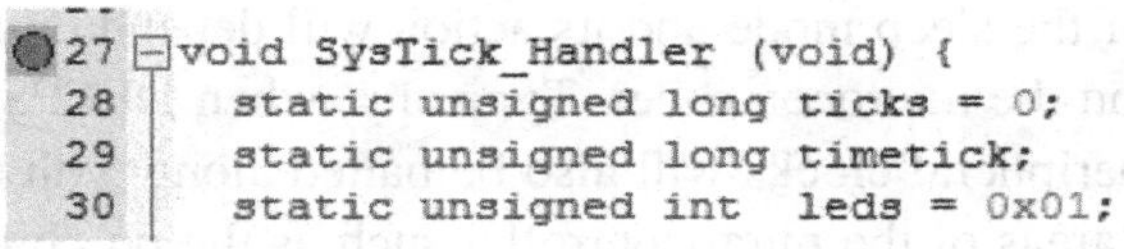

```
27 □ void SysTick_Handler (void) {
28       static unsigned long ticks = 0;
29       static unsigned long timetick;
30       static unsigned int  leds = 0x01;
```

Figure 3.57
Set a breakpoint at the start of the SysTick interrupt handler.

Run the Code.

When the SysTick timer raises, its interrupt the handler address will be fetched from the Blinky vector table and not the default address at the start of the memory and the correct SysTick handler will be executed. Now the Blinky program is running happily at its offset address. If you need the application code to call the bootloader then set a pattern in memory and force a reset within the application code by writing to the "System Control Block Application Interrupt and Reset Control" register.

```
Shared_memory = USER_UPDATE_COMMAND
NVIC_SystemReset();
```

When the bootloader restarts part of its startup checks will be to test the shared_memory location and if it is set run the required code.

Power Management

While the Cortex-M0 and Cortex-M0+ are specifically designed for low-power operation the Cortex-M3 and Cortex-M4 still have remarkably low-power consumption. While the actual power consumption will depend on the manufacturing process used by the Silicon Vendor the figures below give an indication of expected power consumption (Table 3.12).

Table 3.12: Power consumption figures by processor variant in 90 nm LP (Low Power) process

Processor	Dynamic Power Consumption (μW/MHz)	Details
Cortex-M0 +	11	Excludes debug units
Cortex-M0	16	Excludes debug units
Cortex-M3	32	Excludes MPU and debug units
Cortex-M4	33	Excludes FPU, MPU, and debug units

The Cortex-M processors are capable of entering low-power modes called SLEEP and DEEPSLEEP. When the processor in placed in SLEEP mode, the main CPU clock signal is stopped which halts the Cortex-M processor. The rest of the microcontroller clocks and peripherals will still be running and can be used to wake up the CPU. The DEEPSLEEP mode is an extension of the sleep mode and its action will depend on the specific implementation made on the microcontroller. Typically, when DEEPSLEEP mode is entered the processor peripheral clocks will also be halted along with the Cortex-M processor clock. Other areas of the microcontroller such as the on-chip SRAM and power to the Flash memory may also be switched off depending on the microcontroller configuration. (Fig. 3.58).

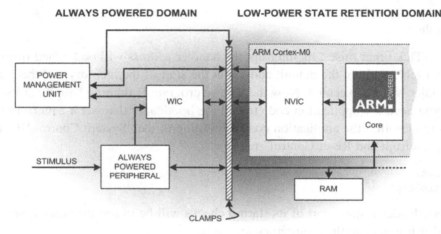

Figure 3.58
The wakesup controller is a small area of gates which do not require a clock source. The WIC can be located on a different power domain to the Cortex-M processor. This allows all the processor clocks to be halted. The range of available power modes is defined by the silicon manufacturer.

When a Cortex-M processor has entered a low-power sleep mode it can be woken up by a microcontroller peripheral raising an interrupt to the NVIC. However, the NVIC needs a clock to operate, so if all the clocks are stopped we need another hardware unit to tell the microcontroller Power Management Unit (PMU) to restore the clocks before NVIC can respond. The Cortex-M processors can be fitted with an optional unit called the wakeup interrupt controller (WIC). The WIC handles interrupt detection when all clocks are stopped and allows all of the Cortex processor to enter low-power modes. The wake-up controller consists of a minimal number of gates and does not need a system clock. The wake-up controller can be placed on a different power domain to the main Cortex-M processor. This allows the microcontroller manufacturers to design a device that can have low-power modes where most of the chip is switched off while keeping key peripherals alive to wake up the processor.

Entering Low-Power Modes

The Thumb-2 instruction set contains two dedicated instructions that will place the Cortex-M processor into SLEEP or DEEPSLEEP mode (Table 3.13).

Table 3.13: Low-power entry instructions

Instruction	Description	CMSIS-Core Intrinsic
WFI	Wait for Interrupt	__WFI()
WFE	Wait for Event	__WFE

As its name implies, the Wait for Interrupt (WFI) instruction will place the Cortex-M processor in the selected low-power mode. When an interrupt is received from one of the microcontroller peripherals, the processor will exit low-power mode and resume processing the interrupt as normal. The wait for event (WFE) instruction is also used to enter the low-power modes but has some configuration options as we shall see next.

Configuring the Low-Power Modes

The system control register is used to configure the Cortex-M processor's low-power options (Fig. 3.59).

Figure 3.59
The system control register contains the Cortex-M processor's low-power configuration bits.

The Cortex-M processor has two external sleep signals which are connected to the power management system designed by the manufacturer. By default, the SLEEPING signal is activated when the WFI or WFE instructions are executed. If the SLEEPDEEP bit is set, the second sleep signal SLEEPDEEP is activated when the Cortex-M processor enters a sleeping mode. The two sleep signals are used by the microcontroller manufacturer to provide a broader power management scheme for the microcontroller. Setting the SLEEPONEXIT bit will force the microcontroller to enter its sleep mode when it reached the end of an interrupt. This allows you to design a system that wakes up in response to an interrupt, runs the required code, and will then automatically return to sleep. In such a system no stack management is required (except in the case of preempted interrupts) during interrupt entry/exit sequence and no background code will be executed. The WFE instruction places the Cortex-M processor into its sleeping mode and the processor may be woken by an interrupt in the same manner as the WFI instruction. However, the WFE instruction has an internal event latch. If the event latch is set to one, the processor will clear the latch but does not enter low-power mode. If the latch is zero, it will enter low-power mode. On a typical microcontroller, the events are the peripheral interrupt signals. So pending interrupts will prevent the processor from sleeping. The SEVONPEND bit is used to change the behavior of the WFE instruction. If this bit is set, the peripheral interrupt lines can be used to wake the processor even if the interrupt is disabled in the NVIC. This allows you to place the processor into its sleep mode, when a peripheral interrupt occurs the processor will wake and resume execution of the instruction following the WFE instruction rather than jumping to an interrupt routine (Fig. 3.60).

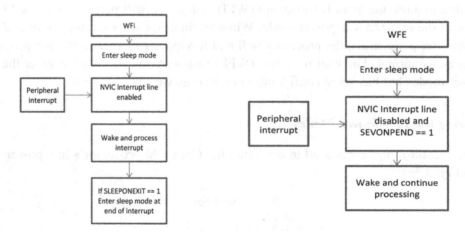

Figure 3.60
The two low-power entry instructions place the Cortex processor into its low-power mode. Both modes use a peripheral interrupt to wake the processor but their wake-up behavior is different.

An interrupt disabled in the NVIC cannot be used to exit a sleep mode entered by the WFI instruction. However, the WFE instruction will respond to activity on any interrupt line even if it is disabled or temporarily disabled by the processor mask registers (BASEPRI, PRIMASK, and FAULTMASK).

Exercise 3.3 Low-Power Modes

In this exercise, we will use the exception project to experiment with the Cortex-M low-power modes.

Select the Boards tab and "The Designers Guide Tutorial Examples." Select the Example tab and Copy "Ex 3.5 Low-Power Modes"

```
while(1)
{
  SLEEP = 1;
  BACKGROUND = 0;
  __wfe();
  BACKGROUND = 1;
  SLEEP = 0;
}
```

This project uses the SysTick and ADC interrupts we saw in the last example. This time we have added an additional SLEEP variable to monitor the processor operating state. The WFI instruction has been added in the main while loop.

Build the project and start the debugger.

Open the Logic Analyzer window and start the code running (Fig. 3.61).

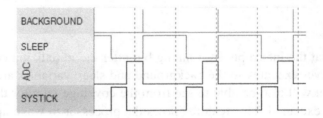

Figure 3.61

In our simple project, the background code only executes the __wfi() instruction forcing the CPU to enter sleep state.

Here, we can see that the background code executes the __wfi() instruction and then goes to sleep until an interrupt is raised. When the interrupts have completed we return to the background code, which will immediately place the processor in its low-power mode.

Exit the debugger and change the code to match the lines below:

```
SCB->SCR = 0x2;
SLEEP = 1;
BACKGROUND = 0;
__wfi();
while(1){
```

Add the code to set bit two of the system control register. This sets the Sleep On Exit flag which forces the processor into a low-power mode when it completes an interrupt. Cut the remaining three lines from inside the while loop and paste them into the initializing section of the main() function.

Build the code, restart the debugger, and observe the execution of the interrupts in the Logic Analyzer window (Fig. 3.62).

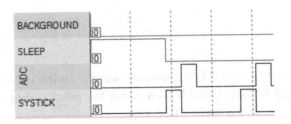

Figure 3.62
Once sleep on exit has been enabled we no longer execute code in the background
(noninterrupt code).

Here, we can see that the interrupts are running but after the initializing code has run the background loop never executes so the background and sleep variables are never updated. In the debugger, you will also be able to see from the coverage monitor that the main() while loop is never executed. This feature allows the processor to wake up, run some critical code, and then sleep with the absolute minimum overhead.

Moving From the Cortex-M3

In this chapter, we have concentrated on learning the Cortex-M3 processor. Now that you are familiar with how the Cortex-M3 works we can examine the differences between the Cortex-M3 and the other Cortex-M variants. As we will see these are mainly architectural differences and if you can use the Cortex-M3 you can easily move up to a Cortex-M4 or down to a Cortex-M0(+)-based microcontroller. Increasingly silicon manufacturers make a microcontroller family were variants have the same package pinout and peripherals buy can be selected with an Cortex-M0 or Cortex-M3 processor allowing you to seamlessly switch devices trading off performance versus cost.

Cortex-M4

The Cortex-M4 is most easily described as a Cortex-M3 with an additional FPU and DSP instructions. We will look at these features in Chapter 8 "Practical DSP for Cortex-M4 and Cortex-M7" but here we will take a tour of the main differences between the Cortex-M3 and Cortex-M4. The Cortex-M4 offers the same processing power of 1.25 DMIPS/MHz as the Cortex-M3 but has a much greater math capability. This is delivered in three ways. The hardware FPU can perform floating point calculations in a little as 1 cycle compared to the hundreds of cycles the same calculation would take on the Cortex-M3. For integer calculations, the Cortex-M4 has a higher performance MAC that improves on the Cortex-M3 MAC to allow single cycle calculations to be performed on 32-bit wide quantities, which yield a 64-bit result. Finally, the Cortex-M4 adds a group of SIMD "Single Instruction Multiple Data" (SIMD) instructions that can perform multiple integer calculations in a single cycle. While the Cortex-M4 has a larger gate count than the Cortex-M3, the FPU contains more gates than the entire Cortex-M0 processor (Table 3.14).

Table 3.14: Additional features in the Cortex-M4

Feature	Comments
Floating point unit DSP SIMD instructions GE field in xPSR	See Chapter 7 "Debugging with CoreSight"
Extended integer MAC unit	The Cortex-M4 extends the integer MAC to support single cycle execution of 32-bit multiplies which yield a 64-bit result

Cortex-M0

The Cortex-M0 is a reduced version of the Cortex-M3; it is intended for low-cost and low-power microcontrollers. The Cortex-M0 has a processing power of 0.84 DMIPS/MHz. While the Cortex-M0 can run at high clock frequencies, it is often designed into low-cost devices with simple memory systems. Hence, a typical Cortex-M0 microcontroller runs with a CPU frequency of 50 DMIPS/MHz, however, its low-power consumption makes it ideal for low-power applications. While it essentially has the same programmer's model as the Cortex-M3, there are some limitations and these are summarized below (Table 3.15).

Table 3.15: Features not included in the Cortex-M0

Feature	Comments
Three-stage pipeline	Same pipeline stages as the Cortex-M3 but no speculative branch target fetch
Von Newman bus interface	Instruction and data use the same bus port which can has a slower performance compared to the Harvard architecture of the M3/M4
No conditional IF THEN blocks	Conditional branches are always used which cause a pipeline flush
No saturated maths instructions	
SYSTICK timer is optional	However, so far every Cortex-M0 microcontroller has it fitted
No memory protection unit	The MPU is covered in Chapter 5 "Advanced Architecture Features"
32 NVIC channels	A limited number of interrupt channels compared to the Cortex-M3/M4, however, in practice this is not a real limitation and the Cortex-M0 is intended for small devices with a limited number of peripherals
Four programmable priority levels	Priority level registers only implemented 2 bits (four levels), and there is no priority group setting
Hard fault exception only	No usage fault, memory management fault, or bus fault exception vectors
16 cycle interrupt latency	Same deterministic interrupt handling as the M3 but with a four more cycle overhead
No priority group	Limited to a fixed four levels of preemption
No BASEPRI register	
No FAULTMASK register	
Reduced debug features	The Cortex-M0 have fewer debug features compared to the Cortex-M3/M4, see Chapter 8 "Practical DSP for Cortex-M4 and Cortex-M7" for more details
No exclusive access instructions	The exclusive access instructions are covered in Chapter 5 "Advanced Architecture Features"
No reverse bit order of count leading zero instructions	
Reduced number of registers in the system control block	See below
Vector table cannot be relocated	The NVIC does not include the vector table offset register
All code executes at privileged level	The Cortex-M0 does not support the unprivileged operating mode

The system control block contains a reduced number of features compared to the Cortex-M3\M4. Also, the SysTick timer registers have been moved from the NVIC to the system control block (Table 3.16).

Table 3.16: Registers in the Cortex-M0 system control block

Register	Size in Words	Description
SysTick control and status	1	Enables the timer and its interrupt
SysTick reload	1	Holds the 24-bit reload value
SysTick current value	1	Holds the current 24-bit timer value
SysTick calibration	1	Allows trimming the input clock frequency
CPU ID	1	Hardwired ID and revision numbers from ARM and the silicon manufacturer
Interrupt control and state	1	Provides pend bits for the SysTick and NMI interrupts and extended interrupt Pending\active information
Application interrupt and reset control	1	Contains the same fields as the Cortex-M3 minus the PRIGROUP field
Configuration and control	1	
System handler priority	2	These registers hold the 8-bit priority fields for the configurable Processor exceptions

Cortex-M0+

The Cortex-M0+ is an enhanced version of the Cortex-M0. As such it boasts lower power consumption figures combined with greater processing power. The Cortex-M0+ also brings the MPU and real-time debug capability to very low-end devices. The Cortex-M0+ also introduces a fast I/O port which speeds up access to peripheral registers typically GPIO ports to allow fast switching of port pins. As we will see in Chapter 7 "Debugging with CoreSight," the debug system is fitted with a new trace unit called the micro trace buffer (MTB) which allows you to capture a history of executed code with a low-cost development tool (Table 3.17).

Table 3.17: Cortex-M0+ features

Feature	Comments
Code compatible with the Cortex-M0	The Cortex-M0+ is code compatible with the Cortex-M0 and provides higher performance with lower power consumption
Two-stage pipeline	This reduces the number of Flash accesses and hence power consumption
I/O port	The I/O port provides single cycle access to GPIO and peripheral registers
Vector table can be relocated	This supports more sophisticated software designs. Typically a Bootloader and separate application
Supports 16-bit Flash memory accesses	This allows devices featuring the Cortex-M0+ to use low-cost memory
Code can execute at privileged and unprivileged levels	The Cortex-M0+ has the same operating modes as the Cortex-M3/M4
Memory protection unit	The Cortex-M0+ has a similar MPU to the Cortex-M3/M4
Micro trace buffer	This is a "snapshot" trace unit which can be accessed by low-cost debug units

Conclusion

In this chapter we have covered the key features of the Cortex-M processor family. To develop successfully with a Cortex-M-based device, you will need to be completely familiar with all the topics covered in this chapter and Chapter 2 "Developing Software for the Cortex-M family". Now that we have a basic understanding of the Cortex-M processor, we will look at the more advanced processor features plus a range of software development methods and techniques.

Cortex Microcontroller Software Interface Standard

Introduction

The widespread adoption of the Cortex-M processor into general purpose microcontrollers has led to two rising trends within the electronics industry. First of all the same processor is available from a wide range of vendors each with their own family of microcontrollers. In most cased each vendor creates a family of microcontrollers that span a range of requirements for embedded systems developers. This proliferation of devices means that as a developer you can select a suitable microcontroller from several thousand devices while still using the same tools and skills regardless of the silicon vendor. This explosive growth in Cortex-M-based microcontrollers has made the Cortex-M processor the de facto industry standard for 32-bit microcontrollers and there are currently no real challengers (Fig. 4.1).

Figure 4.1
CMSIS-compliant software development tools and middleware stacks are allowed to carry the CMSIS logo.

The flip side of the coin is differentiation. It would be possible for a microcontroller vendor to design their own proprietary 32-bit processor. However, this is expensive to do and also requires an ecosystem of affordable tools and software to get mass adoption. It is more cost-effective to license the Cortex-M processor from ARM and then use their own expertise to create a microcontroller with innovative peripherals. There are now more than 17 (up from 10 when I first wrote this book 3 years ago) silicon vendors shipping Cortex-M-based microcontrollers. While in each device the Cortex-M processor is the same but each manufacturer of the final silicon seeks to offer a unique set of user

The Designer's Guide to the Cortex-M Processor Family.
DOI: http://dx.doi.org/10.1016/B978-0-08-100629-0.00004-9

peripherals for a given range of applications. This can be a microcontroller designed for low power applications, motor control, communications, or graphics. This way a silicon vendor can offer a microcontroller with a state-of-the-art processor which has wide development tools support while at the same time using their skill and knowledge to develop a microcontroller featuring an innovative set of peripherals (Fig. 4.2).

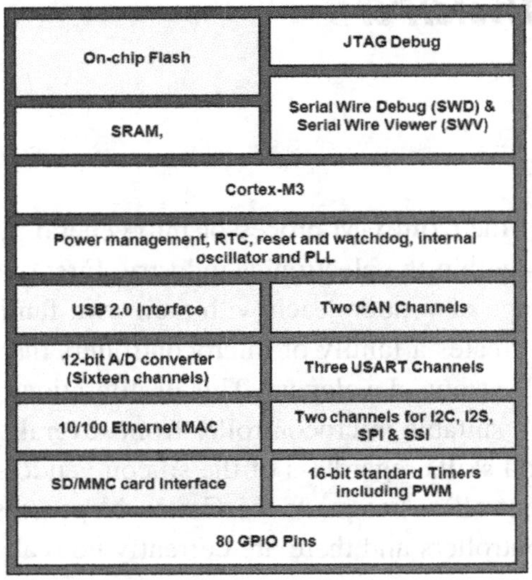

Figure 4.2
Cortex-based microcontrollers can have a number of complex peripherals on a single chip. To make these work you will need to use some form of third-party code. CMSIS is intended to allow stacks from different sources to integrate together easily.

These twin factors have led to a vast "cloud" of standard microcontrollers with increasingly complex peripherals. As well as typical microcontroller peripherals such as USART, I2C, ADC, and DAC, a modern high-end microcontroller could well have a Host/Device USB controller, Ethernet MAC, SDIO controller, and LCD interface. The software to drive any of these peripherals is effectively a project in itself, so gone are the days of a developer using an 8/16 bit microcontroller and writing all of the application code from the reset vector. To release any kind of sophisticated product it is almost certain that you will be using some form of third-party code in order to meet project deadlines. The third-party code may take the form of example code, an open source or commercial stack or a library provided by the silicon vendor. Both of these trends have created a need to make "C" level code more portable between different development tools and different microcontrollers. There is also a need to be able to easily integrate code taken from a variety of sources into a single project.

In order to address these issues a consortium of silicon vendors and tools vendors has developed the "Cortex Microcontroller Software Interface Standard" or CMSIS (Pronounced seeMsys) for short.

CMSIS Specifications

The main aim of CMSIS is to improve software portability and reusability across different microcontrollers and toolchains. This allows software from different sources to integrate seamlessly together. Once learnt CMSIS helps to speed up software development through the use of standardized software functions.

At this point it is worth being clear about exactly what CMSIS is. CMSIS consists of seven interlocking specifications that support code development across all Cortex-M-based microcontrollers. The seven specifications are as follows; CMSIS-Core, CMSIS-RTOS, CMSIS-DSP, CMSIS-Driver, CMSIS-Pack, CMSIS-SVD and CMSIS-DAP (Fig. 4.3).

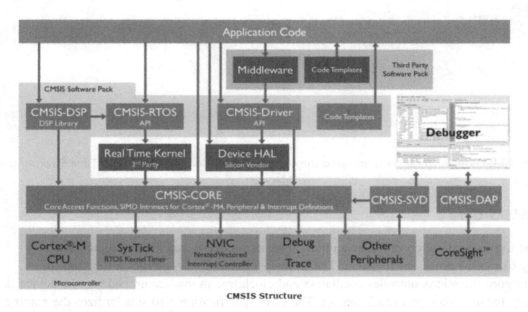

Figure 4.3

CMSIS consists of a several separate specifications (CORE, DSP, RTOS, SVD, DRIVER, DAP, and PACK) that make source code more portable between tools and devices.

It is also worth being clear what CMSIS is not. CMSIS is not a complex abstraction layer that forces you to use a complex and bulky library. Rather the CMSIS-Core specification takes a very small amount of resources about 1 k of code and just 4 bytes of RAM and just standardizes the way you access the Cortex-M processor and microcontroller registers. Furthermore, CMSIS does not really affect the way you develop code or force you to adopt a particular methodology. It simply provides a framework that helps you to integrate third-party code and reuse code on future projects. Each of the CMSIS specifications are not that complicated and can be learnt easily.

The full documentation for each of the CMSIS specifications can be downloaded from the URL www.keil.com/cmsis. Each of the CMSIS specifications are integrated into the MDK-ARM toolchain and the CMSIS documentation is available by opening the Run-Time Environment and clicking on the CMSIS link in the description column (Fig. 4.4).

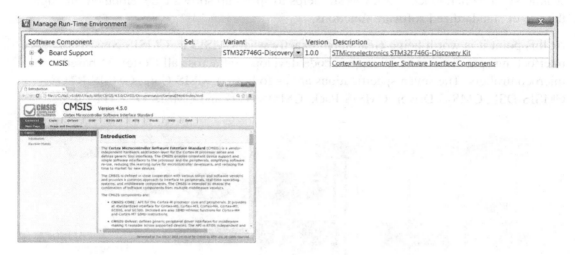

Figure 4.4
The CMSIS documentation is accessed through the description link in the Run-Time Environment Manager.

CMSIS-Core

The Core specification provides a minimal set of functions and macros to access the key Cortex-M processor registers. The Core specification also defines a function to configure the microcontroller oscillators and clock tree in the startup code so the device is ready for use when you reach main(). The Core specification also standardizes the naming convention for the device peripheral registers. The CMSIS-Core specification also includes support for the Instrumentation Trace during debug sessions.

CMSIS-RTOS

The CMSIS-RTOS specification provides a standard API for a Real Time Operating System. This is in effect a set of wrapper functions that translate the CMSIS-RTOS API to the API of the specific RTOS that you are using. We will look at the use of an RTOS in general and the CMSIS-RTOS API in Chapter 9 "CMSIS-RTOS." The Keil RTX RTOS was the first RTOS to support the CMSIS-RTOS API and it has been released as an open source reference implementation. RTX can be compiled with the Keil/ARM, GCC, and IAR compilers. It is licensed with a three clause Berkeley Software Distribution (BSD) license which allows its unrestricted use in commercial and noncommercial applications.

CMSIS-DSP

As we have seen in Chapter 3 "Cortex-M Architecture" the Cortex-M4 is a "Digital Signal Controller" with a number of enhancements to support DSP algorithms. Developing a real time DSP system is best described as a "nontrivial pass time" and can be quite daunting for all but the simplest systems. To help mere mortals include DSP algorithms in Cortex-M4/M7 and Cortex-M3 projects CMSIS includes a DSP library that provides over 60 of the most commonly used DSP mathematical functions. These functions are optimized to run on the Cortex-M4 and Cortex-M7 but can also be compiled to run on the Cortex-M3. We will have a look at using this library in Chapter 8 "Practical DSP for Cortex-M4 and Cortex-M7."

CMSIS-Driver

The CMSIS-Driver specification defines a standard API for a range of peripherals which are common to most microcontroller families. This includes peripherals such as USART, SPI, and I2C as well as more complex peripherals such as Ethernet MAC and USB. The CMSIS-Drivers are intended to provide a standard target for middleware libraries. For example, this would allow a third-party developer to create a USB library that used the CMSIS-USB driver. Such a library could then be deployed on any device that has a CMSIS-USB driver. This greatly speeds up support for new devices and allows library developers to concentrate on adding features to their products rather than continually having to spend time developing support for new devices.

CMSIS-SVD and DAP

One of the key problems for tools vendors is to be able to provide debug support for new devices as soon as they are released. One of the main areas that must be customized in the debugger is the "Peripheral View" windows that show the developer the current state of the microcontroller peripherals. With the growth in both the number of Cortex-M vendors and also the rising number and complexity of on-chip peripherals it is becoming all but impossible for any given tools vendor to maintain support for all possible microcontrollers. To overcome this hurdle the CMSIS-SVD specification defines a "System Viewer Description" file. This file is provided and maintained by the silicon vendor and contains a complete description of the microcontroller peripheral registers in an XML format. This file is then imported by the development tool which uses it to automatically construct the peripheral debug windows for the microcontroller. This approach allows full debugger support to be available as soon as new microcontrollers are released.

The CMSIS-DAP specification defines the interface protocol for a hardware debug unit that sits between the host PC and the Debug Access Port of the microcontroller (Fig. 4.5). This allows any software debugger that supports CMSIS-DAP to connect to any hardware debug unit that also supports CMSIS-DAP. There are an increasing number of very low cost evaluation boards that contain an integral debugger that connects to a PC using USB. In many cases this hardware debugger supports the CMSIS-DAP protocol so that it can be connected to any compliant toolchain.

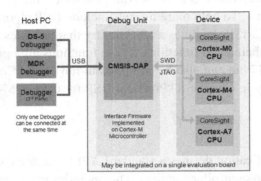

Figure 4.5
CMSIS-DAP allows for interoperability between different vendors—software and hardware debuggers.

CMSIS-Pack

The CMSIS-Core and Driver specifications can be used to develop reusable software components that are portable across device families. The CMSIS-Pack specification defines a method of bundling all of the component elements (software files, examples, help files, templates) into a software Pack. Such a Pack can be downloaded and installed into your toolchain. The Pack also contains information on the software component dependencies, that is, the other files that need to be present when the component is used. This allows you to quickly integrate software from many sources to build a platform on which to develop your application code. As the complexity of Cortex-M microcontrollers is ever increasing this is a very important technology to increase a developer's productivity and code reliability.

In Chapter 12 "Software Components" we will look at using the CMSIS-Driver and CMSIS-Pack specifications to make a software component that can be reused across multiple microcontrollers.

Foundations of CMSIS

The CMSIS-Core specification provides a standard set of low level functions, macros, and peripheral register definitions that allow your application code to easily access the

Cortex-M processor and microcontroller peripheral registers. This framework needs to be added to your code at the start of a project. This is actually very easy to do as the CMSIS-Core functions are very much part of the compiler toolchain.

Coding Rules

While CMSIS is important for providing a standardized software interface for all Cortex-M microcontrollers, it is also interesting for embedded developers because it is based on a consistent set of "C" coding rules called MISRA-C. When applied these coding rules generate clear unambiguous "C" code, and this approach is worth studying as it embodies many of the best practices that should be adopted when writing the "C" source code for your own application software.

MISRA-C

The MISRA-C coding standard is maintained and published by MIRA. MIRA, stands for "Motor Industry Research Agency," is located near Rugby in England and is responsible for many of the industry standards used by the UK motor industry. In 1998 its software division released the first version of its coding rules formally called "MISRA guidelines for the use of C in vehicle electronics" (Fig. 4.6).

Figure 4.6
The CMSIS source code has been developed using MISRA-C as a coding standard.

The original MISRA-C specification contained 127 rules which attempted prevent common coding mistakes and resolve gray areas of the ANSI C specification when applied to embedded systems. Although originally intended for the automotive industry, MISRA-C has found acceptance in the wider embedded systems community. In 2004 a revised edition of MISRA-C was released with the title "MISRA-C Guidelines for the use of C in critical systems." This change in the title reflects the growing adoption of MISRA-C as a coding standard for general embedded systems. There have been two further updates to the MISRA-C standard in 2008 and 2012. One of the other key attractions of MISRA-C is that it was written by engineers and not computer scientists. This has resulted in a clear, compact, and easy to understand set of rules. Each rule is clearly explained with examples of good coding practice. This means that the entire coding standard is contained in a book of just 106 pages which can easily be read in an evening. A typical example of a MISRA-C rule is shown below:

> **Rule 13.6(required) Numeric variables being used within a for loop for iteration counting shall not be modified in the body of the loop**
>
> *Loop counters shall not be modified in the body of the loop. However other loop control variables representing logical values may be modified in the loop. For example a flag to indicate that something has been completed, which is then tested in the for statement.*

```
Flag = 1;
For ((I = 0;(i<5) && (flag==1);i++)
{
/*.......*/
Flag = 0;    /* Compliant — allows early termination of the loop */
i = i + 3;   /* Not Compliant — altering the loop counter */
}
```

Where possible the MISRA-C rules have been designed so that they can be statically checked either manually or by a dedicated tool. The MISRA-C standard is not an open standard and is published in paper and electronic form on the MIRA website. Full details of how to obtain the MISRA Standard are available in Appendix A.

In addition to the MISRA-C guidelines CMSIS enforces some additional coding rules. To prevent any ambiguity in the compiler implementation of standard "C" types, CMSIS uses the data types defined in the ANSI C header file stdint.h (Table 4.1).

Table 4.1: CMSIS variable types

Standard ANSI C Type	MISRA C Type
Signed char	int8_t
Signed short	int16_t
Signed int	int32_t
Signed __int64	int64_t
Unsigned char	unit8_t
Unsigned short	uint16_t
Unsigned int	uint32_t
Unsigned __int64	uint64_t

The type defines ensure that the expected data size is mapped to the correct ANSI type for a given compiler. Using typedefs like this is good practice as it avoids any ambiguity about the underlying variable size which may vary between compilers particularly if you are migrating code between different processor architectures and compiler tools.

CMSIS also specifies IO type qualifiers for accessing peripheral variables. These are typedefs that make clear the type of access each peripheral register has (Table 4.2).

Table 4.2: CMSIS IO qualifiers

MISRA-C IO Qualifier	ANSI C Type	Description
#define __I	Volatile const	Read Only
#define __O	Volatile	Write Only
#define __IO	Volatile	Read and Write

While this does not provide any extra functionality for your code but it provides a common mechanism that can be used by static checking tools to ensure that the correct access is made to each peripheral register.

Much of the CMSIS documentation is auto-generated using a tool called Doxygen. This is a free download released under a GPL license. While Doxygen cannot actually write the documentation for you it does do much of the dull boring stuff for you (leaving you to do the exciting documentation work). Doxygen works by analyzing your source code and extracting declarations and specific source code comments to build up a comprehensive "object dictionary" for your project. The default output format for Doxygen is a browsable HTML but this can be converted to other formats if desired. The CMSIS source code comments contain specific tags prefixed by the @ symbol, for example, @brief. These tags are used by Doxygen to annotate descriptions of the CMSIS functions.

```
/**
 * @brief Enable Interrupt in NVIC Interrupt Controller
 * @param IRQn interrupt number that specifies the interrupt
 * @return none.
 * Enable the specified interrupt in the NVIC Interrupt Controller.
 * Other settings of the interrupt such as priority are not affected.
 */
```

When the Doxygen tool is run it analyzes your source code and generates a report containing a dictionary of you functions and variables based on the comments and source code declarations.

CMSIS-Core Structure

The CMSIS-Core functions can be included in your project through the addition of three files. These include the default startup code with the CMSIS standard vector table.

The second file is the system_<device>.c file which contains the necessary code to initialize the microcontroller system peripherals. Finally, the device includes file which imports the CMSIS header files that contain the CMSIS-Core functions and macros (Fig. 4.7).

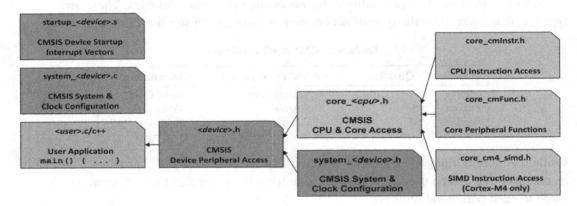

Figure 4.7
The CMSIS-Core standard consists of the device startup, system C code, and a device header. The device header defines the device peripheral registers and pulls in the CMSIS header files. The CMSIS header files contain all of the CMSIS-Core functions.

Startup Code

The startup code provides the reset vector, initial stack pointer value, and a symbol for each of the interrupt vectors.

```
__Vectors   DCD   __initial_sp    ; Top of Stack
   DCD   Reset_Handler             ; Reset Handler
   DCD   NMI_Handler               ; NMI Handler
   DCD   HardFault_Handler         ; Hard Fault Handler
   DCD   MemManage_Handler         ; MPU Fault Handler
```

When the processor starts it will initialize the main stack pointer by loading the value stored in the first 4 bytes of the vector table. Then it will jump to the reset handler;

```
Reset_Handler PROC
  EXPORT Reset_Handler [WEAK]
  IMPORT __main
  IMPORT SystemInit
    LDR R0, =SystemInit
    BLX R0
    LDR R0, =__main
    BX R0
    ENDP
```

System Code

The reset handler calls the SystemInit() function which is located in the CMSIS system_ < device >.c file. This code is delivered by the silicon manufacturer and it provides all the necessary code to configure the microcontroller after it leaves the reset vector. Typically this includes setting up the internal phase locked loops, configuring the microcontroller clock tree and internal buss structure, enabling the external buss if required. The configuration of the initializing functions is controlled by a set of #defines located at the start of the module. This allows you to customize the basic configuration of the microcontroller system peripherals. Since the SystemInit() function is run when the microcontroller leaves reset the microcontroller system peripherals and the Cortex-M processor will be in a fully configured state when the program reaches main(). In the past this system initializing code is something you would have had to write or crib from example code. On a new microcontroller this would have been a few days' work, so the SystemInit() function does save you a lot of time and effort. The SystemInit() function also sets the CMSIS global variable SystemCoreClock to the CPU frequency. This variable can then be used by the application code as a reference value when configuring the microcontroller peripherals. In addition to the SystemInit() function the CMSIS system file contains an additional function to update the SystemCoreClock variable if the CPU clock frequency is changed on the fly. The function SystemCoreClockUpdate(); is a void function which must be called if the CPU clock frequency is changed. This function is tailored to each microcontroller and will evaluate the clock tree registers to calculate the new CPU operating frequency and change the SystemCoreClock variable accordingly.

Once the SystemInit() function has run and we reach the application code we will need to access the CMSIS-Core functions. This framework is added to the application modules through the microcontroller-specific header file.

Device Header File

The header file first defines all of the microcontroller special function registers in a CMSIS standard format.

A typedef structure is defined for each group of special function registers on the supported microcontroller. In the code below, a general GPIO typedef is declared for the group of GPIO reregisters. This is a standard typedef but we are using the IO qualifiers to designate the type of access granted to a given register.

```
typedef struct
{
__IO uint32_t MODER;    /*!< GPIO port mode register, Address offset: 0x00 */
__IO uint32_t OTYPER;   /*!< GPIO port output type register, Address offset: 0x04 */
__IO uint32_t OSPEEDR;  /*!< GPIO port output speed register, Address offset: 0x08 */
```

```
__IO uint32_t PUPDR;    /*!< GPIO port pull-up/pull-down register, Address offset: 0x0C */
__IO uint32_t IDR;      /*!< GPIO port input data register, Address offset: 0x10 */
__IO uint32_t ODR;      /*!< GPIO port output data register, Address offset: 0x14 */
__IO uint16_t BSRRL;    /*!< GPIO port bit set/reset low register, Address offset: 0x18 */
__IO uint16_t BSRRH;    /*!< GPIO port bit set/reset high register, Address offset: 0x1A */
__IO uint32_t LCKR;     /*!< GPIO port configuration lock register, Address offset: 0x1C */
__IO uint32_t AFR[2];   /*!< GPIO alternate function registers, Address offset: 0x24-0x28 */
} GPIO_TypeDef;
```

Next #defines are used to layout the microcontroller memory map. First the base address of the peripheral special function registers is declared and then offset addresses to each of the peripheral busses and finally an offset to the base address of each GPIO port.

```
#define PERIPH_BASE        ((uint32_t)0x40000000)
#define APB1PERIPH_BASE    PERIPH_BASE
#define GPIOA_BASE         (AHB1PERIPH_BASE + 0x0000)
#define GPIOB_BASE         (AHB1PERIPH_BASE + 0x0400)
#define GPIOC_BASE         (AHB1PERIPH_BASE + 0x0800)
#define GPIOD_BASE         (AHB1PERIPH_BASE + 0x0C00)
```

Then the register symbols for each GPIO port can be declared.

```
#define GPIOA  ((GPIO_TypeDef *) GPIOA_BASE)
#define GPIOB  ((GPIO_TypeDef *) GPIOB_BASE)
#define GPIOC  ((GPIO_TypeDef *) GPIOC_BASE)
#define GPIOD  ((GPIO_TypeDef *) GPIOD_BASE)
```

Then in the application code we can program the peripheral special function registers by accessing the structure elements.

```
void LED_Init (void) {
   RCC->AHB1ENR |= ((1UL << 3) );       /* Enable GPIOD clock */
   GPIOD->MODER &= ~((3UL << 2*12) |
                    (3UL << 2*13) |
                    (3UL << 2*14) |
                    (3UL << 2*15) );   /* PD.12..15 is output */
   GPIOD->MODER |= ((1UL << 2*12) |
                   (1UL << 2*13) |
                   (1UL << 2*14) |
                   (1UL << 2*15) );
```

The microcontroller <device>.h include file provides similar definitions for all of the on-chip peripheral special function registers. These definitions are created and maintained by the silicon manufacturer and as they do not use any non ANSI keywords in the include file may be used with any "C" compiler. This means that any peripheral driver code written to the CMSIS specification is fully portable between CMSIS-compliant tools.

The microcontroller include file also provides definitions of the interrupt channel number for each peripheral interrupt source.

```
WWDG_IRQn = 0,         /*!< Window WatchDog Interrupt */
PVD_IRQn = 1,          /*!< PVD through EXTI Line detection Interrupt */
TAMP_STAMP_IRQn = 2,   /*!< Tamper and TimeStamp interrupts through the EXTI line */
RTC_WKUP_IRQn = 3,     /*!< RTC Wakeup interrupt through the EXTI line */
FLASH_IRQn = 4,        /*!< FLASH global Interrupt */
RCC_IRQn = 5,          /*!< RCC global Interrupt */
EXTIO_IRQn = 6,        /*!< EXTI Line0 Interrupt */
EXTI1_IRQn = 7,        /*!< EXTI Line1 Interrupt */
EXTI2_IRQn = 8,        /*!< EXTI Line2 Interrupt */
EXTI3_IRQn = 9,        /*!< EXTI Line3 Interrupt */
EXTI4_IRQn = 10,       /*!< EXTI Line4 Interrupt */
```

In addition to the register and interrupt definitions, the silicon vendor may also provide a library of peripheral driver functions. Again as this code is written to the CMSIS standard it will compile with any suitable development tool. Often these libraries are very useful for getting a project working quickly and minimize the amount of time you have to spend writing low level code. However they are often very general libraries that do not yield the most optimized code. So if you need to get the maximum performance or minimal code size you will need to rewrite the driver functions to suit your specific application. The microcontroller include file also imports up to five further include files. These are "stdint.h", a "CMSIS-Core" file for the Cortex-M processor you are using. A header file "system_<device>.h" is also included to give access to the functions in the system file. The CMSIS instruction intrinsic and helper functions are contained in two further files "core_cminstr.h" and "core_cmfunc.h". If you are using the Cortex-M4 or Cortex-M7 an additional file "core_CM4_simd.h" is added to provide support for the Cortex-M4 SIMD instructions. As discussed earlier the "stdint.h" file provides the MISRA-C types which are used in the CMSIS definitions and should be used through your application code.

CMSIS-Core Header Files

Within the CMSIS-Core specification there are a small number of defines that are setup for a given microcontroller. These can be found in the <device>.h processor include file (Table 4.3).

Table 4.3: CMSIS configuration values

CMSIS Define	Description
__CMx_REV	Core revision number
__NVIC_PRIO_BITS	Number of priority bits implemented in the NVIC priority registers
__MPU_PRESENT	Defines if an MPU is present (see Chapter: 5 Advanced Architecture Features)
__FPU_PRESENT	Defines if an FPU is present (see Chapter: 7 Debugging with CoreSight)
__Vendor_SysTickConfig	Defines if there is a vendor specific SysTick Configuration

The processor include file also imports the CMSIS header files which contain the CMSIS-Core helper functions. The helper functions are split into the groups as shown in Table 4.4:

Table 4.4: CMSIS function groups

CMSIS-Core Function Groups
NVIC access functions
SysTick configuration
CPU register Access
CPU instruction intrinsics
Cortex-M4-SIMD intrinsics
ITM debug functions
FPU Functions (Cortex-M7 only)
Cache Functions (Cortex-M7 only)

The NVIC group provides all the functions necessary to configure the Cortex-M interrupts and exceptions. A similar function is provided to configure the SysTick timer and interrupt. The CPU register group allows you to easily read and write to the CPU registers using the Move Register to Special Register (MRS) and Move Special Registers to Register (MSR) instructions. Any instructions that are not reachable by the "C" language are supported by dedicated intrinsic functions and are contained in the CPU instructions group. An extended set of intrinsics are also provided for the Cortex-M4 and Cortex-M7 to access the SIMD instructions. Finally some standard functions are provided to access the debug Instrumentation Trace.

Interrupts and Exceptions

Management of the NVIC registers may be done by the functions provided in the interrupt and exception group. These functions allow you to setup an NVIC interrupt channel and manage its priority as well as interrogate the NVIC registers during run time (Table 4.5).

Table 4.5: CMSIS interrupt and exception group

CMSIS Function	Description
NVIC_SetPriorityGrouping	Set the priority grouping
NVIC_GetPriorityGrouping	Read the priority grouping
NVIC_EnableIRQ	Enable a peripheral interrupt channel
NVIC_DisableIRQ	Disable a peripheral interrupt channel
NVIC_GetPendingIRQ	Read the pending status of an interrupt channel
NVIC_SetPendingIRQ	Set the pending status of an interrupt channel
NVIC_ClearPendingIRQ	Clear the pending status of an interrupt channel
NVIC_GetActive	Get the active status of an interrupt channel
NVIC_SetPriority	Set the active status of an interrupt channel
NVIC_GetPriority	Get the priority of an interrupt channel
NVIC_EncodePriority	Encodes the priority group
NVIC_DecodePriority	Decodes the priority group
NVIC_SystemReset	Forces a system reset

A configuration function is also provided for the SysTick timer (Table 4.6).

Table 4.6: CMSIS-SysTick function

CMSIS Function	Description
SysTick_Config	Configures the timer and enables the interrupt

So for example to configure an external interrupt line we first need to find the name for the external interrupt vector used in the startup code vector table.

```
    DCD FLASH_IRQHandler          ; FLASH
DCD RCC_IRQHandler                ; RCC
DCD EXTI0_IRQHandler              ; EXTI Line0
DCD EXTI1_IRQHandler              ; EXTI Line1
DCD EXTI2_IRQHandler              ; EXTI Line2
DCD EXTI3_IRQHandler              ; EXTI Line3
DCD EXTI4_IRQHandler              ; EXTI Line4
DCD DMA1_Stream0_IRQHandler       ; DMA1 Stream 0
```

So for external interrupt line 0 we simply need to create a void function duplicating the name used in the vector table;

```
void EXTI0_IRQHandler (void);
```

This now becomes our interrupt service routine. In addition we must configure the microcontroller peripheral and NVIC to enable the interrupt channel. In the case of the external interrupt line the following code will setup Port A pin 0 to generate an interrupt to the NVIC on a falling edge.

```
AFIO->EXTICR[0]   & = ~AFIO_EXTICR1_EXTI0;      /* clear used pin */
AFIO->EXTICR[0]   | = AFIO_EXTICR1_EXTI0_PA;    /* set PA.0 to use */
EXTI->IMR         | = EXTI_IMR_MR0;             /* unmask interrupt */
EXTI->EMR         & = ~EXTI_EMR_MR0;            /* no event */
EXTI->RTSR        & = ~EXTI_RTSR_TR0;           /* no rising edge trigger */
EXTI->FTSR        | = EXTI_FTSR_TR0;            /* set falling edge trigger */
```

Next we can use the CMSIS functions to enable the interrupt channel.

```
NVIC_EnableIRQ(EXTI0_IRQn);
```

Here we are using the defined enumerated type for the interrupt channel number. This is declared in the microcontroller header file <device>.h. Once you get a bit familiar with the CMSIS-Core functions it becomes easy to intuitively work out the name rather than having to look it up or look up the NVIC channel number.

We can also add a second interrupt source by using the SysTick configuration function which is the only function in the SysTick group.

```
uint32_t SysTick_Config(uint32_t ticks)
```

This function configures the countdown value of the SysTick timer and enables its interrupt so an exception will be raised when its count reaches zero. Since the systemInit() function sets the global variable SystemCoreClock with the CPU frequency that is also used you the SysTick timer we can easily setup the SysTick timer to generate a desired periodic interrupt. So a one millisecond interrupt can be generated as follows.

```
SysTick_Config(SystemCoreClock/1000);
```

Again we can look up the exception handler from the vector table:

```
DCD  0 ;              Reserved
DCD  PendSV_Handler;  PendSV Handler
DCD  SysTick_Handler; SysTick Handler
```

and create a matching "C" function;

```
void SysTick_Handler (void);
```

Now that we have two interrupt sources we can use other CMSIS interrupt and exception functions to manage the priority levels. The number of priority levels will depend on how many priority bits have been implemented by the silicon manufacturer. For all of the Cortex-M processors we can use a simple "flat" priority scheme where zero is the highest priority. The priority level is set by:

```
NVIC_SetPriority(IRQn_Type IRQn,uint32_t priority);
```

The NVIC_SetPriority() function is a bit more intelligent than a simple macro. It uses the IRQn NVIC channel number to differentiate between user peripherals and the Cortex-M processor exceptions. This allows it to program either the system handler priority registers in the system control block or the interrupt priority registers in the NVIC itself. The NVIC_SetPriority() function also uses NVIC_PRIO_BITS definition to shift the priority value into the active priority bits which have been implemented by the silicon vendor.

```
_STATIC_INLINE void NVIC_SetPriority(IRQn_Type IRQn, uint32_t priority)
{
  if(IRQn<0) {
   SCB->SHP[((uint32_t)(IRQn) & 0xF)-4] = ((priority << (8 - __NVIC_PRIO_BITS)) & 0xff);
   }                         /* set Priority for Cortex-M System Interrupts */
  else {
   NVIC->IP[(uint32_t)(IRQn)] = ((priority << (8 - __NVIC_PRIO_BITS)) & 0xff); } /* set
   Priority for device specific Interrupts */
}
```

However for Cortex-M3/M4 and Cortex-M7 we have the option to set priority groups and subgroups as discussed in Chapter 3 "Cortex-M Architecture." Depending on the number of priority bits defined by the manufacturer we can configure priority groups and subgroups.

```
NVIC_SetPriorityGrouping();
```

To set the NVIC priority grouping you must write to the "Application Interrupt and Reset Control" Register. As discussed in Chapter 3 "Cortex-M Architecture" this register is protected by its VECTKEY field. In order to update this register you must write "0x5FA" to the VECTKEY field. The SetPriorityGrouping() function provides all the necessary code to do this.

```
__STATIC_INLINE void NVIC_SetPriorityGrouping(uint32_t PriorityGroup)
{
  uint32_t reg_value;
  uint32_t PriorityGroupTmp=(PriorityGroup&(uint32_t)0x07); /* only values 0..7 are used */
  reg_value = SCB->AIRCR;                            /* read old register configuration */
  reg_value &= ~(SCB_AIRCR_VECTKEY_Msk|SCB_AIRCR_PRIGROUP_Msk);/* clear bits to change */
  reg_value = (reg_value | ((uint32_t)0x5FA << SCB_AIRCR_VECTKEY_Pos) |/* Insert write
  key and priority group */
    (PriorityGroupTmp << 8));
SCB->AIRCR = reg_value;
}
```

The "Interrupt and Exception" group also provides a system reset function that will generate a hard reset of the whole microcontroller.

```
NVIC_SystemReset(void);
```

This function writes to bit two of the "Application Interrupt Reset Control" Register. This strobes a logic line out of the Cortex-M Core to the microcontroller reset circuitry which resets the microcontroller peripherals and the Cortex-M processor. However, you should be a little careful here as the implementation of this feature is down to the microcontroller manufacturer and may not be fully implemented. So if you are going to use this feature you need to test it first. Bit zero of the same register will do a warm reset of the Cortex-M processor. That is force a reset of the Cortex-M processor but leave the microcontroller registers configured.

Exercise 4.1 CMSIS and User Code Comparison

In this exercise we will revisit the multiple interrupts example and examine a rewrite of the code using the CMSIS-Core functions.

Open the Pack Installer.

Select the Boards::Designers Guide Tutorial.

Select the example tab and Copy "EX 4.1 CMSIS Multiple Interrupt".

Select the example tab and Copy "EX 3.3 Multiple Interrupts".

Open main.c in both projects and compare the initializing code.

The SysTick timer and ADC interrupts can be initialized with the following CMSIS functions.

```
SysTick_Config(SystemCoreClock / 100);
NVIC_EnableIRQ          (ADC1_2_IRQn);
NVIC_SetPriorityGrouping (5);
NVIC_SetPriority          (SysTick_IRQn,4);
NVIC_SetPriority          (ADC1_2_IRQn,4);
```

Or you can use the equivalent non-CMSIS code. . ..

```
SysTick->VAL = 0x9000;          //Start value for the sys Tick counter
SysTick->LOAD = 0x9000;         //Reload value
SysTick->CTRL = SYSTICK_INTERRUPT_ENABLE      |SYSTICK_COUNT_ENABLE;     //Start and
enable interrupt
NVIC->ISER[0] = ( 1UL ≪ 18);       /* enable ADC Interrupt */
NVIC->IP[18] = (2≪6 | 2≪4);
SCB->SHP[11] = (1≪6 | 3≪4);
Temp = SCB->AIRC;
Temp & = ~0x
Temp = Temp|(0xAF0≪ )|(0x05≪);
```

Although both blocks of code achieve the same thing, the CMSIS version is much faster to write more readable and far less prone to coding mistakes.

Build both projects and compare the size of the code produced.

The CMSIS functions introduce a small overhead but this is an acceptable trade off against ease of use and maintainability.

CMSIS-Core Register Access

The next group of CMSIS functions gives you direct access to the processor CPU registers (Table 4.7).

Table 4.7: CMSIS CPU register functions

Core Function	Description
__get_Control	Read the control register
__set_Control	Write to the control register
__get_IPSR	Read the IPSR register
__get_APSR	Read the APSR register
__get_xPSR	Read the xPSR register
__get_PSP	Read the process stack pointer
__set_PSP	Write to the process stack pointer
__get_MSP	Read the main stack pointer
__set_MSP	Write to the main stack pointer
__get_PRIMASK	Read the PRIMASK
__set_PRIMASK	Write to the PRIMASK
__get_BASEPRI	Read the BASEPRI register
__set_BASEPRI	Write to the BASEPRI register
__get_FAULTMASK	Read the FAULTMASK
__set_FAULTMASK	Write to the FAULTMASK
__get_FPSCR	Read the FPSCR
__set_FPSCR	Write to the FPSCR
__enable_irq	Enable interrupts and configurable fault exceptions
__disable_irq	Disable interrupts and configurable fault exceptions
__enable_fault_irq	Enables interrupts and all fault handlers
__disable_fault_irq	Disables interrupts and all fault handlers

These functions provide you with the ability to globally control the NVIC interrupts and set the configuration of the Cortex-M processor into its more advanced operating mode. First we can globally enable and disable the microcontroller interrupts with the following functions.

```
__set_PRIMASK(void);
__set_FAULTMASK (void);
__enable-IRQ
__enable_Fault-irq
__set_BASEPRI()
```

While all of these functions are enabling and disabling interrupt sources, they all have slightly different effects. The __set_PRIMASK() function and the enable_IRQ/Disable_IRQ functions have the same effect in that they set and clear the PRIMASK bit which enables and disables all interrupt sources except the Hard Fault Handler and the Non-Maskable interrupt. The __set_FAULTMASK() function can be used to disable all interrupts except the Non-Maskable interrupt. We will see later how this can be useful when we want to bypass the Memory protection unit. Finally the __set_BASEPRI() function sets the minimum active priority level for user peripheral interrupts. When the Base priority register is set to a nonzero level and interrupt at the same priority level or lower will be disabled.

These functions allow you to read the program status register and its aliases. You can also access the control register to enable the advanced operating modes of the Cortex-M processor as well as explicitly setting the stack pointer values. A dedicated function is also provided to access the "Floating Point tatus and Control" register if you are using the Cortex-M4 or Cortex-M7. We will have a closer look at the more advanced operating modes of the Cortex-M processor in Chapter 5 "Advanced Architecture Features".

CMSIS-Core CPU Intrinsic Instructions

The CMSIS-Core header also provides two groups of standardized intrinsic functions. The first group is common to all Cortex-M processors and the second provides standard intrinsic for the Cortex-M4 SIMD instructions (Table 4.8).

Table 4.8: CMSIS instruction intrinsics

CMSIS Function	Description	More Information
__NOP	No operation	
__WFI	Wait for interrupt	
__WFE	Wait for event	
__SEV	Send event	See Chapter 3 "Cortex-M Architecture"
__ISB	Instruction synchronization barrier	
__DSB	Data synchronization barrier	
__DMD	Data memory synchronization barrier	

(Continued)

Table 4.8: (Continued)

CMSIS Function	Description	More Information
__REV	Reverse byte order (32 bit)	See Chapter 4 "Cortex Microcontroller Software Interface Standard" for rotation instructions
__REV16	Reverse byte order (16 bit)	
__REVSH	Reverse byte order, signed short	
__RBIT	Reverse bit order (not for Cortex-M0)	
__ROR	Rotate right by *n* bits	
__LDREXB	Load exclusive (8 bits)	
__LDREXH	Load exclusive (16 bits)	
__LDREXW	Load exclusive (32 bits)	See Chapter 5 "Advanced Architecture Features" for exclusive access instructions
__STREXB	Store exclusive (8 bits)	
__STREXH	Store exclusive (16 bits)	
__STREXW	Store exclusive (32 bits)	
__CLREX	Remove exclusive lock	
__SSAT	Signed saturate	See Chapter 3 "Cortex-M Architecture"
__USAT	Unsigned saturate	
__CLZ	Count leading zeros	See Chapter 4 "Cortex Microcontroller Software Interface Standard"

The CPU intrinsics provide direct access to Cortex-M processor instructions that are not directly reachable from the "C" language. Using an intrinsic will allow a dedicated single cycle instruction to replace multiple instructions generated by standard "C" code.

With the CPU intrinsic we can enter the low power modes using the __WFI() and __WFE() instructions. The CPU intrinsics also provide access to the saturated math's instructions that we met in Chapter 3 "Cortex-M Architecture". The intrinsic functions also give access to the execution barrier instructions that ensure completion of a data write or instruction execution before continuing with the next instruction. The next group of instruction intrinsics is used to guarantee exclusive access to a memory region by one region of code. We will have a look at these in Chapter 5 "Advanced Architecture Features." The remainder of the CPU intrinsics supports single cycle data manipulation functions such as the rotate and reverse bit order instructions.

Exercise 4.2 Intrinsic Bit Manipulation

In this exercise we will look at the data manipulation intrinsic supported in CMSIS.

Open the Pack Installer.

Select the Boards::Designers Guide Tutorial.

Select the example tab and Copy "EX 4.2 CMSIS-Core Intrinsic".

The exercise declares an input variable and a group of output variables and then uses each of the intrinsic data manipulation functions.

```
outputREV    = __REV(input);
outputREV16  = __REV16(input);
outputREVSH  = __REVSH(input);
outputRBIT   = __RBIT(input);
outputROR    = __ROR(input,8);
outputCLZ    = __CLZ(input);
```

Build the project and start the debugger.

Add the input and each of the output variables to the watch window.

Step through the code and count the cycles taken for each function.

While each intrinsic instruction takes a single cycle, some surrounding instructions are required so the intrinsic functions take between 9 and 18 cycles.

Examine the values in the output variables to familiarize yourself with the action of each intrinsic (Fig. 4.8).

Name	Value	Type
input	0x00112233	unsigned int
outputREV	0x33221100	unsigned int
outputREV16	0x11003322	unsigned int
outputREVSH	0x00003322	int
outputRBIT	0xCC448800	unsigned int
outputROR	0x33001122	unsigned int
outputCLZ	0x0000000B	unsigned int

Figure 4.8
Results of the intrinsics operations.

Consider how you would code each intrinsic using standard "C" instructions.

CMSIS-SIMD Intrinsics

The next group of CMSIS intrinsics provides direct access to the Cortex-M4 and Cortex-M7 SIMD instructions.

The SIMD instructions provide simultaneous calculations for two 16-bit operations or four 8-bit operations. This greatly enhances any form of repetitive calculation over a data set, as in a digital filter and we will take a close look at these instructions in Chapter 8 "Practical DSP for Cortex-M4 and Cortex-M7."

CMSIS-Core Debug Functions

The CMSIS-Core functions also provide enhanced debug support through the CoreSight Instrumentation Trace. The CMSIS standard has two dedicated debug specifications CMSIS-SVD and CMSIS-DAP which we will look at in Chapter 7 "Debugging with CoreSight." However, the CMSIS-Core specification contains some useful debug support.

Hardware Breakpoint

First of all there is a dedicated intrinsic to add a hardware breakpoint to your code.

```
__BKPT(uint8_t value)
```

Using this intrinsic will place a hardware breakpoint instruction at this location in your code. When this point is reached execution will be halted and the "value" will be passed to the debugger. During development the __BKPT() intrinsic can be used to trap error conditions and halt the debugger.

Instrumentation Trace

As part of its hardware debug system, the Cortex-M3, Cortex-M4, and Cortex-M7 provide an "Instrumentation Trace" unit (ITM). This can be thought of as a debug UART which is connected to a console window in the debugger. By adding debug hooks (Instrumenting) into your code it is possible to read and write data to and from the debugger while the code is running. We will look at using the Instrumentation Trace for additional debug and software testing in Chapter 7 "Debugging with CoreSight." For now there are a number of CMSIS functions that standardize communication with the ITM (Table 4.9).

Table 4.9: CMSIS debug functions

CMSIS Debug Function	Description
volatile int ITM_RxBuffer = ITM_RXBUFFER_EMPTY;	Declare one word of storage for receive flag
ITM_SendChar(c);	Send one character to the ITM
ITM_CheckChar()	Check if any data has been received
ITM_ReceiveChar()	Read one character from the ITM

CMSIS-Core Functions for Corex-M7

With the release of the Cortex-M7 processor at the end of 2014 the CMSIS-Core specification was extended to provide some additional functions to support the new features introduced by the Corex-M7 (Table 4.10).

Table 4.10: CMSIS Cortex-M7 function

CMSIS Cortex-M7 Function	Description
Cache functions	Eleven functions to support the Instruction and Data caches
FPU function	One function to support the FPU

As we will see in Chapter 6 "Cortex-M7 Processor," the Cortex-M7 introduces Data and Instruction caches to the Cortex-M processor family. The CMSIS cache functions allow you to enable and disable the caches and manage them as your code executes. We will look at these functions in Chapter 6 "Cortex-M7 Processor."

Conclusion

A good understanding of each CMSIS specification is key to effectively developing applications for any Cortex-M-based microcontroller. In this chapter we have introduced each CMSIS specification and taken a detailed look at the CMSIS-Core specification. We will look at the remaining CMSIS specifications through the rest of this book.

CHAPTER 5

Advanced Architecture Features

Introduction

In the last few chapters, we have covered most of what you need to know to develop with a Cortex-M-based microcontroller. In this chapter, we will look at some of the more advanced features of the Cortex-M processor. All of the features discussed in this chapter are included in the Cortex-M0+, -M3, -M4, and -M7. In this chapter, we will look at the different operating modes built into each of the Cortex-M processors and some additional instructions that are designed to support the use of a real-time operating System (RTOS). We will also have a look at the optional Memory Protection Unit (MPU) which can be fitted to the Cortex-M0+, -M3, -M4, and -M7 and how this can partition the memory map. This provides controlled access to different regions of memory depending on the running processor mode. To round the chapter off, we will have a look at the bus interface between the Cortex-M processor and the microcontroller system.

Cortex Processor Operating Modes

When the Cortex-M processor comes out of reset, it is running in a simple "flat" mode where all of the application code has access to the full processor address space and unrestricted access to the CPU and NVIC registers. While this is ok for many applications, the Cortex-M processor has a number of features that let you place the processor into a more advanced operating mode that is suitable for high-integrity software and also supports a RTOS.

As a first step to understanding the more advanced operating modes of the Cortex-M processor, we need to understand its operating modes. The CPU can be running it two different modes, Thread mode and Handler mode. When the processor is executing background code (ie, noninterrupt code) it is running in Thread mode. When the processor is executing interrupt code, it is running in Handler mode (Fig. 5.1).

The Designer's Guide to the Cortex-M Processor Family.
DOI: http://dx.doi.org/10.1016/B978-0-08-100629-0.00005-0

		Operations (privilege out of reset)	Stacks (Main out of reset)
Modes (Thread out of reset)	Handler - Processing of exceptions	**Privileged execution Full control**	**Main Stack Used by OS and Exceptions**
	Thread - No exception is being processed - Normal code execution	**Privileged or Unprivileged**	**Main or Process**

Figure 5.1
Each Cortex-M processor has two execution modes, Handler (interrupt) and Thread (background). It is possible to configure these modes to have privileged and unprivileged access to memory regions. It is also possible to configure a two-stack operating mode.

When the processor starts to run out of reset, there is no operating difference between Thread and Handler mode. Both modes have full access to all features of the CPU, this is known as privileged mode. By programming the Cortex-M processor CONTROL register, it is possible to place the Thread mode in unprivileged mode by setting the thread privilege level (TPL) bit (Fig. 5.2).

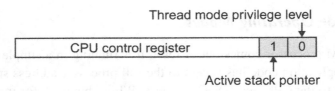

Figure 5.2
The Control register is a CPU register which can only be accessed by the MRS and MSR instructions. It contains two bits which configure the Thread mode privilege level and activation of the process stack pointer.

In unprivileged mode, the Move Register to Special Register (MRS), Move Special Registers to Register (MSR), and Change Processor State (CPS) instructions are disabled for all special CPU registers, except for APSR. This prevents the Cortex-M processor from accessing the CONTROL, FAULTMASK, and PRIMASK registers, and the PROGRAM STATUS register (except the APSR). In unprivileged mode, it is also not possible to access the SysTick timer registers, NVIC, or the System Control Block. This limits the possibility of unprivileged code accidentally disturbing the operation of the Cortex-M processor. If Thread mode has been limited to unprivileged access, it is not possible to clear the Thread mode privilege level bit even if the CPU is running in Handler mode with privilege access. Once the TLP bit has been set, the application code running in Thread mode can no longer influence the operation of the Cortex-M processor. When the processor responds to an exception or interrupt, it moves into Handler mode, which always executes code in privileged mode regardless of the contents of the CONTROL register. The CONTROL register also contains an additional bit, the Active Stack Pointer Selection (ASPEL). Setting this bit enables an additional stack called the Process Stack Pointer (PSP). The CONTROL register is a CPU register rather than a memory-mapped register and can only be accessed by the MRS and MSR instructions. The CMSIS-core specification provides dedicated functions to read and write to the CONRTOL register.

```
void __set_CONTROL(uint32_t value);
uint32_t   __get_CONTROL(void);
```

The Process Stack Pointer is a banked R13 stack pointer which is used by code running in Thread mode. When the Cortex-M processor responds to an exception it enters Handler mode. This causes the CPU to switch stack pointers. This means that the Handler mode will use the Main Stack Pointer (MSP) while Thread mode uses the Process Stack Pointer (Fig. 5.3).

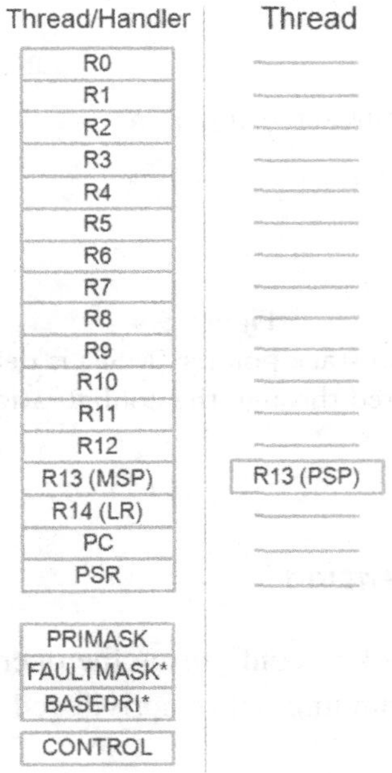

Figure 5.3
At reset, R13 is the main stack pointer and is automatically loaded with the initial stack value. The CPU control register can be used to enable a second banked R13 register. This is the Process Stack which is used in Thread mode. The application code must load an initial stack value into this register.

As we have seen in Chapter 3 "Cortex-M Architecture" at reset, the MSP will be loaded with the value stored in the first 4 bytes of memory. However, the Process Stack Pointer is not automatically initialized and must be set up by the application code before it is enabled. Fortunately the CMSIS-Core specification contains functions to configure the Process Stack.

```
void __set_PSP(uint32_t TopOfProcStack);
uint32_t __get_PSP(void);
```

So, if you have to manually set the initial value of the Process Stack, what should it be? There is not an easy way to answer this, but the compiler produces a report file that details the static calling tree for the project. This file is created each time the project is built and is called <project name>. htm. The report file includes a value for the maximum stack usage and a calling tree for the longest call chain.

This calling tree is likely to be for background functions and will be the maximum value for the Process Stack Pointer. This value can also be used as a starting point for the MSP (Fig. 5.4).

```
Stack_Size    EQU    0x00000400
              AREA    STACK, NOINIT, READWRITE, ALIGN = 3
Stack_Mem    SPACE    Stack_Size
__initial_sp
```

Option	Value
⊟ Stack Configuration	
Stack Size (in Bytes)	0x400
⊟ Heap Configuration	
Heap Size (in Bytes)	0x0

Figure 5.4
The stack size allocated to the main stack pointer (MSP) is defined in the startup code and can be configured through the configuration wizard.

Exercise 5.1 Stack Configuration

In this exercise, we will have a look at configuring the operating mode of the Cortex-M processor so the Thread mode is running with unprivileged access and uses the Process Stack Pointer.

Open the Pack Installer.

Select the Boards::Designers Guide Tutorial.

Select the Example tab and Copy "EX 5.1 Process Stack Configuration."

Build the code, start the debugger, and run to main().

This is a version of the Blinky project we used earlier with some code added to configure the processor operating mode. The new code includes a set of #defines.

```
#define USE_PSP_IN_THREAD_MODE  (1<<1)
#define THREAD_MODE_IS_UNPRIVILIGED 1
#define PSP_STACK_SIZE      0x200
```

The first two declarations define the location of the bits which need to be set in the control register to enable the Process Stack Pointer and switch the Thread mode into unprivileged access. Then, we define the size of the Process Stack space. At the start of main, we can use the CMSIS functions to configure and enable the Process Stack Pointer. We can also examine the operating modes of the processor in the Register window (Fig. 5.5).

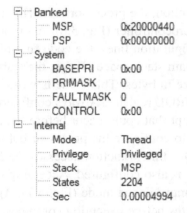

Figure 5.5
The Cortex-M processor is in Thread/privileged mode using the main stack. The PSP is not initialized.

```
 _initalPSPValue = __get_MSP() + PSP_STACK_SIZE;
_set_PSP(initalPSPValue);
__set_CONTROL(USE_PSP_IN_THREAD_MODE);
__ISB();
__ISB();
```

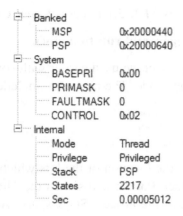

Figure 5.6
Now the processor is in Thread/privileged mode but is using the Process Stack Pointer which has been initialized with a stack space of 200H bytes.

When you reach the main() function, the processor is in Thread mode with full privileged access to all features of the microcontroller (Fig. 5.6). Also, only the MSP is being used. If you step through the three configuration lines, the code first reads the contents of the MSP. This will be at the top of the main stack space. To get the start address for the PSP, we simply add the desired stack size in bytes. This value is written to the Process Stack Pointer before enabling it in the CONTROL register. Always configure the stack before enabling it in case there is an active interrupt that could occur before the stack is ready. Next, we need to execute any code that needs to configure the processor before switching the Thread mode to unprivileged access. The ADC_Init() function accesses the NVIC to configure an interrupt and the SysTick timer is also configured. Accessing these registers will be prohibited when we switch to unprivileged mode (Fig. 5.7). Again, an instruction barrier is used to ensure the code completes before execution continues.

```
ADC_Init();
    SysTick_Config(SystemCoreClock / 100);
__set_CONTROL(USE_PSP_IN_THREAD_MODE
        |THREAD_MODE_IS_UNPRIVILIGED);
__ISB();
```

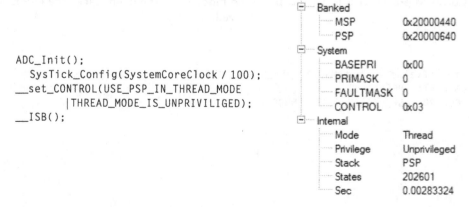

Figure 5.7
Now the processor has been set into Thread/unprivileged mode.

Set a breakpoint in the IRQ.c module line 32.

This is in the SysTick interrupt handler routine. Now, run the code and it will hit the breakpoint when the SysTick handler interrupt is raised (Fig. 5.8).

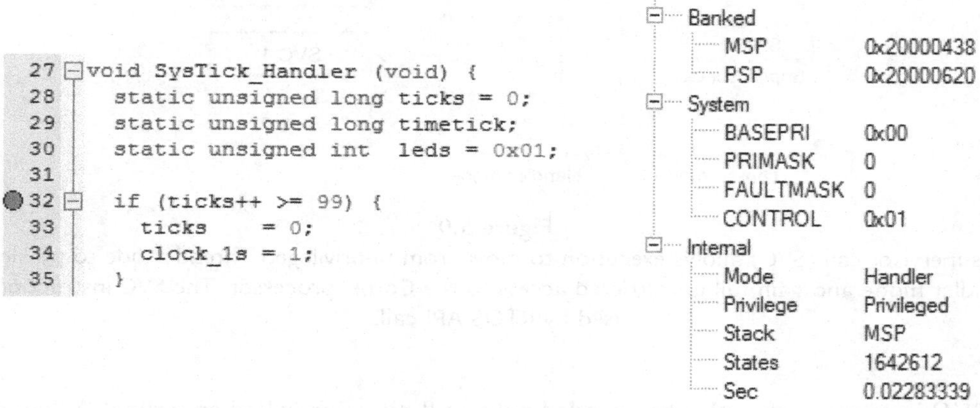

```
27 □ void SysTick_Handler (void) {
28       static unsigned long ticks = 0;
29       static unsigned long timetick;
30       static unsigned int  leds = 0x01;
31
● 32 □    if (ticks++ >= 99) {
33          ticks   = 0;
34          clock_1s = 1;
35       }
```

```
□ Banked
    MSP        0x20000438
    PSP        0x20000620
□ System
    BASEPRI    0x00
    PRIMASK    0
    FAULTMASK  0
    CONTROL    0x01
□ Internal
    Mode       Handler
    Privilege  Privileged
    Stack      MSP
    States     1642612
    Sec        0.02283339
```

Figure 5.8
During the exception the processor enters Handler/privileged mode.

Now that the processor is serving an interrupt it has moved into interrupt Handler mode with privileged access to the Cortex-M processor and is using the main stack.

Supervisor Call

Once configured, this more advanced operating mode provides a partition between the exception/interrupt code running in Handler mode and the background application code running in Thread mode. Each operating mode can have its own code region, RAM region, and stack. This allows the interrupt Handler code full access to the chip without the risk that it may be corrupted by the application code. However, at some point, the application code will need to access features of the Cortex-M processor that are only available in the Handler mode with its full privileged access. To allow this to happen, the Thumb-2 instruction set has an instruction called supervisor call (SVC). When this instruction is executed, it raises a supervisor exception which moves the processor from executing the application code in thread/unprivileged mode to an exception routine in handler/privileged mode. The SVC has its own location within the vector table and behaves like any other exception (Fig. 5.9).

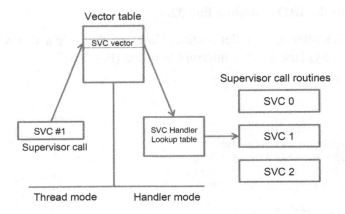

Figure 5.9
The supervisor call (SVC) allows execution to move from unprivileged Thread mode to privileged Handler mode and gain full unrestricted access to the Cortex processor. The SVC instruction is used by RTOS API calls.

The SVC instruction may also be encoded with an 8-bit value called an ordinal. When the SVC call is executed, this ordinal value can be read and used as an index to call one of 256 different supervisor functions (Fig. 5.10).

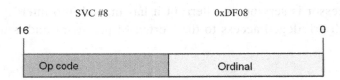

Figure 5.10
The unused portion of the SVC instruction can be encoded with an ordinal number. On entry to the SVC handler, this number can be read to determine which SVC functions to execute.

The compiler toolchain provides a dedicated SVC support function that is used to extract the ordinal value and then call the appropriate function. First, the SVC support function reads the link register to determine the operating mode, then it reads the value of the saved PC from the appropriate stack. We can then read the memory location holding the SVC instruction and extract the ordinal value. This number is then used as an index into a lookup table to load the address of the function which is being called. The function is then called and is executed in privileged mode before we return back to the application code running in unprivileged thread mode. This mechanism may seem an overly complicated way of calling a function, but it provides the basis of a supervisor/user split where an operating system (OS) is running in privileged mode and acts as a supervisor to application threads running in unprivileged thread mode. This way the individual threads do not have access to critical processor features except by making API calls to the OS.

Exercise 5.2 Supervisor Call

In this exercise, we will look at calling some functions with the SVC instruction rather than branching to the routine as in a standard function call.

Open the Pack Installer.

Select the Boards::Designers Guide Tutorial.

Select the Example tab and Copy "EX 5.2 Supervisor Call"

First, let us have a look at the project structure (Fig. 5.11).

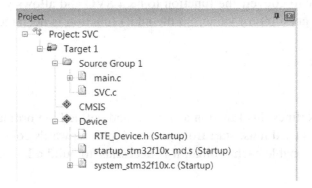

Figure 5.11
SVC instructions are supported by adding the additional SVC.c module. This module provides code to "decode" the SVC instruction + ordinal.

The project consists of the standard project startup file and the initializing system file. The application source code is in the file main.c. There is an additional source file SVC.c that provides support for handling SVC exceptions. The SVC.c file contains the SVC exception handler, this is a standard support file that is provided with the ARM compiler. We will have a closer look at its operation later. The application code in main.c is calling two simple functions that in turn call routines to perform basic arithmetic operations.

```
int main (void) {
   test_a();
   test_t();
while(1);
}
void test_a (void) {
   res  = add (74, 27);
   res + = mul4(res);
}
void test_t (void) {
   res  = div (res, 10);
   res  = mod (res, 3);
}
```

Each of the arithmetic functions is designed to be called with a SVC instruction so that all of these functions run in handler mode rather than thread mode. In order to convert the arithmetic functions from standard functions to software interrupt functions, we need to change the way the function prototype is declared. The way this is done will vary between compilers but in the ARM compiler there is a function qualifier __svc. This is used as shown below to convert the function to be a SVC and allows you to pass up to four parameters and get a return value. So the add() function is declared as follows:

```
int __svc(0) add (int i1, int i2);
int __SVC_0    (int i1, int i2) {
   return (i1 + i2);
}
```

The __svc qualifier defines this function as a SVC and defined the ordinal number of the function. The ordinals used must start from zero and grow upwards contiguously to a maximum of 256. To enable each ordinal, it is necessary to build a lookup table in the SVC.c file.

```
; Import user SVC functions here.
         IMPORT __SVC_0
         IMPORT __SVC_1
         IMPORT __SVC_2
         IMPORT __SVC_3
SVC_Table
; Insert user SVC functions here
         DCD    __SVC_0
         DCD    __SVC_1      ;
         DCD    __SVC_2      ;
         DCD    __SVC_3      ;
```

You must import the label used for each supervisor function and then add the function labels to the SVC table. When the code is compiled, the labels will be replaced by the entry address of each function.

In the project build the code and start the simulator. Step the code until you reach line 61, the call to the add function.

The following code is displayed in the disassembly window, and in the register window we can see that the processor is running in thread mode (Fig. 5.12).

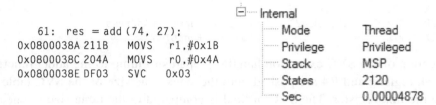

```
    61:  res = add (74, 27);
0x0800038A 211B   MOVS   r1,#0x1B
0x0800038C 204A   MOVS   r0,#0x4A
0x0800038E DF03   SVC    0x03
```

⊟ Internal	
Mode	Thread
Privilege	Privileged
Stack	MSP
States	2120
Sec	0.00004878

Figure 5.12
Prior to the SVC instruction, the processor is running in Thread mode.

The function parameters are loaded into the parameter passing registers R0 and R1, and the normal branch instruction is replaced by an SVC instruction. The SVC instruction is encoded with an ordinal value of 3. If you make the disassembly window the active window and step through these instructions, the SVC exception will be raised and you will enter the SVC handler in SVC.c. In the Registers window, you can also see that the processor is now running in handler mode (Fig. 5.13).

```
__asm void SVC_Handler (void) {
    PRESERVE8
        TST    LR,#4        ; Called from Handler Mode?
        MRSNE  R12,PSP      ; Yes, use PSP
        MOVEQ  R12,SP       ; No, use MSP
        LDR    R12,[R12,#24]   ; Read Saved PC from Stack
        LDRH   R12,[R12,#-2]   ; Load Halfword
        BICS   R12,R12,#0xFF00 ; Extract SVC Number
```

⊟ Internal	
Mode	Handler
Privilege	Privileged
Stack	MSP
States	2150
Sec	0.00004919

Figure 5.13
Once the SVC instruction has been executed, the provesor will be running in Handler mode.

The first section of the SVC_Handler code works out which stack is in use and then reads the value of the program counter saved on the stack. The program counter value is the return address, so the load instruction deducts 2 to get the address of the SVC instruction.

This will be the address of the SVC instruction that raised the exception. The SVC instruction is then loaded into R12 and the ordinal value is extracted. The code is using R12 because the ARM binary interface standard defines R12 as the "Intra Procedure Call Scratch Register"; this means it will not contain any program data and is free for use.

```
PUSH    {R4,LR}                     ; Save Registers
        LDR     LR,=SVC_Count
        LDR     LR,[LR]
        CMP     R12,LR
        BHS     SVC_Dead            ; Overflow
        LDR     LR,=SVC_Table
        LDR     R12,[LR,R12,LSL #2]  ; Load SVC Function Address
        BLX     R12                 ; Call SVC Function
```

The next section of the SVC exception handler prepares to jump to the add() function. First, the link register and R4 are pushed onto the stack. The size of the SVC table is loaded into the link register. The SVC ordinal is compared to the table size to check it is less than the SVC table size and hence a valid number. If it is valid, the function address is loaded into R12 from the SVC table and the function is called. If the ordinal number has not been added to the table, the code will jump to a trap called SVC_DEAD. Although R4 is not used in this example, it is preserved on the stack as it is possible for the called function to use it.

```
        POP     {R4,LR}
        TST     LR,#4
        MRSNE   R12,PSP
        MOVEQ   R12,SP
        STM     R12,{R0-R3}         ; Function return values
        BX      LR                  ; RETI
```

Once the SVC function has been executed, it will return back to the SVC exception handler to clean up before returning to the background code in handler mode.

PEND_SV Exception

The Cortex-M0+, -M3, -M4, and -M7 have an additional processor exception called the PENDSV exception. The PENDSV exception has been added to the Cortex-M processor primarily to support a RTOS. We will take a closer look at how an RTOS uses the PENDSV exception in Chapter 9 "CMSIS-RTOS" but for now we will look at how it works. The PENDSV exception can be thought of as an NVIC interrupt channel that is connected to a processor register rather than a microcontroller peripheral. A PENDSV exception can be raised by the application software writing to the PENDSV register, the PENDSV exception is then handled in the same way as any other exception.

Example Pend_SV

In this example, we will examine how to use the PENDSV system service call interrupt.

Open the Pack Installer.

Select the Boards::Designers Guide Tutorial.

Select the Example tab and Copy "EX 5.3 PENDSV Exception"

Build the project and start the debugger.

The code initializes the ADC and enables its "End of Conversion Interrupt." It also changes the PENDSV and ADC interrupt priority from their default options (Fig. 5.14). The SVC has priority zero (highest) while the ADC has priority 1, and the PENDSV interrupt has priority 2 (lowest). The systemCode routine uses an SVC instruction to raise an exception and move to Handler mode (Fig. 5.15).

```
NVIC_SetPriority(PendSV_IRQn,2);     //set interrupt priorities
NVIC_SetPriority(ADC1_2_IRQn,1);
NVIC_EnableIRQ(ADC1_2_IRQn);     //enable the ADC interrupt
ADC1->CR1  |= (1UL << 5);         //switch on the ADC and start a conversion
ADC1->CR2  |= (1UL << 0);
ADC1->CR2  |= (1UL << 22);
systemCode();        //call some system code with an SVC interrupt
```

Set a breakpoint on the systemCode() function and run the code.

```
58 | ADC1->CR2    |=  (1UL << 22);
59 | systemCode();
60 |
```

Figure 5.14
Execute the code up to the systemCode() function.

Open the peripherals/core peripherals/Nested vector interrupt controller window and check the priority levels of the SVC, PENDSV, and ADC interrupts.

Idx	Source	Name	E	P	A	Priority
11	System Service Call	SVCALL	1	0	0	0
14	Pend System Service	PENDSV	1	0	0	2
34	ADC Global Interrupt	ADC	1	0	0	1

Figure 5.15
The Peripherals/NVIC window shows that the interrupts are enabled, none are active or pending. We can also see their priority levels.

Now step into the systemCode routine (F11) until you reach the C function.

```
void __svc(0) systemCode (void);
void __SVC_0    (void) {
unsigned int i, pending;
   for(i = 0;i<100;i++);
pending = NVIC_GetPendingIRQ(ADC1_2_IRQn);
if(pending ==1){
     SCB->ICSR |= 1<<28;        //set the pend pend
}else{
     Do_System_Code();
}
     }
```

Inside the systemCode() routine, there is a short loop which represents the critical section of code that must be run. While this loop is running, the ADC will finish conversion and, as it has a lower priority than the SCV interrupt it will enter a pending state. When we exit the loop, we test the state of any critical interrupts by reading their pending bits. If a critical interrupt is pending, then the remainder of the system code routine can be delayed. To do this, we set the PENDSVSET bit in the Interrupt Control and State register and quit the SVC handler.

Set a breakpoint on the exit brace (}) of the systemCode() routine and run the code (Fig. 5.16).

```
  38 │    Do_System_Code();
  39 ├ }
○ 40 │  }
```

Figure 5.16
Run the Do_System_Code() routine.

Now use the NVIC debug window to examine the state of the interrupts (Fig. 5.17).

Idx	Source	Name	E	P	A	Priority
11	System Service Call	SVCALL	1	0	1	0
14	Pend System Service	PENDSV	1	1	0	2
34	ADC Global Interrupt	ADC	1	1	0	1

Figure 5.17
Now the SVC exception is active with the ADC and PENDSV exceptions pending.

Now the SVC is active with the ADC and PENDSV system service call in a pending state.

Single step out of the System Service Call until you enter the next interrupt.

Both of the pending interrupts will be tail chained onto the end of the system service call. The ADC has the highest priority so it will be served next (Fig. 5.18).

Idx	Source	Name	E	P	A	Priority
11	System Service Call	SVCALL	1	0	0	0
14	Pend System Service	PENDSV	1	1	0	2
34	ADC Global Interrupt	ADC	1	0	1	1

Figure 5.18
Now the ADC interrupt is active when this ends, the PENDSV routine will be served and the system code routine will resume.

Step out of the ADC handler and you will immediately enter the PENDSV system service interrupt which allows you to resume execution of the system code that was requested to be executed in the System Service Call interrupt.

Interprocessor Events

The Cortex-M processors are designed so that it is possible to build multiprocessor devices. An example would be to have a Cortex-M4 and a Cortex-M0 within the same microcontroller. The Cortex-M0 will typically manage the user peripherals, while the Cortex-M4 runs the intensive portions of the application code. Alternatively, there are devices which have a Cortex-A9 which can run Linux and manage a complex user interface; on the same chip there are two Cortex-M4 which manage the real-time code. These more complex system-on-chip designs require methods of signaling activity between the different processors. The Cortex-M processors can be chained together by an external event signal. The event signal is set by using a set event instruction. This instruction can be added to your C code using the __SEV() intrinsic provided by the CMSIS-Core specification. When a __SEV() instruction is issued it will wake up the target processor if it has entered a low-power mode using the __WFE() instruction. If the target processor is running the event latch will be set so that when the target processor executes the __WFE() instruction it will reset the event latch and keep running without entering the low-power mode.

Exclusive Access

One of the key features of an RTOS is multitasking support. As we will see in the next chapter this allows you to develop your code as independent threads that conceptually are running in parallel on the Cortex-M processor. As your code develops, the program threads will often need to access common resources be it SRAM or peripherals. An RTOS provides mechanisms called semaphores and mutexes which are used to control access to peripherals and common memory objects (Fig. 5.19).

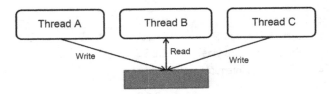

Figure 5.19
In a multiprocessor or multithread environment it is necessary to control access to shared resources or errors such as read before write can occur.

While it is possible to design "memory lock" routines on any processor, the Cortex-M3, -M4, and -M7 provide a set of instructions that can be used to optimize exclusive access routines (Table 5.1).

Table 5.1: Exclusive access instructions

__LDREXB	Load exclusive (8 bits)
__LDREXH	Load exclusive (16 bits)
__LDREXW	Load exclusive (32 bits)
__STREXB	Store exclusive (8 bits)
__STREXH	Store exclusive (16 bits)
__STREXW	Store exclusive (32 bits)
__CLREX	Remove exclusive lock

In earlier ARM processors like the ARM7 and ARM9, the problem of exclusive access was answered by a swap instruction that could be used to exchange the contents of two registers. This instruction took 4 cycles but it was an atomic instruction meaning that once started it could not be interrupted and was guaranteed exclusive access to the CPU to carry out its operation. As Cortex-M processors have multiple busses, it is possible for read and write accesses to be carried out on different busses and even by different bus masters which may themselves be additional Cortex-M processors. On the Cortex-M processor, the new technique of exclusive access instructions has been introduced to support multitasking and multiprocessor environments (Fig. 5.20).

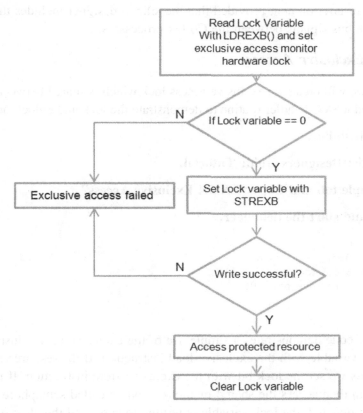

Figure 5.20
The Load and store exclusive instructions can be used to control access to a memory recourse.
They are designed to work with single and multiprocessor devices.

The exclusive access system works by defining a lock variable to protect the shared resource. Before the shared resource can be accessed, the locked variable is checked using the exclusive read instruction; if it is zero, then the shared resource is not currently being accessed. Before we access the shared resource, the lock variable must be set using the exclusive store instruction. Once the lock variable has been set, we now have control of the shared resource and can write to it. If our process is preempted by an interrupt or another thread that also performs an exclusive access read, then a hardware lock in the exclusive access monitor is set, preventing the original exclusive store instruction from writing to the lock variable. This gives exclusive control to the preempting process.

When we are finished with the shared resource, the lock variable must be written to zero; this clears the variable and also removes the lock. If your code starts the exclusive access process but needs to abandon it, there is a clear exclusive (CLREX) instruction that can be used to remove the lock. The exclusive access instructions control access between different processes running on a single Cortex-M processor, but the same technique can be extended to a multiprocessor environment provided that the silicon designer includes the additional monitor hardware bus signals between the Cortex processors.

Exercise 5.4 Exclusive Access

In this exercise, we will create an exclusive access lock which is shared between a background thread process and a SVC handler routine to demonstrate the lock and unlock process.

Open the Pack Installer.

Select the Boards::Designers Guide Tutorial.

Select the Example tab and Copy "EX 5.4 Exclusive Access"

Build the code and start the debugger.

```
int main (void) {
  if(__LDREXB( &lock_bit) = = 0){
    if (!__STREXB(1,& lock_bit = = 0)){
      semaphore + + ; lock_bit = 0;
    }
  }
```

The first block of code demonstrates a simple use of the exclusive access instructions. We first test the lock variable with the exclusive load instruction. If the resource is not locked by another process we set the lock bit with the exclusive store instruction. If this is successful, we can then access the shared memory resource called semaphore. Once this variable has been updated, the lock variable is written to zero, and the clear exclusive instruction releases the hardware lock.

Step through the code to observe its behaviour.

```
if(__LDREXB( &lock_bit) ==0){
  thread_lock();
  if (!__STREXB(1,&lock_bit)){
    semaphore++;
  }
}
```

The second block of code does exactly the same thing except between the exclusive load and exclusive store the function thread lock is called. This is an SVC routine that will enter handler mode and jump to the SVC0 routine.

```
void __svc(0) thread_lock (void);
void __SVC_0 (void) {
  __LDREXB( &lock_bit);
}
```

The SVC routine simply does another exclusive read of the lock variable that will set the hardware lock in the exclusive access monitor. When we return to the original routine and try to execute the exclusive store instruction, it will fail because any exception that happens between LDREX and STREX will cause the STREX to fail. The local exclusive access monitor is cleared automatically at exception entry/exit.

Step through the second block of code and observe the lock process.

Memory Protection Unit

The Cortex-M0+, -M3, -M4, and -M7 processors have an optional MPU which may be included in the processor core by the silicon manufacturer when the microcontroller is designed. The MPU allows you to extend the privileged/unprivileged code model. If it is fitted, the MPU allows you to define regions within the memory map of the Cortex-M processor and grant privileged or unprivileged access to these regions. If the processor is running in unprivileged mode and tries to access an address within a privileged region, a memory protection exception will be raised and the processor will vector to the memory protection ISR. This allows you to detect and correct run-time memory errors. When used with the Cortex-M7, the MPU is used to configure the Instruction and Data Caches present in the Cortex-M7. In the next section, we will look at how to use of the MPU with the Cortex-M0+, -M3, and -M4. We will look at the cache configuration features in Chapter 6 "Cortex-M7 Processor" (Fig. 5.21).

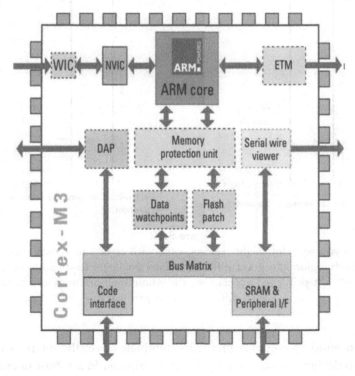

Figure 5.21
The memory protection unit is available on the Cortex-M0+, M3, M4, and M7. It allows you to place a protection template over the processor memory map.

In practice, the MPU allows you to define eight memory regions within the Cortex-M processor address space and grant privileged or unprivileged access to each region. These regions can then be further subdivided into eight equally sized subregions which in turn can be granted privileged or unprivileged access. There is also a default background region which covers the entire 4 GB address space. When this region is enabled, it makes access to all memory locations privileged. To further complicate things, memory regions may be overlapped with the highest numbered region taking precedent. Also any area of memory that is not covered by an MPU region may not have any kind of memory access (Fig. 5.22).

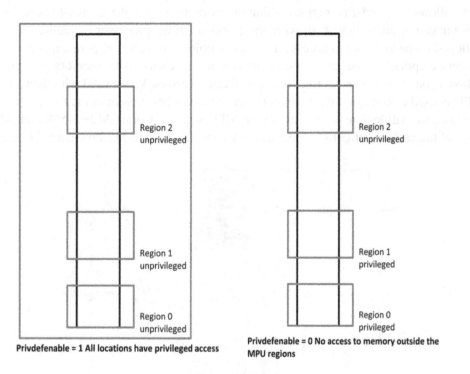

Figure 5.22
The MPU allows you to define eight regions, each with eight subregions over the processor memory map. Each region can grant different access privileges to its address range. It is also possible to set a default privileged access over the whole memory map and then create "holes" with different access privileges.

So it is possible to build up complex protection templates over the memory address space. This allows you to design a protection regime that helps build a robust operating environment but also gives you enough rope to hang yourself.

Configuring the MPU

The MPU is configured through a group of memory-mapped registers located in the Cortex-M processor system block. These registers may only be accessed when the Cortex processor is operating in privileged mode (Fig. 5.23).

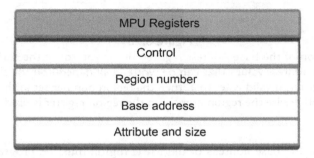

Figure 5.23
Each MPU region is configured through the region, base address, and attribute registers. Once each region is configured, the control register makes the MPU regions active.

The CONTROL register contains three active bits that affect the overall operation of the MPU. These are the PRIVDEFENABLE bit which enables privileged access over the whole 4 GB memory map. The next bit is the HFNMIENA bit; when set, this bit enables the operation of the MPU during a Hard Fault, NMI, or FAULTMASK exception. The final bit is the MPU ENABLE bit; when set, this enables the operation of the MPU. Typically, when configuring the MPU, the last operation performed is to set this bit. After reset, all of these bits are cleared to zero (Fig. 5.24).

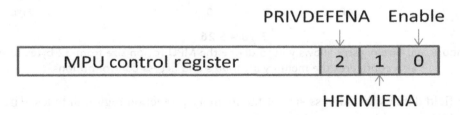

Figure 5.24
The control register allows you to enable the global privileged region (PRIVDEFENA). The enable bit is used to make the configured MPU regions active and HFMIENA is used to enable the regions when the Hard Fault, NMI, or FAULTMASK exceptions are active.

The remaining registers are used to configure the eight MPU regions. To configure a given region (0–7) first select the region by writing its region number into the region number register. Once a region has been selected, it can then be configured by the base address and the attribute and size register. The base address register a 27-bit address field, a valid bit and also a repeat of the MPU region number (Fig. 5.25).

Figure 5.25
The address region of the Base Address register allows you to set the start address of an MPU region. The address values that can be written will depend on the size setting in the Attribute and Size register. If valid is set to 1, then the region number set in the region field is used; otherwise the region number in the region register is used.

As you might expect, the base address of the MPU region must be programmed into the address field. However, the available base addresses that may be used depend on the size of the defined for the region. The minimum size for a region is from 32 bytes upto 4 GB. The base address of a MPU region must be a multiple of the region size. Programming the address field sets the selected regions base address. You do not need to set the valid bit. If you write a new region number into the base address register region field, set the valid bit and write a new address; you can start to configure a new region without the need to update the region number register. Programming the attribute and size register finishes configuration of an MPU region (Fig. 5.26).

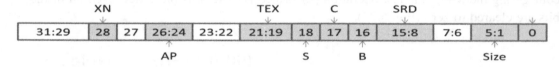

Figure 5.26
The Attribute and Size register allows you to define the MPU region size from 32 bytes to 4 GB. It also configures the memory attributes and access control options.

The size field defines the address size of the memory protection region in bytes. The region size is calculated using the formula:

$$\text{MPU Region memory size} = 2\,\text{POW}(\text{SIZE} + 1)$$

This gives us a minimum size starting at just 32 bytes. As noted above, the selected size also defines the range of possible base addresses. Next, it is possible to set the region attributes and access privileges. Like the Cortex-M processor, the MPU is designed to support multiprocessor systems. Consequently, it is possible to define regions as being shared between Cortex-M processors or as being exclusive to the given processor. It is also possible to define the cache policy for the area of memory covered by the MPU region. Currently, the vast majority of microcontrollers only have a single Cortex-M processor, though asymmetrical multiprocessor devices have started to appear (Cortex-M4 and Cortex-M0). The Cortex-M7 introduces Instruction and Data Caches which are configured through the MPU. We will look at this in more detail in Chapter 6 "Cortex-M7."

The MPU attributes are defined by the TEX, C, B, and S bits and suitable settings for most microcontrollers are shown in Table 5.2.

Table 5.2: Memory region attributes

Memory Region	TEX	C	B	S	Attributes
Flash	000	1	0	0	Normal memory, nonshareable, write through
Internal SRAM	000	1	0	1	Normal memory, shareable, write through
External SRAM	000	1	1	1	Normal memory, shareable, write back write allocate
Peripherals	000	0	1	1	Device memory, shareable

When working with the MPU, we are more interested in defining the access permissions. These are defined for each region in the AP field (Table 5.3).

Table 5.3: Memory access rights

AP	Privileged	Unprivileged	Description
000	No access	No access	All accesses generate a permission fault
001	RW	No access	Access form privileged code only
010	RW	RO	Unprivileged writes cause a permission fault
011	RW	RW	Full access
100	Unpredictable	Unpredictable	Reserved
101	RO	No access	Reads by privileged code only
110	RO	RO	Read only for privileged and unprivileged code
111	RO	RO	

Once the size, attributes, and access permissions are defined, the enable bit can be set to make the region active. When each of the required regions has been defined, the MPU can be activated by setting the global enable bit in the CONTROL register. When the MPU is active and the application code makes an access that violates the permissions of a region, an MPU exception will be raised. Once you enter an MPU exception, there are a couple of registers that provide information to help diagnose the problem. The first byte of the Configurable Fault Status register located in the system control block is called the Memory Manager Fault Status register (Fig. 5.27).

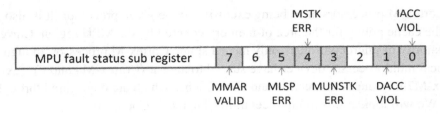

Figure 5.27

The MPU fault status register is a subsection of the configurable fault status register in the system control block. It contains error flags which are set when an MPU exception occurs. The purpose of each flag is shown in Table 5.4.

Table 5.4: Fault status register flag descriptions

Flag	Description
IACCVIOL	Instruction access violation status flag
DACCVIOL	Data access violation status flag
MUNSTKERR	Memory manager fault on unstacking
MSTKERR	Memory manager fault on stacking
MLSPERR	Memory manager FPU lazy stacking error Cortex-M4 only (see Chapter 5: "Advanced Architectural Features")
MMARVALID	Memory manager fault address valid

Depending on the status of the fault conditions, the address of the instruction that caused the memory fault may be written to a second register, the memory manager fault address register. If this address is valid it may be used to help diagnose the fault.

Exercise 5.5 MPU Configuration

In this exercise, we will configure the MPU to work with the Blinky project.

Open the Pack Installer.

Select the Boards::Designers Guide Tutorial.

Select the Example tab and Copy "EX 5.5 MPU Configuration."

This time, the microcontroller used is an NXP LPC1768 which has a Cortex-M3 processor fitted with the MPU. First, we have to configure the project so that there are distinct regions of code and data that will be used by the processor in thread and handler modes. When doing this, it is useful to sketch out the memory map of the application code and then define a matching MPU template (Fig. 5.28).

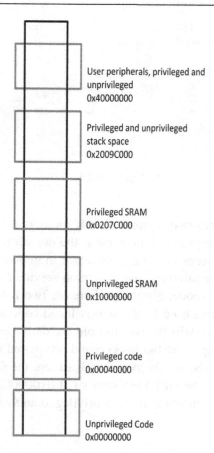

User peripherals, privileged and
unprivileged
0x40000000

Privileged and unprivileged
stack space
0x2009C000

Privileged SRAM
0x0207C000

Unprivileged SRAM
0x10000000

Privileged code
0x00040000

Unprivileged Code
0x00000000

Figure 5.28
The memory map of the Blinky project can be split into six regions.

Region 0   unprivileged application code
Region 1   Privileged system code
Region 2   Unprivileged SRAM
Region 3   Privileged SRAM
Region 4   Privileged and unprivileged stack space
Region 5   Privileged and unprivileged user peripherals

Now open the Options for Target/Target menu. Here, we can set up the memory regions
to match the proposed MPU protection template (Fig. 5.29).

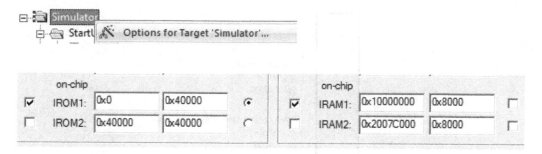

Figure 5.29

The target memory map defines two regions of code memory 0-0x3FFF and 0x4000−0x7FFF. The lower region will be used as the default region to hold the application code and will be accessed by the processor in unprivileged mode. The upper region will be used to hold the interrupt and exception service routines and will be accessed by the processor in privileged mode. Similarly, there are two RAM regions 0x10000000 and 0x100008000 will hold data used by the unprivileged code and the system stacks, while the upper region 0x0207C000—will be used to hold the data used by the privileged code. In this example, we are not going to set the background privileged region, so we must map MPU regions for the peripherals. All the peripherals except the GPIO are in one contiguous block from 0x40000000 while the GPIO registers sit at 0x000002C9. The peripherals will be accessed by the processor while it is in both privileged and unprivileged modes (Fig. 5.30).

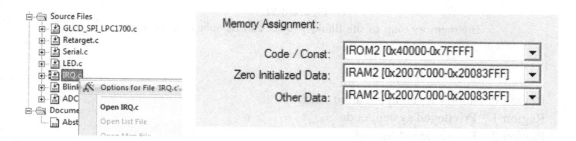

Figure 5.30
Use the local memory assignment options to configure the memory regions used by the interrupt code.

To prepare the code, we need to force the interrupt handler code into the regions which will be granted privileged access. In this example, all the code which will run in Handler mode has been placed in one module. In the local options for this module, we can select the code and data regions which will be given privileged access rights by the MPU. All of the other code and data will be placed in the default memory regions which will run in unprivileged mode. Once the project memory layout has been defined, we can add code to the project to set up the MPU protection template.

```
#define SIZE_FIELD                      1
#define ATTRIBUTE_FIELD                 16
#define ACCESS_FIELD                    24
#define ENABLE                          1
#define ATTRIBUTE_FLASH                 0x4
#define ATTRIBUTE_SRAM                  0x5
#define ATTRIBUTE_PERIPHERAL            0x3
#define PRIV_RW_UPRIV_RW                3
#define PRIV_RO_UPRIV_NONE              5
#define PRIV_RO_UPRIV_RO                6
#define PRIV_RW_UPRIV_RO                2
#define USE_PSP_IN_THREAD_MODE          2
#define THREAD_MODE_IS_UNPRIVILIGED     1
#define PSP_STACK_SIZE                  0x200
#define TOP_OF_THREAD_RAM               0x10007FF0
MPU->RNR     =     0x00000000;
MPU->RBAR    =     0x00000000;
MPU->RASR    =     (PRIV_RO_UPRIV_RO<<ACCESS_FIELD)
                   |(ATTRIBUTE_FLASH<<ATTRIBUTE_FIELD)
                   | (17<<SIZE_FIELD)|ENABLE;
```

The code shown above is used to set the MPU region for the unprivileged thread code at the start of memory. First, we need to set a region number followed by the base address of the region. Since this will be FLASH memory, we can use the standard attribute for this memory type. Next, we can define its access type. In this case, we can grant Read Only access for both privileged and unprivileged modes. Next, we can set the size of the region which is 256K which must equal 2POW(SIZE + 1) which equate to 17. The enable bit is set to activate this region when the MPU is fully enabled. Each of the other regions are programmed in a similar fashion. Finally, the memory management exception and the MPU are enabled.

```
NVIC_EnableIRQ (MemoryManagement_IRQn);
MPU->CTRL = ENABLE;
```

Start the debugger, set a breakpoint on line 82, and run the code.

When the breakpoint is reached, we can view the MPU configuration via the peripherals/ core peripherals/MPU (Fig. 5.31).

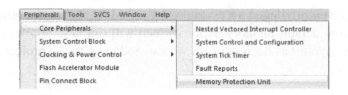

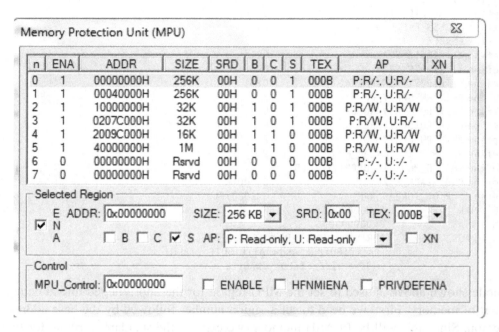

Figure 5.31
The debugger Peripheral windows provide a detailed view of the configured MPU.

Here, we can easily see the regions defined and the access rights that have been granted.

Now run the code for a few seconds and then halt the processor.

An MPU exception has been raised and execution has jumped to the MemManager_Handler (Fig. 5.32).

```
147    MemManage_Handler\
148                      PROC
149                      EXPORT   MemManage_Handler        [WEAK]
150                      B         .
151                      ENDP
```

Figure 5.32
When a MPU exception occurs, the code will vector to the default MemManager_Handler in the startup code.

The question is now, what caused the MPU exception? We can find this out by looking at the memory manager fault and status register (Fig. 5.33).

Open the Peripherals/Core Peripherals/Fault Reports window.

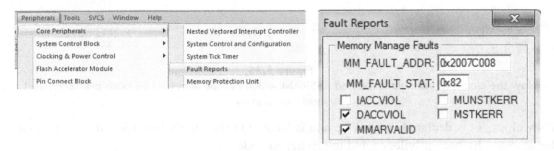

Figure 5.33
The debugger also provides a condensed view of all the fault registers.

Here, we can see that the fault was caused by a data access violation to address 0x2007C008. If we now open the map file produced by the linker, we can search for this address and find what variable is placed at this location.

Highlight the project name in the project window and double click. This will open the map file. Now use the Edit/Find dialog to search the map file for the address 0x2007C008 (Fig. 5.34).

Figure 5.34
You can view the linker MAP file by double clicking on the project root node in the Project window. Then search through this file to find the symbol located at 0x2007C008.

This shows that the variable clock_1s is at 0x2007C008 and that it is declared in irq.c.

Clock_1s is a global variable that is also accessed from the main loop running in unprivileged mode. However, this variable is located in the privileged RAM region so accessing it while the processor is running in unprivileged mode will cause an MPU fault (Fig. 5.35).

Find the declaration of clock_1s in blinky.c and remove the extern keyword.

Now find the declaration for clock_1s in irq.c and add the keyword extern.

Build the code and view the updated map file.

```
__stdin                          0x10000008   Data    4   retarget.o(.data)
clock_1s                         0x1000000c   Data    1   blinky.o(.data)
AD_done                          0x1000000e   Data    1   adc.o(.data)
```

Figure 5.35
Now the clock_1s variable is located in SRAM which can be accessed by both privileged and unprivileged code.

Now clock_1s is declared in blinky.c and is located in the unprivileged RAM region so can be accessed by both privileged and unprivileged code.

Restart the debugger and the code will run without any raising any MPU exception.

MPU Subregions

As we have seen, the MPU has a maximum of eight regions that can be individually configured with location size and access type. Any region which is configured with a size of 256 bytes or more will contain eight equally space subregions. When the region is configured, each of the subregions is enabled and has the default region attributes and access settings. It is possible to disable a subregion by setting a matching subregion bit in the SRD field of the MPU attribute and size register. When a subregion is disabled, a "hole" is created in the region, this "hole" inherits the attributes and access permission of any overlapped region. If there is no overlapped region, then the global privileged background region will be used. If the background region is not enabled then no access rights will be granted and a MPU exception will be raised if an access is made to an address in the subregion "hole" (Fig. 5.36).

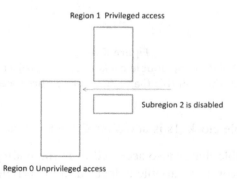

Figure 5.36
Each region has eight subregions. If a subregion is disabled, it inherits the access rights from an overlapped region or the global background region.

If we have two overlapped regions, the region with the highest region number will take precedence. In the case above, an unprivileged region is overlapped by a privileged region. The overlapped section will have privileged access. If a subregion is disabled in region 1, then the access rights in region 0 will be inherited and grant unprivileged access to the subregion range of addresses.

MPU Limitations

When designing your application software to use the MPU it is necessary to realize that the MPU only monitors the activity of the Cortex-M processor. Many, if not most, Cortex−M-based microcontrollers have other peripherals, such as DMA units, which are capable of autonomously accessing memory and peripheral registers. These units are additional "Bus Masters" which arbitrate with the Cortex-M processor to gain access to the microcontroller resources. If such a unit makes an access to a prohibited region of memory, it will not trigger an MPU exception. This is important to remember as the Cortex-M processor has a bus structure that is designed to support multiple independent "Bus Master" devices.

AHB Lite Bus Interface

The Cortex-M processor family has a final important architectural improvement over the earlier generation of ARM7- and ARM9-based microcontrollers. In these first generation of ARM-based microcontrollers the CPU was interfaced to the microcontroller through two types of busses. These were the advanced high-speed bus (AHB) and the advanced peripheral bus (APB) (Fig. 5.37).

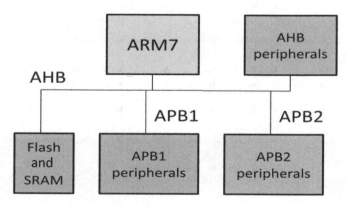

Figure 5.37
The first generation of ARM-based microcontrollers had an internal bus system based on the advanced high speed bus and the advanced peripheral bus. As multiple bus masters (CPU, DMA) were introduced, a bus arbitration phase had to be completed before a transfer could be made across the bus.

The high-speed bus connected the CPU to the Flash and SRAM memory while the microcontroller peripherals were connected to one or more APB busses. The AHB bus also supported additional bus masters such as DMA units to sit alongside the ARM7 processor. While this system worked, the bus structure started to become a bottleneck particularly as more complex peripherals such as Ethernet MAC and USB were added. These peripherals contained their own DMA units which also needed to act as a bus master. This meant that there could be several devices (ARM7 CPU, general-purpose DMA, and Ethernet MAC DMA) arbitrating for the AHB bus at any given point in time. As more and more complex peripherals are added, the overall throughput and deterministic performance became difficult to predict.

The Cortex-M processor family overcomes this problem by using an AHB bus matrix (Fig. 5.38). The AHB bus matrix consists of a number of parallel AHB busses that are connected to different device resources such as a block of RAM or a group of peripherals on an APB bus. The mix of AHB busses and layout of the device resources assigned by the manufacturer when the chip is designed. Each region of device resources is a slave device. Each of these regions is then connected back to each of the bus masters through additional AHB busses to form the bus matrix. This allows manufacturers to design complex devices with multiple Cortex-M processors, DMA units, and advanced peripherals, each with parallel paths to the different device resources. The bus matrix is hardwired into the

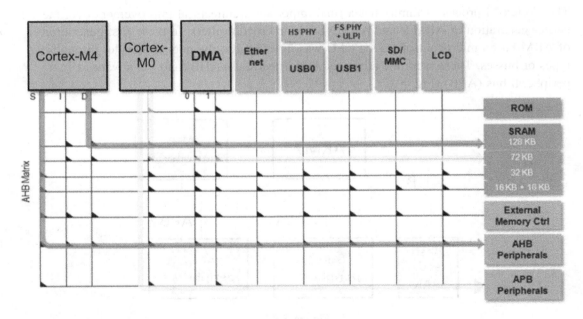

Figure 5.38

The Cortex-M processor family replaces the single AHB bus with a bus matrix that provides parallel paths for each bus master to each block of slave devices.

microcontroller and does not need any configuration by your application code. However, when you are designing the application code, you should pay attention to where different memory objects are located. For example, the memory used by the Ethernet controller should be placed in one block of SRAM while the USB memory is located in a separate SRAM block. This allows the Ethernet and USB DMA unit to work in parallel while the Cortex-M processor is accessing the FLASH and user peripherals. So by structuring the memory map of your application code you can exploit this degree of parallelism and gain an extra boost in performance.

Conclusion

If you are intending to write bare metal code then the bulk of information in this chapter can be ignored. However, these features are critically important if you want to use a more advanced development framework such as an RTOS as we will see in Chapter 9 "CMSIS-RTOS" While the Cortex-M processors are simple to use, they do include features to support safety, critical, and security applications as well as multicore systems.

CHAPTER 6

Cortex-M7 Processor

In the first few chapters of this book we have looked at the Cortex-M0, -M0+, -M3, and -M4 processors. Next we are going to examine the latest Cortex-M processor, the -M7. The Cortex-M7 is currently the highest performing Cortex-M processor. It retains the same programmer's model as the other members of the family so everything we have learned so far can be applied to the Cortex-M7. However, its internal architecture has some radical differences to the earlier Cortex-M processors that dramatically improve its performance. The Cortex-M7 also has a more complex memory system which we need to understand and manage (Fig. 6.1).

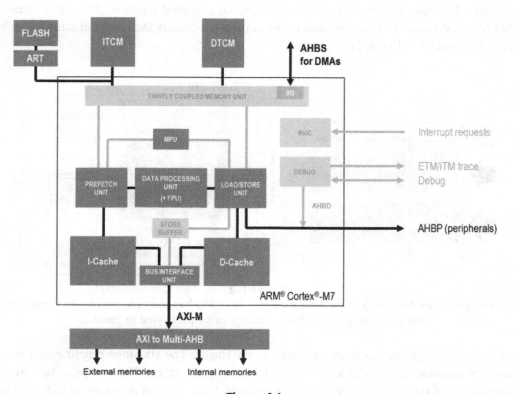

Figure 6.1

The Cortex-M7 has a more complex bus and memory structure but it retains the Cortex-M programmer's model.

The Designer's Guide to the Cortex-M Processor Family.
DOI: http://dx.doi.org/10.1016/B978-0-08-100629-0.00006-2

To a developer the Cortex-M7 processor programmer's model looks the same as the other Cortex-M processors. However, its performance is improved by a number of architectural features that are very distinct from the rest of the family. These include a six-stage dual issue pipeline, an enhanced branch prediction unit, and a double precision floating point maths unit. While these features boost the performance of the Cortex-M7 they are largely transparent to a developer. For a developer the Cortex-M7 really starts to differ from the other Cortex-M processors in terms of its memory model. In order to achieve sustained high performance processing the Cortex-M7 introduces a memory hierarchy which consists of "tightly coupled memories" and a pair of on-chip caches in addition to the main microcontroller FLASH and SRAM. The Cortex-M7 also introduces a new type of bus interface, the AXI-M bus. The AXI-M bus allows 64-bit data transfers and also supports multiple transactions.

Superscalar Architecture

The Cortex-M7 has a superscalar architecture and an extended pipeline. This means that it is able to "dual issue" instructions and under certain conditions the two instructions can be processed in parallel (Fig. 6.2).

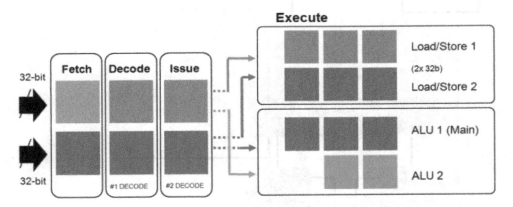

Figure 6.2
The cortex-M7 has a six-stage dual issue pipeline. The CPU has multiple processing "pipes" that allow different groups of instructions to be processed in parallel.

The Cortex-M7 pipeline has been increased to six stages. The six-stage pipeline acts as two three-stage pipelines each with a fetch decode and execute stage. The two pipelines are connected to the memory store via a 64-bit bus which allows two instructions to be fetched and injected into the pipelines in parallel. Both pipelines are in turn connected to several "pipes" within the processing unit. Each of the processor pipes is capable of executing a subset of instructions. This means that there are separate processing paths for Load and Store instructions, MAC instructions, ALU, and floating point instructions. This makes it possible to parallel process instructions if we can issue the two instructions in the pipeline to different processing pipes. While this increases the complexity of the processor, for the developer it happens "under the hood" and is a challenge for the compiler to order the program instructions as interleaved memory accesses and data processing instructions.

Branch Prediction

The performance of the Cortex-M7 is further improved by adding a branch cache to the branch prediction unit (Fig. 6.3). This is a dedicated cache unit with 64 entries each 64 bits wide. During runtime execution, the Branch Target Address Cache (BTAC) will provide the correct branch address. If correctly predicted the BTAC allows loop branches to be processed in a single cycle. The branch instructions can also be processed in parallel with another dual issued instruction further minimizing the branch overhead. This is a critical feature as it allows a general purpose microcontroller to come close to matching the loop performance of a dedicated DSP device.

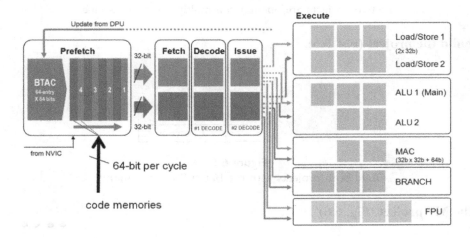

Figure 6.3
The BTAC ensures branch instructions are single cycle. This greatly reduces any overhead in program loops.

Exercise 6.1 Simple Loop

In this exercise we are going to use some real hardware in the form of the STM32F7 Discovery board and the STM32F4 Discovery board in place of the software simulator.

In the remainder of this chapter's exercises we will continue to use the STM32F7 Discovery board.

In this exercise we will measure the execution time of a simple loop on both the Corex-M4 and the Cortex-M7.

Open the Pack Installer.

Select the Boards::Designers Guide Tutorial.

Select the example tab and Copy "Ex 6.1 Cortex-M7 Simple Loop".

This is a multiproject workspace. The same code is built in a Cortex-M4 project and a Cortex-M7 project (Fig. 6.4).

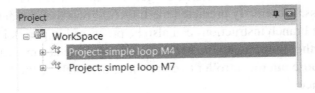

Figure 6.4
The two projects are opened in a multiproject workspace.

Batch build the project (Fig. 6.5).

Figure 6.5
Build both projects with the Batch Build command.

Select the M7 project (Fig. 6.6).

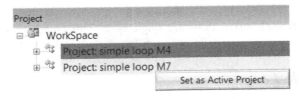

Figure 6.6
Select the M7 as the active project.

Start the debugger and set a breakpoint.

Run the debugger until it hits the breakpoint instruction.

Take the cycle count in the register window (Fig. 6.7).

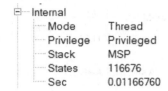

Figure 6.7
Read the cycle count for the M7.

Repeat for the -M4 and compare (Fig. 6.8).

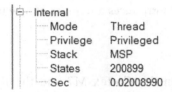

Figure 6.8
Read the cycle count for the -M4.

With this simple example we can see that code compiled for the Cortex-M7 runs much more efficiently than the Cortex-M4 even though this example does not use the caches and tightly coupled memory (TCM) in the Cortex-M7. Combine these architectural improvements with a higher clock rate than a typical Cortex-M4 device then the Cortex-M7 offers a massive performance upgrade. In the remainder of this chapter we will look at optimizing the Cortex-M7 memory system to achieve even greater performance boosts.

Bus Structure

The Cortex-M7 also has a more complex bus structure than the earlier Cortex-M processors while still maintaining a linear memory map. The Cortex-M7 has an AXI-M bus interface, an AHB peripheral interface bus, an AHB slave bus, and also dedicated local busses for the Instruction-TCM and Data-TCM (Fig. 6.9).

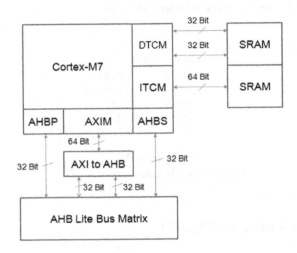

Figure 6.9
The Cortex-M7 introduces a number of new bus interfaces.

The main bus interface between the Cortex-M7 and the microcontroller is the ARM AXI-M bus. This is a 64-bit wide bus capable of supporting multiple transactions. The AXI bus provides a number of slave ports which are bridged on to an AHB lite bus. The AHB lite bus is the same as is found on the other members of the Cortex-M family; this helps silicon vendors reuse existing microcontroller designs with the Cortex-M7 processor. The Cortex-M7 also has an AHB peripheral bus (AHBP) port which is used to access the microcontroller peripherals via the AHB bus matrix. The AHBP bus is optimized for peripheral access so peripheral register access will have a lower latency compared to other Cortex-M processors. The Cortex-M7 also has a second AHB port; this is the AHB slave (AHBS) port. This port allows other bus masters, typically DMA units, access to the Instruction-TCM and Data-TCM. A typical implementation of these busses within a microcontroller is shown in (Fig. 6.10).

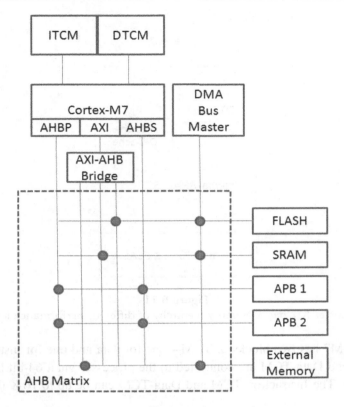

Figure 6.10
A typical bus interface configuration for a Cortex-M7 microcontroller.

In this example the AXI-M bus is connected to the AHB bus matrix via a bridge. However the AHB bus only connects the AXI-M bus ports to the microcontroller system memory. The AHB bus peripheral busses are connected back to the Cortex-M7 through the AHBP ports. Each of the DMA units can act as bus masters and have access to all of the system memory and peripherals and can use the ADB Slave bus to access the Instruction and Data TCM. Although the AHBS bus is shown as being routed through the Cortex-M7 this bus will stay awake if the Cortex-M7 processor is placed into a low power mode. This makes it possible for other bus masters to access the TCM even when the Cortex-M7 is fully asleep.

Memory Hierarchy

The more complex bus structure of the Cortex-M7 supports a memory map which contains a hierarchy of memory regions, which needs to be understood and controlled by a designer (Fig. 6.11).

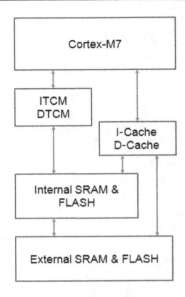

Figure 6.11
The Cortex-M7 has a memory hierarchy of different performance levels.

Firstly the Cortex-M7 has two blocks of TCM—one for data and one for instructions. The Instruction-TCM and Data-TCM are connected to the processor via a 64-bit bus and are zero wait state memory. The Instruction-TCM and Data-TCM can be up to 16 K (Fig. 6.12).

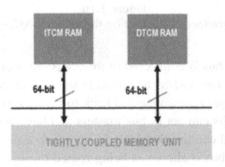

Figure 6.12
The TCMs are zero wait state memories interfaced directly to the CPU through dedicated 64-bit busses.

Code located in the Instruction-TCM will run at zero wait states and will be highly deterministic. So this region of memory should be home to any critical routines and interrupt service routines. The Data-TCM is the fastest data RAM available to the processor and should be home of any frequently used data and is also a good location for stack and heap memory.

Exercise 6.2 Locating Code and Data into the TCM

This exercise demonstrates how to load code into the Instruction-TCM and data into the Data-TCM and compare its execution to code located in the system FLASH memory.

Open the Pack Installer.

Select the Boards::Designers Guide Tutorial.

Select the example tab and Copy "Ex 6.2−6.4 Cortex-M7 TCM and I-Cache" (Fig. 6.13).

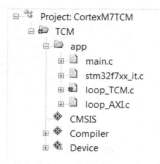

Figure 6.13
The project has two loop modules which have their code located in the ITCM and the AXI
FLASH memory.

Open the options for target menu.

In this project we have two sets of loop functions that accumulate the values stored in a thousand-element array. By default the project will place the code in the flash memory which is located on the AXI-M bus at 0x8000000 (Fig. 6.14).

Read/Only Memory Areas					Read/Write Memory Areas				
default	off-chip	Start	Size	Startup	default	off-chip	Start	Size	NoInit
☐	ROM1:			○	☐	RAM1:	0x00000000	0x8000	☐
☐	ROM2:			○	☐	RAM2:			☐
☐	ROM3:			○	☐	RAM3:			☐
	on-chip					on-chip			
☑	IROM1:	0x8000000	0x100000	◉	☑	IRAM1:	0x20020000	0x30000	☐
☐	IROM2:	0x200000	0x100000	○	☐	IRAM2:	0x20000000	0x10000	☐

Figure 6.14
Memory layout with an additional region (RAM1) created to cover the ITCM.

However we have also defined a RAM (RAM1) region at 0x0000000 for 32 K which is the location of the Instruction-TCM on this device. The default tick box is unchecked so the linker will not place any objects in this region unless we specifically tell it to.

Open the "Options for file loop_TCM.c" (Fig. 6.15).

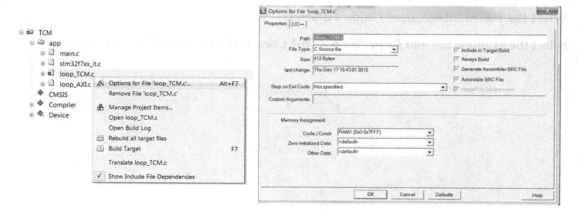

Figure 6.15
The module local options are used to map the code/const data into the RAM1 region. The Code located into the I-TCM RAM region will be automatically booted out of the FLASH into the I-TCM by the startup code.

In the memory assignment options we can select the RAM1 region for the Code/Const segments in this file. This means that the code in the loop_TCM.c file will be stored in the flash memory but will be copied into the RAM1 region (The Instruction TCM SRAM) at startup. As a bonus the debug symbols will also match the load address, this allows us to debug code loaded into the Instruction-TCM without having to make a special build.

Open main.c.

The arrays which hold the data to be accumulated are defined at the start of main().

```
uint32_t array_DTCM[NUM_SAMPLES]  __attribute__((at(0x20000000)));  //Locate in DTCM
                                                                     fastest data memory
uint32_t array_SDRAM[NUM_SAMPLES]  __attribute__((at(0xC0000000))); //Locate in SDRAM
                                                                     slowest data memory
```

The __attribute directive is used to force the linker to locate the array at an absolute address. In this case we are forcing the first array into the D-TCM region and the second into the SRAM located on the AXI bus. In the main() application we are configuring the SysTick timer to act as a stopwatch timer. Main() then calls identical loop functions which are executing from either the Instruction-TCM or the AXI FLASH. The loop functions are also accessing an instance of the array which is located either in the Data-TCM or the AXI-M SRAM.

Build the code and start the debugger.

Open the view/serial windows/debug printf window.

Run the code and observe the different run times displayed in the serial window.

```
Measure Loop Timing in AXI FLASH and ITCM using the DTCM for data
AXI FLASH loop timing using DTCM RAM = 79156
ITCM loop timing using DTCM RAM = 24432
AXI FLASH loop timing using SDRAM = 89827
ITCM loop timing using SDRAM = 37306
```

Cache Units

Outside of the TCMs, all program code and data will be held in the system memory. This will be the microcontroller internal FLASH and SRAM. The microcontroller may also have an external bus interface to access additional RAM and ROM memory. As we have seen with the Cortex-M3 and -M4 it is necessary to provide some form of "Flash Memory Accelerator" to deliver instructions from the flash memory to the CPU in order to maintain processing at the full CPU clock speed. The Cortex-M7 is both a higher performance processor and is designed to run at higher clock frequencies and this form of memory acceleration is no longer fully effective. In order to meet both its code and data throughput requirements the Cortex-M7 can be implemented with both Data and Instruction Caches (Fig. 6.16).

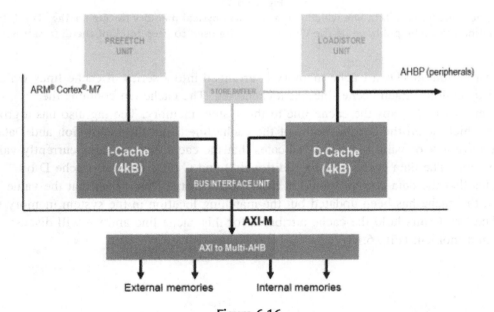

Figure 6.16
The I-Cache and D-Cache are used to cache memory located on the AXI-M bus.

Both of the caches are used to hold copies of instructions and data fetched from memory accessed over the AXI-M bus. Once the instruction or data value has been loaded into the cache it may be accessed more faster than the system FLASH and SRAM.

Cache Operation

Like the TCMs, the Instruction and Data Caches are blocks of memory internal to the Cortex-M7 processor which are capable of being accessed with zero wait states. The Instruction Cache can be up to 64 Kb in size and the Data Cache may be up to 16 Kb (Fig. 6.17).

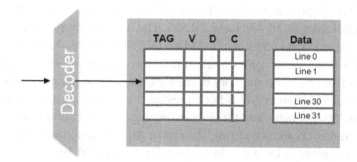

Figure 6.17
A Cache consists of a data line which is mapped to physical memory through a Tag. The C bits define the cache policy while the V and D bits are used to keep track of the data status.

The Instruction and Data Cache memory is arranged into a series of cache lines which are 32 bytes long. Each cache line has a cache tag. The cache tag contains the information which maps the cache line to the system memory. The tag also has a group of bits which detail the current status of the cache line. Both the Instruction and Data Cache have a V or valid bit which indicates that the cache line contains currently valid information. The data cache has the additional D and C bits. The data cache D bit indicates that the data currently held in the cache is dirty. This means that the value held in the cache has been updated but the matching location in the system memory has not. The two C bits hold the cache attributes for this cache line and we will discuss these in a moment (Fig. 6.18).

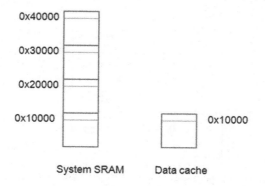

Figure 6.18
A cache line is mirrored to memory locations through the physical memory space.

The cache is mapped as a page of memory that is reflected up through the address space of the Cortex-M7 processor. This means that each cache line has a number of matching memory locations evenly spaced through the system memory.

When the system memory is accessed the value stored in the system memory is loaded into the CPU and also into the matching cache location. The cache tag will be set to decode the data address in the system memory and the valid bit will be set. Once loaded into the cache further accesses to this location will read the value from the cache rather than the system memory. Depending on the cache policy which is set by the application code, writes to a memory location will cause a write to the matching cache line. Again the cache tag will be updated with the decode address of the system memory location. The valid bit will be set to indicate that the cache holds the correct current data value. The point at which the system memory is updated will depend on the cache policy which is set by the C bits in the cache tag. We will look at the cache policy settings later. If the cache memory holds a value which has not yet been written to the system memory the D bit will be set. As the application code executes, data and instructions will be moved into the caches. If data has been loaded into a cache line and an access is made to another location in system memory that is mapped to the same cache line then the old data will be "evicted." The new data will be copied into the cache line and the cache tag will be updated (Fig. 6.19).

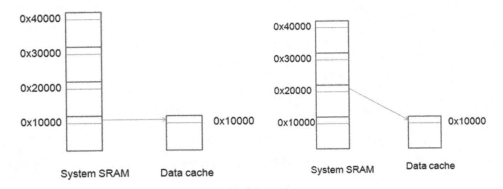

Figure 6.19
During execution data will be loaded into the cache. It can also be evicted if the mirror cache location is accessed.

If several locations mapped to a single cache line are frequently accessed they will continually be evicting each other. This is known as cache thrashing and a lot of thrashing will reduce the effectiveness of the cache (Fig. 6.20).

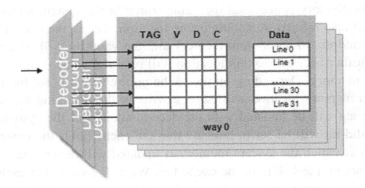

Figure 6.20
To improve cache efficiency the cache I is arranged as a number of parallel pages. This provides multiple locations to store cached data.

To reduce the impact of thrashing the cache is arranged as a set of parallel pages called "ways" rather than one contiguous block of memory. The Instruction Cache is arranged as two ways and the data cache is arranged as four ways. Now the cache memory is mapped up through the system at the granularity of each cache way. This creates a number of parallel locations within the cache that a given memory location can be mapped into. This reduces the probability of cache thrashing and leads to a more efficient use of the cache memory.

Instruction Cache

The Instruction Cache is controlled by a small group of functions provided by the CMSIS-Core specification. Functions are provided to enable and disable the Instruction Cache. There is also a function to invalidate the cache contents. When the Instruction Cache is invalidated all of the valid bits are cleared effectively emptying the cache of any loaded instructions. As execution continues fetches to instruction address will start to reload the cache (Table 6.1).

Table 6.1: CMSIS-Core I-Cache management functions

CMSIS-I-Cache Management Functions
void SCB_EnableICache(void)
void SCB_DisableICache(void)
void SCB_InvalidateICache(void)

After reset both of the caches are disabled. To correctly start the Instruction Cache you must first invalidate its contents before enabling it. Once enabled it is self-managing and in most applications will not require any further attention from the application code.

Exercise 6.3 Instruction Cache

This exercise measures the performance increase achieved by using the Instruction cache.

Open the Pack Installer.

Select the Boards::Designers Guide Tutorial.

Reopen the last exercise "Ex 6.2−6.4 Cortex-M7 TCM and I-Cache".

Open main.c.

Remove the comments at lines 75 and 91.

This adds a new block of code to the last example. Here we enable the Instruction Cache and rerun the AXI loop functions to see the change in timing.

Open the view/serial windows/debug window.

Run the code and observe the difference in run time compared to the original example.

```
Measure Loop Timing in AXI FLASH with I-Cache Enabled
AXI loop timing with I-Cache and DTCM RAM = 62395
AXI loop timing with I-Cache and SDRAM = 76300
```

The Instruction Cache is a big improvement over executing from the FLASH but not as good as the Instruction-TCM.

Data Cache

The Data Cache is also located on the AXI-M bus to cache any system RAM accesses. While this speeds up access to program data it can introduce unexpected side effects due to data coherency issues between the system memory and the Data Cache. The Data Cache is in effect shadowing the system memory so that a program variable could be living in two places, within the cache and within the SRAM memory. This would be fine if the CPU were the only bus master in the system. In a typical Cortex-M7 microcontroller there may be several other bus masters in the form of peripheral DMA units or even other Cortex-M processors. In such a multi master system the cache may now introduce the problem of data coherency (Fig. 6.21).

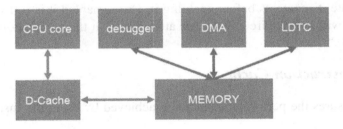

Figure 6.21
The D-Cache can introduce data coherency problems. The System memory can be updated by other bus masters. The D-Cache can "hide" these changes from the CPU.

Since our data may now be held is several locations within the memory hierarchy (CPU register, Data Cache, microcontroller SRAM) the slower system memory may be holding a historical value which is different to the current data located in the Data Cache, A DMA unit will only have access to the system memory so will use the older historical value rather than the current value held in the cache Alternatively a DMA update to the microcontroller SRAM may be masked from the CPU by an outdated value held in the data cache. This means that the Data Cache, does require more management than the Instruction Cache. In the CMSIS-Core specification there are eight Data Cache management functions (Table 6.2).

Table 6.2: CMSIS-Core data cache functions

CMSIS-Core Data Cache Functions
SCB_EnableDCache(void)
SCB_DisableDCache(void)
SCB_InvalidateDCache(void)
SCB_CleanDCache(void)
SCB_CleanInvalidateDCache(void)
SCB_InvalidateDCache_by_Addr(uint32_t *addr,int32_t dsize)
SCB_CleanDCache_by_Addr(uint32_t *addr,int32_t dsize)
SCB_CleanInvalidateDCache_by_Addr(uint32_t *addr,int32_t dsize)

Like the Instruction Cache there are functions to enable, disable, and invalidate the data cache. In addition we can clean the cache. This forces all the data held in the cache to be written to the system memory making the whole memory store coherent. It is also possible to clean and invalidate regions of the system memory which may be held in the cache.

Data Cache Invalidate	—	Flush what is in the cache memory and force a reload from system memory
Data Cache Clean	—	Write the data held in cache memory to the system memory
Data Cache Clean and Invalidate	—	Write cache memory to system memory and flush the cache memory

As with the Instruction Cache, the Data Cache is disabled after reset and must first be invalidated before it is enabled. The CMSIS-Core cache enable functions perform an invalidation before enabling the either cache.

Memory Barriers

In Chapter 5 "Advanced Architecture Features" we saw that the Cortex-M thumb-2 instruction set contains a group of memory barrier instructions. The memory barrier instructions are used to maintain data and instruction coherency within a Cortex-M microcontroller. These instructions are rarely used in most Cortex-M projects and when they are it is mainly for defensive programming. However, when working with the Cortex-M7 Instruction and Data Caches they are used to ensure that the cache operations have finished before allowing the CPU to continue processing. The CMSIS-Core cache functions include the necessary memory barrier instructions to ensure that all cache operations have completed when the function returns.

```
__STATIC_INLINE void SCB_EnableICache (void)

  {
    #if (__ICACHE_PRESENT == 1U)
      __DSB();
```

```
    __ISB();
    SCB->ICIALLU = 0UL;          /* invalidate I-Cache */
    SCB->CCR |= (uint32_t)SCB_CCR_IC_Msk; /* enable I-Cache */
    __DSB();
    __ISB();
  #endif
}
```

Exercise 6.4 Example Data Cache

This exercise measures the performance increase achieved by enabling the Data Cache.

Open the Pack Installer.

Select the Boards::Designers Guide Tutorial.

Reopen the last exercise "Ex 6.2−6.4 Cortex-M7 TCM and I-CACHE".

Open main.c.

Remove the comments at lines 92 and 106.

This adds a new block of code to the last example. Here we enable the Instruction Cache and rerun the AXI loop functions to see the change in timing.

Open the view/serial windows/debug window.

Run the code and observe the difference in run time compared to the original example.

```
Measure Loop Timing in AXI FLASH with I-Cache and D-Cache Enabled
AXI FLASH loop timing with I-Cache and D-Cache = 33305
```

So switching on the Data Cache gives another performance boost when data are located in the internal or external RAM, but as expected the best level of performance comes from locating code and data into the TCMs.

Memory Protection Unit and Cache Configuration

The Memory Protection Unit (MPU) is common across all members of the Cortex-M processor family except for the Cortex-M0. As we saw in Chapter 5 "Advanced Architecture Features" the MPU can be used to control access to different regions of the memory map within the Cortex-M0+\-M3 and -M4. On the Cortex-M7 the MPU offers the same functionality but in addition it is used to configure the cache regions and cache policy.

Cache Policy

The cache policy for the different regions of memory within the microcontroller can be defined through the MPU regions with the cache configuration bits in the MPU "Attribute and Size" register (Fig. 6.22; Table 6.3).

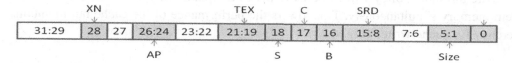

Figure 6.22
The MPU "Attribute and Size" register.

Table 6.3: The MPU D-Cache configuration options

Cache Configuration Field	Description
TEX	Type Extension
C	Cacheable
B	Bufferable
S	Sharable

The sharable option in the MPU "Attribute and Size" register defines if a memory region can be shared between two bus masters. If a region is declared as shareable its data will be kept coherent between the cache and the physical memory. This means that any bus master accessing the data will always get the correct current value. However, in order to do this the CPU will have to write through the Data Cache when it updates a RAM value, if the RAM is updated by another bus master it must be loaded into the cache. So making the SRAM shareable removes coherency problems but reduces the overall Data Cache performance.

The cache policy also defines how data is written between the cache memory and the system memory. First we can define when data are allocated into the cache. There are two options for the allocate policy, we can force data to be loaded into the cache when a RAM location is read or we can select to load a RAM location into the cache when it is read or written to. This policy is set by configuring the Type Extension bits (TEX) (Table 6.4).

Table 6.4: Cache allocate policy settings

TEX	Cache Allocate Policy
000	Read allocate, load memory location into cache when it is read from
001	Read and write allocate, load memory location into cache when it is read from or written to

The remaining TEX values are there to support more complex memory systems which may have a second outer (L2) cache.

Once a memory location has been loaded into the cache we can define when the system RAM will be updated. There are two possible update policies "write through" and "write back."

The "write through" policy will force CPU writes to update the cache memory and the system memory simultaneously. This lowers the performance of the cache but maintains coherency for cache writes (Fig. 6.23).

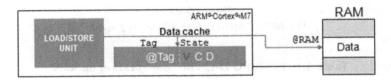

Figure 6.23
Write through cache policy.

The write back policy will force writes to the cache memory only and the data will only be written to the system memory when the cache data is evicted or a cache clean instruction is issued. This maximizes the performance of the cache for data writes at the expense of cache coherency (Fig. 6.24).

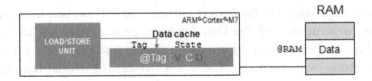

Figure 6.24
Write back cache policy.

The cache update policy is configured through the C and B bits in the MPU register (Table 6.5).

Table 6.5: Cache update policy

C	B	Cache Update Policy
1	1	Write Back—cacheable and bufferable
1	0	Write Through—cacheable but no buffering

Managing the Data Cache

As we have seen the Instruction Cache will look after itself and in most cases we just need to enable it. However, we need to pay more attention to the Data Cache and the overall system coherency.

Switch Off the Cache

The simplest thing to do is switch off the Data Cache. While this will remove any coherency problem it will reduce the performance of the Cortex-M7. However, it can be a useful option at the beginning of a project in order to simplify the system.

Disable Caching over a Region of System Memory

This is similar to the first option except we can program an MPU region to prevent data caching on a region of the system memory. We can then locate the data that is common to the Cortex-M7 and other bus masters into this region. Since all access will be to the system memory there will be no coherency problems but accesses to this memory region with the Corex-M7 will be at their lowest.

Change the Cache Policy for a Region of System Memory

If the cache policy for a region is set to write through the system memory will always be updated. So this is more limited option where the application code is writing into memory shared by bus masters.

Use the Cache Management Functions to Guarantee Coherency

The cache management functions can be used to clean or invalidate regions of system memory before they are accessed. This will cause dirty data to be written to the system memory prior to a second bus master accessing the data. We can also invalidate data causing the Cortex-M7 to reload data from system memory shared with another bus master.

Exercise 6.5 Data Cache Configuration

This exercise demonstrates the behavior of various cache configurations. If you are planning to use a Cortex-M7 you should experiment with this example to get a thorough understanding of the Data Cache.

```
Initilise_Device();
MPU_Config();
//SCB_EnableDCache();  //cache is disabled at the start of the project
```

```
Initilise_Arrays();     //load data into array_sdram1 and zero array_sdram2
//Cache_Clean();
Update_Source_Array(); //Write new values to array_sdram1
//Cache_Clean();
DMA_Transfer();         //copy sdram1 to array_sdram2 using a DMA bus master
//Cache_Invalidate();
Test_Coherence();       //Test the coherency between array_sdram1 and array_sdram2
Test_Coherence();       //Test the coherency between array_sdram1 and array_sdram2
```

In this exercise the MPU configures the Data Cache to cover the first 128 Kb of external SDRAM which is populated with two data arrays. The Application code fills an array with some ordered data and writes zero into a second. Then a DMA unit (a bus master) is used to copy the data from the first array into the second. Finally we test for data coherence between the contents of the two arrays. We can use this simple program to see the effect of different cache settings. The MPU code is contained in a single source file and header file so it can be reused in other projects.

Open the Pack Installer.

Select the Boards::Designers Guide Tutorial.

Reopen the last exercise "Ex 6.5 Data Cache".

Comment out line 90 so the Data Cache is not enabled.

Build the project.

Start the debugger.

Open the view/serial windows/debug printf window.

Run the code.

During this first run the Data Cache is disabled to give us a baseline for different cache configurations. The cycles taken for each operation are as follows:

```
->CPU cycles spent for arrays initialization: 5217
->CPU cycles spent re initializing source array 1065
->CPU cycles spent on DMA Transfer: 11985
->CPU cycles spent for comparison: 13912
There were 0 different numbers
```

Open main.c and uncomment line 90 to enable the data cache.

Open mpu.h.

This is a templated configuration file for the code in mpu.c. The initial configuration maps the Data Cache region at the start of the external SRAM at 0xC0000000 for 128 Kb. In the Data Cache configuration options we are enabling the region as sharable with other bus masters within the microcontroller (Fig. 6.25).

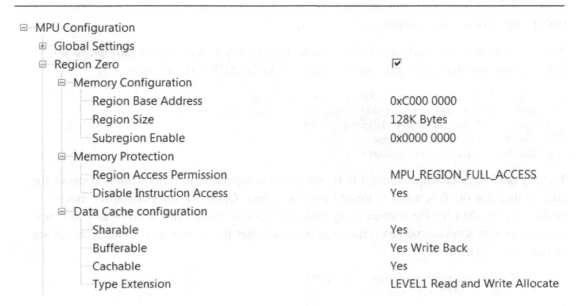

Figure 6.25
MPU configuration options.

Rebuild the code.

Start the debugger and run the code.

```
->CPU cycles spent for arrays initialization: 5184
->CPU cycles spent re initializing source array 1064
->CPU cycles spent on DMA Transfer: 11985
->CPU cycles spent for comparison: 13910
There were 0 different numbers
```

While the Data Cache is now operating it does not have a significant effect because it has write and read from the system memory in order to maintain coherency.

In mpu.h change the sharable option from Yes to NO (Fig. 6.26).

Figure 6.26
Set the D-Cache region to be nonsharable.

Build and rerun the example.

Now the region is not shareable but the cache load policy is set to write through so array data written into the cache will also be written into the SDRAM memory.

```
->CPU cycles spent for arrays initialization: 3388
  ->CPU cycles spent re initializing source array 1057
  ->CPU cycles spent on DMA Transfer: 11981
  ->CPU cycles spent for comparison: 4418
  There were 0 different numbers
```

The big performance improvement is found in the comparison routine. This is because the current data for the first array is loaded into the cache. Once the comparison has been performed the data for the second array will now also be loaded (Read Allocate) so if we run the comparison routine again the time is even faster because now all data accesses are to the Data Cache.

```
  ->CPU cycles spent for comparison: 2083
  There were 0 different numbers
```

Open mpu.h.

Change Bufferable to Yes Write back.

Change Type extension to Level 1 Read and Write Allocate (Fig. 6.27).

Data Cache configuration	
Sharable	No
Bufferable	Yes Write Back
Cachable	Yes
Type Extension	LEVEL1 Read and Write Allocate

Figure 6.27
Set the D-Cache policy to be "Write Back" and "Read and Write Allocate."

Build and run the code.

This time the code fails because the array data values are held in the Data Cache. The correct values are not copied by the DMA. Also since we have selected read and Write Allocate, the array_sdram2[] is loaded into the Data Cache when it is initialized. This means that the DMA updates the SDRAM but the Cortex-M7 will only see the old values stored in the cache.

Uncomment line 98 Invalidate_Cache();

Uncomment line 94 Clean_Cache();

Build and run the code.

Again the code fails but in a more interesting way. This time the Invalidate_Cache() function has loaded the values in the SDRAM into the cache but they are the wrong values. We have initialized both arrays and then called the Clean_Cache() function. This places the initial values in the SDRAM. However, we then update the array_sdram1[] values which will only be changed in the cache.

Uncomment Line 96 Clean_Cache();

Build and run the code.

```
->CPU cycles spent for arrays initialization: 3051
->CPU cycles spent on cache clean: 1067
->CPU cycles spent re initializing source array 1062
->CPU cycles spent on cache clean: 1095
->CPU cycles spent on DMA Transfer: 11999
->CPU cycles spent on cache invalidate: 371
->CPU cycles spent for comparison: 2696
There were 0 different numbers
->CPU cycles spent for comparison: 2081
There were 0 different numbers
```

This time the code runs correctly because we have maintained coherency between the Data Cache and the SDRAM.

Double Precision Floating Point Unit

Like the Cortex-M4, the Cortex-M7 has a Hardware Floating Point Unit (FPU). This unit is tightly integrated into the processor pipeline to allow very fast processing of both single and double precision floating point values. The Cortex-M7 FPU is enabled through the CPARC register in the Cortex-M7 processor System Control Block, this is typically done by the system startup code. Floating point calculations will now make use of the hardware unit rather than software libraries provided the compiler options have been set correctly. Because the Cortex-M7 may be fitted with a single or double precision FPU the CMSIS-Core specification provides a function to query the processor and return the type of FPU fitted (Fig. 6.28).

```
__STATIC_INLINE uint32_t SCB_GetFPUType ( void )
```

Returns

- **0**: No FPU
- **1**: Single precision FPU
- **2**: Double + Single precision FPU

The function returns the implemented FPU type.

Figure 6.28
CMSIS-Core gets FPU type function.

Once the FPU has been enabled it can be used in the same way we saw in Chapter 3 "Cortex-M Architecture."

Functional Safety

Many design sectors require compliance with some form of safety standard. Some obvious sectors are Medical, Automotive, and Aerospace. Increasingly other sectors such as Industrial Control, Robotics, and Consumer Electronics all require compliance to safety standards. As a rule of thumb design processes required for functional safety today are adopted by mainstream designers tomorrow. Developing a functional safety project is a lot more than writing and testing the code thoroughly. The underlying silicon device must be the product of an equally rigorous development process with available documentation to back this up. The microcontroller must also incorporate features (error correction codes on bus interfaces, built in self-test) that allow the software to validate that the processor is running correctly and be able to detect and correct faults. An IEC 61508-3 (SIL3) level system will often consist of dual processors running either in lock step or with an application and supervisor arrangement. This allows the system to detect and manage processor faults.

The Cortex-M7 has been designed with both hardware and process/safety documentation that allows it to be used in high integrity systems. At the time of writing no microcontrollers using these features have yet been released.

Cortex-M7 Safety Features

The Cortex-M7 introduces a range of safety features that allow a silicon vendor to design a device suitable for a high reliability system. The safety features shared with other Cortex-M processors are as follows (Table 6.6):

Table 6.6: Common Cortex-M safety features

Memory protection unit
Instruction and Data Trace
Fault exceptions

The Cortex-M7 extends the available safety features with the additional features (Table 6.7).

Table 6.7: Corex-M7 safety features

Safety Feature	Description
Error Correction Codes	The Cortex-M7 may be configured to detect and correct hard and soft errors in Cache RAMS using ECC The TCM interfaces support interfacing to memory with ECC hardware
Dual Core Lock Step	The Cortex-M7 may be implemented as a dual core design were both processors are operating in Lock Step
MBIST	The Cortex-M7 may be synthesized with a Memory Built in self-test interface that supports memory validation during production and run time

Safety Documentation

To enable silicon vendors to design a new microcontroller which is suitable for safety use, ARM also provides extensive documentation in the form of a safety package (Table 6.8).

Table 6.8: Cortex-M7 safety documentation

Document	Description
Safety manual	A safety manual describing in detail the processor's fault detection and control features and information about integration aspects in Silicon Partner's device implementations
FMEA manual	A Failure Modes and Effects Analysis with a qualitative analysis of failure modes within the processor logic, failure effects on the processor's behavior, and an example of quantitative hardware metrics
Development interface report	Report making clear how the Silicon Partner's engineers should manage the ARM deliverables and what to expect from them

These reports are only of interest to Silicon Designers and some companies specializing in safety testing and tools. As a Designer you will be more interested in the software tools rather than the Silicon Design process.

Development Tools

There are a number of toolchains that have been certified for safety use. This mainly applies to the compiler and linker. The ARM compiler used in MDK-ARM is certified as follows (Table 6.9):

Table 6.9: Compiler safety documentation

Safety Feature	Description
Safety certification	TÜV SUD certification
	ISO 26262 (ASILD)
	IEC 61508-3 (SIL3)
Maintenance	Five years extended maintenance
	Critical defects fixes
	Maintained compiler branch
Documentation	Safety manual
	Development process documentation
	Test report
	Defect report

Conclusion

In this chapter we have seen that the Cortex-M7 retains the Cortex-M programmer's model allowing us for easy transition to this new processor. However, it does take a while to realize the big performance increase over the Cortex-M4. This will become more marked a Silicon Vendors move the smaller process technologies and achieve even higher clock frequencies. The Cortex-M7 is also the first Cortex-M processor with a full suite of safety documentation which allows silicon vendors to design devices suitable for safety critical applications.

CHAPTER 7

Debugging with CoreSight

Many developers who start work with a Cortex-M microcontroller assume that its debug system is a form of "JTAG (Joint Test Action Group)" interface. In fact a Cortex-M processor has a debug architecture called "CoreSight" which is considerably more powerful. In addition to the run control and memory access features provided by "JTAG," the "CoreSight" debug system includes a number of real-time trace units that provide you with a detailed debug view of the processor as it runs. In this chapter, we will see what features are available and how to configure them.

Introduction

Going back to the dawn of modern times, microcontroller development tools were quite primitive. The application code was written in assembler and tested on an erasable programmable read-only memory (EPROM) version of the target microcontroller. Each EPROM had to be erased by UV light before it could be reprogrammed for the next test run (Fig. 7.1).

Figure 7.1
The electrically erasable programmable read-only memory was the forerunner of today's FLASH memory.

To help debugging, the code was "instrumented" by adding additional lines of code to write debug information out of the UART or to toggle an I/O pin. Monitor debugger programs were developed to run on the target microcontroller and control execution of the application code. While monitor debuggers were a big step forward, they consumed resources on the microcontroller and any bug that was likely to crash the application code would also crash the monitor program just at the point you needed it (Fig. 7.2).

The Designer's Guide to the Cortex-M Processor Family.
DOI: http://dx.doi.org/10.1016/B978-0-08-100629-0.00007-4

Figure 7.2
An in-circuit emulator provides nonintrusive real-time debug for older embedded microcontrollers.

If you had the money, an alternative was to use an in-circuit emulator. This was a sophisticated piece of hardware that replaced the target microcontroller and allowed full control of the program execution without any intrusion on the CPU runtime or resources. In the late 1990s the more advanced microcontrollers began to feature various forms of on-chip debug unit. One of the most popular on-chip debug units was specified by the "Joint Test Action Group" and is known by the initials JTAG. The JTAG debug interface provides a basic debug connection between the microcontroller CPU and the PC debugger via a low-cost debug adapter (Fig. 7.3).

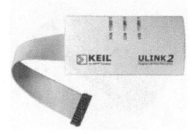

Figure 7.3
Today low cost debug hardware is available to program all Cortex-M devices.

JTAG allows you to start and stop the CPU running. It also allows you to read and write to memory locations and insert instructions into the CPU. This allows the debugger designer to halt the CPU, save the state of the processor, run a series of debug commands and then restore the state of the CPU, and restart execution of the application program. While this process is transparent to the user, it means that the PC debugger program has run control of the CPU (reset, run, halt, and set breakpoint) and memory access (read/write to user memory and peripherals). The key advantage of JTAG is that it provides a core set of debug features with the reliability of an emulator at a much lower cost. The disadvantage of JTAG is that you have to add a hardware socket to the development board and the JTAG interface uses some of the microcontroller pins. Typically, the JTAG interface requires five GPIO pins which may also be multiplexed with other peripherals. More importantly, the JTAG interface needs to halt the CPU before any debug information can be provided to the PC debugger. This run/stop style of debugging becomes very limited when you are dealing with a real-time system such as a communication protocol or motor control. While the JTAG interface was used on ARM7/9-based microcontrollers, a new debug architecture called CoreSight was introduced by ARM for all the Cortex-M/R and A-based processors.

CoreSight Hardware

When you first look at the datasheet of a Cortex-M-based microcontroller, it is easy to miss the debug features available or assume it has a form of JTAG interface. However, the CoreSight debug architecture provides a very powerful set of debug features that go way beyond what can be offered by JTAG. First of all on the practical side a basic CoreSight debug connection only requires two pins, Serial In and Serial Out (Fig. 7.4).

Figure 7.4
The CoreSight debug architecture replaces the JTAG berg connector with two styles of subminiature connector.

The JTAG hardware socket is a 20-pin berg connector that often has bigger footprint on the PCB than the microcontroller that is being debugged. CoreSight specifies two connectors: a 10-pin connector for the standard debug features and a 20-pin connector for the standard debug features and instruction trace. We will talk about trace options later but if your microcontroller supports instruction trace then it is recommended to fit the larger 20-pin socket so you have access to the trace unit even if you do not initially intend to use it. A complex bug can be sorted out in hours with a trace tool where with basic run/stop debugging it could take weeks (Fig. 7.5) (Table 7.1).

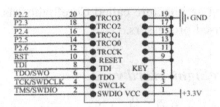

Figure 7.5
The two debug connectors require a minimum number of processor pins for hardware debug.

Table 7.1: CoreSight debug sockets

Socket	Samtec	Don Connex
10-pin standard debug	FTSH-105-01-L-DV-K	C42-10-B-G-1
20-pin standard + ETM trace	FTSH-110-01-L-DV-K	C42-20-B-G-1

This standard debug system uses the 10-pin socket. This requires a minimum of two of the microcontroller pins, serial wire IO (SWIO) and serial wire clock (SWCLK) plus the target Vcc, ground, and reset. As we will see below, the Cortex-M3 and Cortex-M4 are fitted with two trace units which require and extra pin called serial wire out (SWO). Some Cortex-M3 and Cortex-M4 devices are fitted with an additional instruction trace unit. This is supported by the 20-pin connector and uses an additional four processor pins for the instruction trace pipe.

Debugger Hardware

Once you have your board fitted with a suitable CoreSight socket you will need a debug adapter unit that plugs into the socket and is connected to the PC usually through a USB or Ethernet connection. An increasing number of low-cost development boards also incorporate a USB debug interface based on the CMSIS-DAP (Cortex Microcontroller Software Interface Standard-Debug Access Port) specification.

CoreSight Debug Architecture

There are several levels of debug support provided over the Cortex-M family. In all cases, the debug system is independent of the CPU and does not use processor resources or runtime (Fig. 7.6).

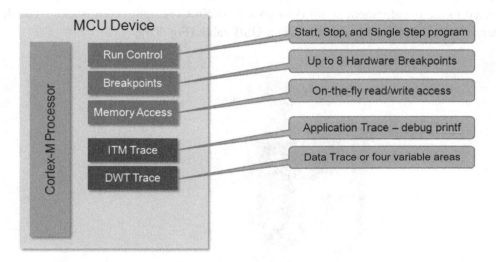

Figure 7.6
In addition to run control, the Cortex-M3 and Cortex-M4 basic debug system includes two trace units. A data trace and an instrumentation trace.

The minimal debug system available on the Cortex-M3 and Cortex-M4 consists of the serial wire (SW) interface connected to a debug control system which consists of a run control unit, breakpoint unit, and a memory access unit. The breakpoint unit support up to eight hardware breakpoints, the total number actually available will depend on the number specified by the Silicon Vendor when the chip is designed. In addition to the debug control units, the standard CoreSight debug support for Cortex-M3 and Cortex-M4 includes two trace units, a data watch trace and an instrumentation trace (ITM). The data watch trace allows you to view internal RAM and peripheral locations "on-the-fly" without using any CPU cycles. The data watch unit allows you to visualize the behaviour of your applications data.

Exercise 7.1 CoreSight Debug

For most of the examples in this book, I have used the simulator debugger which is part of the μVision IDE. In this exercise, however, I will run through setting up the debugger to work with the STM32F7 Discovery board. This evaluation board has its own debug adapter included as part of the board. While there are a plethora of evaluation boards available for different Cortex-M devices the configuration of the hardware debug interface is essentially the same for all boards.

Hardware Configuration

If you are developing your own hardware and software, an external debug adapter is connected to the development board through the 10-pin CoreSight debug socket. The debug adapter is in turn connected to the PC via a USB cable (Fig. 7.7).

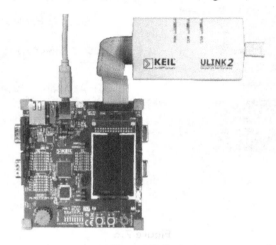

Figure 7.7

A typical evaluation board provides a JTAG/CoreSight connector. The evaluation board must also have its own power supply. Often this is provided via a USB socket.

It is important to note here that the development board also has its own power supply. The debug adapter will sink some current into the development board, enough to switch on LEDs on the board and give the appearance that the hardware is working. However, not enough current is provided to allow the board to run reliably; always ensure that the development board is powered by its usual power supply (Fig. 7.8).

Figure 7.8

STM32F7xx Discovery board.

In this example, we will configure the hardware debugger built into the STM32F7 Discovery board. This board is connected to the development PC using a USB cable. On this board, the USB cable provides power to the Discovery board and also connects to the built-in debug interface.

Software Configuration

Open the Pack Installer.

Select the Boards::Designers Guide Tutorial.

Select the Example tab and Copy "EX 7.1 Hardware Debug."

This is a version of the Blinky example for the ST Discovery board.

Now open the Options for Target\Debug tab

Figure 7.9
The Debug tab allows you to configure the debugger for hardware debug and select the debug adapter.

It is possible to switch from using the simulator to the CoreSight debugger by clicking on the "Use" radio button on the right-hand side of the menu (Fig. 7.9). In the drop-down box make sure ST-Link Debugger is selected. Like the simulator it is possible to run a script when the microvision debugger is started. While the debugger will start successfully without this script, the script is used to program a register within the microcontroller debug system that allows us to configure any unique options for the microcontroller in use (Fig. 7.10).

Debug MCU Configuration
- DBG_SLEEP ☑
- DBG_STOP ☑
- DBG_STANDBY ☑
- TRACE_IOEN ☑
- TRACE_MODE Asynchronous
- DBG_IWDG_STOP ☐
- DBG_WWDG_STOP ☐
- DBG_TIM1_STOP ☐
- DBG_TIM2_STOP ☐
- DBG_TIM3_STOP ☐
- DBG_TIM4_STOP ☐
- DBG_CAN_STOP ☐

Figure 7.10
The debugger script file allows you to configure the CoreSight debug registers. Here, we can freeze timers and control the debugger behaviour during powerdown.

In the case of this microcontroller, the manufacturer has implemented options that allow some of the user peripherals to be halted when the Cortex-M CPU is halted by the debug system. In this case, the user timers, watchdogs, and CAN module may be frozen when the CPU is halted. The script also allows us to enable debug support when the cortex CPU is placed into low-power modes. In this case, the sleep modes designed by the microcontroller manufacturer can be configured to allow the clock source to the CoreSight debug system to be kept running while the rest of the microcontroller enters the requested low-power mode. Finally, we can configure the serial trace pin. It must be enabled by selecting TRACE_IOEN with TRACE_MODE set to Asynchronous. When the debugger starts, the script will write the custom configuration value to the MCU debug register (Fig. 7.11).

```
_WDWORD(0xE0042004, 0x00000027);
```

Now press the settings button.

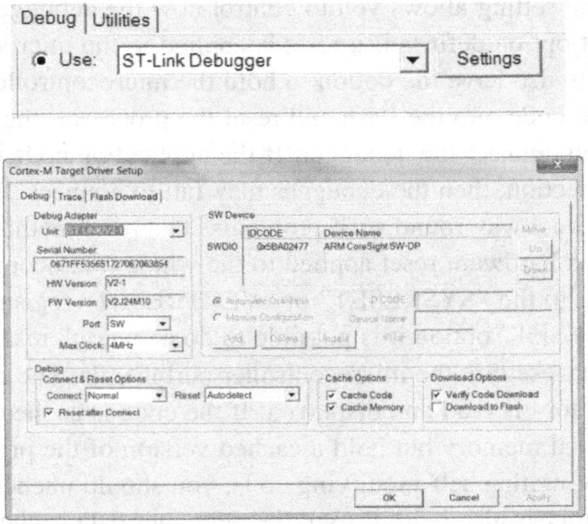

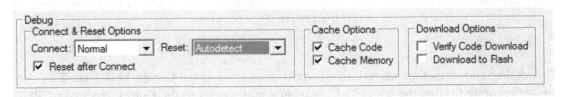

Figure 7.11
The driver setup menu will initially show you that the debug adapter is connected and working.
Successfully reading the coreID proves the microcontroller is running.

This will open the debug adapter configuration window. The two main panes in this dialog show the connection state of the debug adapter to the PC and to the microcontroller. The debug adapter USB pane shows that the debug adapter serial number and firmware version. It also allows you to select which style of debug interface to use when connecting to the microcontroller. For Cortex-M microcontrollers, you should normally always use SW but JTAG is also available should you need it. The SWJ tick box allows the debugger to select the debug interface style dynamically.

When you are connected to a microcontroller, the SW device dialog will display all of the available debug interfaces (on some microcontrollers there may be more than one). Although you normally do not need to do any configuration to select a debug port, this information is useful in that it tells you that the microcontroller is running. When you are bringing up a new board for the first time, it can be useful to check this screen when the board is first connected to get some confidence that the microcontroller is working (Fig. 7.12).

Figure 7.12
The debug section allows you to define the connection and reset method used by the debugger.

The debug dialog in the setting allows you to control how the debugger connects to the hardware. The connect option defines if a reset is applied to the microcontroller when the debugger connects, you also have the option to hold the microcontroller in reset. Remember that when you program code into the flash and reset the processor, the code will run for many cycles before you connect the debugger. If the application code does something to disturb the debug connection, then the debugger may fail to connect. Being able to reset and halt, the code can be a way round such problems. The reset method may also be controlled, this can be a hardware reset applied to the whole microcontroller or a software reset caused by writing to the "SYSRESET" or "VECTRESET" registers in the NVIC. In the case of the "SYSRESET" option it is possible to do a "warm" reset that is resetting the Cortex-M CPU without resetting the microcontroller peripherals. The cache options affect when the physical memory is read and displayed. If the code is cached then the debugger will not read the physical memory but hold a cached version of the program image in the PC memory. If you are writing self-modifying code, you should uncheck this option. A cache of the data memory is also held, if used the debugger data cache is only updated once when the code is halted. If you want to see the peripheral registers update while the code is halted, you should uncheck this option (Fig. 7.13).

Now click on the Trace tab.

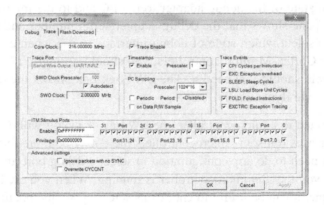

Figure 7.13
The Trace tab is used to configure the Instruction, data, and instrumentation trace units and also enable the performance counters.

This dialog allows us to configure the internal trace units of the Cortex-M device. To ensure accurate timing, the core clock frequency must be set to the Cortex processor CPU clock frequency. The trace port options are configured for the SW interface using the UART communication protocol. In this menu, we can also enable and configure the data watch trace and enable the various trace event counters. This dialog also configures the ITM and we will have a look at this later in this chapter (Fig. 7.14).

Now click the Flash Download tab.

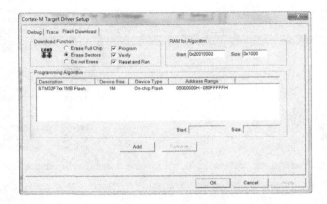

Figure 7.14
The programming algorithm for the internal microcontroller FLASH memory.

This dialog allows you to set the Flash Download algorithm for the microcontroller FLASH memory. This will normally be configured automatically when the project is defined. This menu allows you to update or add algorithms to support additional external parallel or serial FLASH memory.

Once configured, you can start the debugger as normal and it will be connected to the hardware in place of the simulation model. This allows you to exercise the code on the real hardware through the same debugger interface we have been using for the simulator. With the data watch trace enabled, you can see the current state of your variables in the watch and memory windows without having to halt the application code. It is also possible to add global variables to the Logic analyzer to trace the values of a variable over time. This can be incredibly useful when working with real-time data (Fig. 7.15).

Type	Ovf	Num	Address	Data	PC	Dly	Cycles	Time[s]	
Counter Event				01H			22593	0.00010460	
Counter Event				10H		X	57420	0.00026583	
Counter Event				01H		X	57420	0.00026583	
Counter Event	X			11H		X	57420	0.00026583	
Counter Event	X			11H		X	57420	0.00026583	
Counter Event	X			11H		X	57420	0.00026583	
ITM		31		08001569H		X	57420	0.00026583	
ITM		31		0103H		X	57420	0.00026583	
ITM		31		03H		X	57420	0.00026583	
ITM		31		FFH		X	57420	0.00026583	
Exception Return	X	0				X	57420	0.00026583	
Counter Event	X			19H		X	57420	0.00026583	
Counter Event				02H			251504	0.00116437	
Exception Entry		15					251508	0.00116439	
Counter Event				10H			251668	0.00116513	
Counter Event				01H			251719	0.00116537	
Exception Return	X	0				X	256896	0.00118933	
Exception Entry		15					467496	0.00216433	
Exception Exit		15					467595	0.00216479	
Exception Return		0					467604	0.00216483	

Trace Records

Trace Exceptions

☑ EXCTRC: Exception Tracing ☑ Timestamps Enable

#	Name	Count	Total Time	Min Time ...	Max Time...	Min Time ...	Max Time...	First Time...	Last Time...
15	SysTick	10586	5.822 ms	430.556 ns	55.000 us	945.000 us	2.000 ms	0.00116439	10.586164...
16	WWDG	0	0 s						
18	TAMP_STAMP	0	0 s						
19	RTC_WKUP	0	0 s						
20	FLASH	0	0 s						

Figure 7.15
The Exception tracing and Event counters provide useful code execution metrics.

The data watch trace windows give some high-level information about the runtime performance of the application code. The data watch trace windows provide a raw high-level trace exception and data access. An exception trace is also available which provides detailed information of exception and interrupt behaviour (Fig. 7.16).

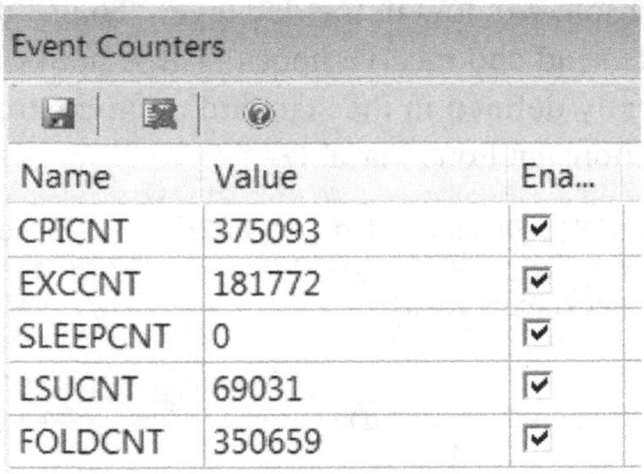

Figure 7.16
The Event counters are a guide to how efficiently the Cortex-M processor is running within the microcontroller.

The CoreSight debug architecture also contains a number of counters that show the performance of the Cortex-M processor. The extra cycles per instruction count is the number of wait states the processor has encountered waiting for the instructions to be delivered from the FLASH memory. This is a good indication at how efficiently the processor is running.

Debug Limitations

When the PC debugger is connected to the real hardware, there are some limitations compared to the simulator. Firstly, you are limited to the number of hardware breakpoints provided by the silicon manufacturer and there will be a maximum of eight breakpoints. This is not normally a limitation but when you are trying to track down a bug or test code, it is easy to run out. The basic trace units do not provide any instruction trace or timing information. This means that the code coverage and performance analysis features are disabled.

Instrumentation Trace

In addition to the data watch trace, the basic debug structure on Cortex-M3 and Cortex-M4 includes a second trace unit called the ITM. You can think of the ITM as serial port which is connected to the debugger. You can then add code to your application that writes custom debug messages to the ITM which are then displayed within the debugger. By instrumenting your code this way, you can send complex debug information to the debugger. This can be used to help locate obscure bugs but it is also especially useful for software testing.

The ITM is slightly different from the other two trace units in that it is intrusive on the CPU, that is, it does use a few CPU cycles. The ITM is best thought of as a debug UART which is connected to a console window in the debugger. To use it, you need to instrument your code by adding simple send and receive hooks. These hooks are part of the CMSIS standard and are automatically defined in the standard microcontroller header file. The hooks consist of three functions and one variable.

```
static __INLINE uint32_t ITM_SendChar (uint32_t ch);    //Send a character to the ITM
volatile int ITM_RxBuffer = ITM_RXBUFFER_EMPTY;         //Receive buffer for the ITM
static __INLINE int ITM_CheckChar (void);               //Check to see if a character has
                                                        //  been sent
                                                        //from the debugger
static __INLINE int ITM_ReceiveChar (void);             //Read a character from the ITM
```

The ITM is actually a bit more complicated in that it has 32 separate channels. Currently, channel 31 is used by the RTOS (Real-Time Operating System) kernel to send messages to the debugger for the kernel aware debug windows. Channel 0 is the user channel which can be used by your application code to send printf() style messages to a console window within the debugger.

Exercise 7.2 Setting Up the ITM

In this exercise, we will look at configuring the ITM to send and receive messages between the microcontroller and the PC debugger.

Open the Pack Installer.

Select the Boards::Designers Guide Tutorial.

Select the Example tab and Copy "EX 7.1 Hardware Debug."

We will configure the Blinky project so that it is possible to send messages through the ITM from the application code. First, we need to configure the debugger to enable the application ITM channel.

Open the "Options for Target\debug" dialog.

Press the ST-Link settings button and select the Trace tab (Fig. 7.17).

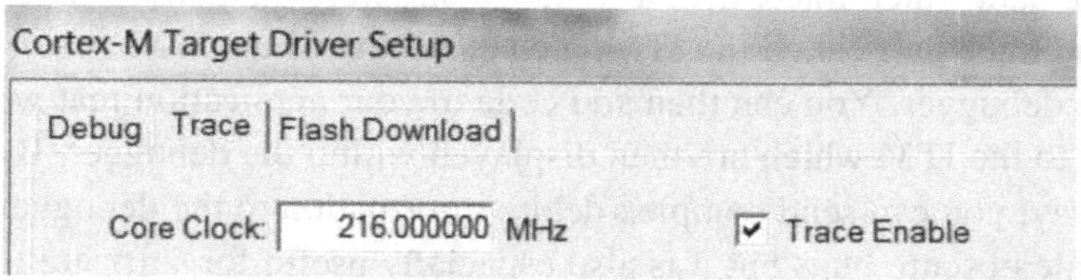

Figure 7.17
The Core Clock frequency must be set correctly for the ITM to work successfully.

To enable the ITM, the Core Clock must be set to the correct frequency as discussed in the last exercise and the Trace must be enabled (Fig. 7.18).

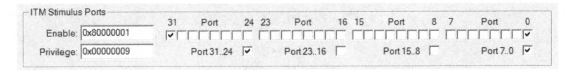

Figure 7.18
The ITM has 32 channels. Currently channel 31 and 0 are used for the RTOS event view and the ITM user debug channel.

In the ITM stimulus ports menu, port 31 will be enabled by default. To enable the application port we must enable port 0. We want to be able to use the Application ITM form privileged and unprivileged mode so the Privileged Port 7..0 should be unchecked.

Once the Trace and ITM settings are configured, click OK and return back to the editor.

Open the Runtime Editor.

Select the Compiler::IO section.

Here, we can configure the channel used as the standard IO (STDIO). The File option provides support for low-level calls made by the stdio.h functions. Leave this unchecked unless you have added a file system (Fig. 7.19).

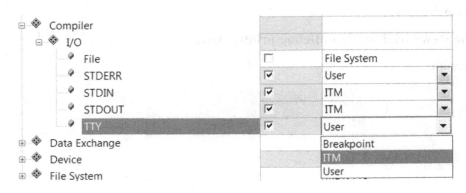

Figure 7.19
The Compiler::I/O options allow you to select the low-level STDIO channel.

Enable STDERR, STDIN, STDOUT, and TTY, and select ITM from the drop-down menu.

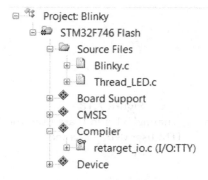

Figure 7.20
Updated project with the ITM support for STDIO in retarget_IO.c.

This adds a "retarget_io.c" file to the project. This file contains the low-level STDIO driver functions which read and write a single character to the ITM (Fig. 7.20).

Open Blinky.c.

Add stdio.h to the list of include files.

```
#include "RTE_Components.h"
#include <stdio.h>
```

Add a printf statement after the kernelStart() API call.

```
osKernelStart();
printf("RTX Started\n");
while (1) {
```

Start the debugger.

Open the view\serial windows\Debug(printf) viewer.

Run the code.

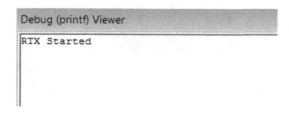

Figure 7.21
Debug message sent from the application code to the debugger console window.

The printf() message is sent to the STDIO channel which is the ITM. This message is read and displayed in the debugger console window (Fig. 7.21).

We will see how to use the ITM to send run-time diagnostic messages in Chapter 10 "RTOS Techniques" and also how to use it as part of a software testing scheme in Chapter 11 "Test-Driven Development."

System Control Block Debug Support

The CoreSight debugger interface allows you to control execution of your application code and examine values in the memory and peripheral registers. Combined with the various trace units, this provides you with a powerful debug system for normal program development. However, as we saw in Chapter 3 "Cortex-M Architecture," the Cortex-M processors have up to four fault exceptions which will be triggered if the application code makes incorrect use of the Cortex-M processor or the microcontroller hardware (Table 7.2).

Table 7.2: Fault exceptions

Fault Exception	Priority	Cortex Processor
Hard fault	−1	Cortex-M0, Cortex-M0+, Cortex-M3, Cortex-M4
Bus fault	Programmable	Cortex-M3, Cortex-M4
Usage fault	Programmable	Cortex-M3, Cortex-M4
Memory manager fault	Programmable	Cortex-M3, Cortex-M4 (Optional)

When this happens, your program will be trapped on the default fault handler in the startup code. It can be very hard to work out how you got there. If you have an instruction trace tool, you can work this out in seconds. If you do not have access to instruction trace then resolving a run-time crash can take a long, long time. In this section, we will look at configuring the fault exceptions and then looking at how to track back to the source of a fault exception.

The behaviour of the fault exceptions can be configured by registers in the system control block. The key registers are listed in Table 7.3.

Table 7.3: Fault exception configuration registers

Register	Processor	Description
Configuration and control	M3, M4	Enable additional fault exception features
System handler control and state	M3, M4	Enable and pending bits for fault exceptions
Configurable fault status register	M3, M4	Detailed fault status bits
Hard fault status	M3, M4	Reports a hard fault or fault escalation
Memory manager fault address	M3, M4	Address of location that caused the memory manager fault
Bus fault address	M3, M4	Address of location that caused the bus fault

When the Cortex-M processor comes out of reset, only the hard fault handler is enabled. If a usage, bus, or memory manager fault is raised and the exception handler for these faults is not enabled, then the fault will "escalate" to a hard fault. The hard fault status register provide two status bits that indicate the source of the hard fault (Table 7.4).

Table 7.4: Hard fault status register

Name	Bit	Use
FORCED	30	Reached the hard fault due to fault escalation
VECTTBL	1	Reached the hard fault due to a faulty read of the vector table

The "System Handler Control and State" register contains enable, pending, and active bits for the bus, usage, and memory manager exception handlers. We can also configure the behaviour of the fault exceptions with the "Configuration and Control" register (Table 7.5).

Table 7.5: Configuration and control register

Name	Bit	Use
STKALIGN	9	Configures 4- or 8-byte stack alignment
BFHFMIGN	8	Disables data bus faults caused by load and store instructions
DIV_0_TRP	4	Enables a usage fault for divide by zero
UNALIGNTRP	3	Enables a usage fault for unaligned memory access

The "divide by zero" exception can be a useful trap to enable, particularly during development. The remaining exceptions should be left disabled unless you have a good reason to switch them on. When a memory manager fault exception occurs, the address of the instruction that attempted to access a prohibited memory region will be stored in the "Memory Fault Address Register," similarly when a bus fault is raised the address of the instruction that caused the fault will be stored in the "Bus Fault Address" register. However, under some conditions, it is not always possible to write the fault addresses to these registers. The configurable "Fault Status" register contains an extensive set of flags that report the Cortex-M processor error conditions that help you track down the cause of a fault exception.

Tracking Faults

If you have arrived at the hard fault handler, first check the "Hard Fault Status" register. This will tell you if you have reached the hard fault due to fault escalation or a vector table read error. If there is a fault escalation, next check the "System Handler Control and State" register to see which other fault exception is active. The next port of call is the

"Configurable Fault Status" register. This has a wide range of flags that report processor error conditions (Table 7.6).

Table 7.6: Configurable fault status register

Configurable fault status register		
Name	Bit	Use
DIVBYZERO	25	Divide by zero error
UNALIGNED	24	Unaligned memory access
NOCP	19	No CoProcessor present
INVPC	18	Invalid PC load
INVSTATE	17	Illegal access the execution program status register EPSR
UNDEFINSTR	16	Attempted execution of an undefined instruction
BFARVALID	15	Address in bus fault address register is valid
STKERR	12	Bus fault on exception entry stacking.
UNSTKERR	11	Bus fault on exception exit unstacking
IMPRECISERR	10	Data bus error. Error address not stacked
PRECISERR	9	Data bus error. Error address stacked
IBUSERR	8	Instruction bus error
MMARVALID	7	Address in the memory manager fault address register is valid
MSTKERR	4	Stacking on exception entry caused a memory manager fault
MUNSTKERR	3	Stacking on exception exit caused a memory manager fault
DACCVIOL	1	Data access violation flag
IACCVIOL	0	Instruction access violation flag

When the processor fault exception is entered a stack frame will normally be pushed onto the stack. In some cases, the stack frame will not be valid and this will be indicated by the flags in the "Configurable Fault Status" register. When a valid stack frame is pushed, it will contain the PC address of the instruction that generated the fault. By decoding the stack frame, you can retrieve this address and locate the problem instruction. The system control block provides a memory and bus fault address register which depending on the cause of the error may hold the address of the instruction that caused the error exception.

Exercise 7.3 Processor Fault Exceptions

Open the Pack Installer.

Select the Boards::Designers Guide Tutorial.

Select the Example tab and Copy "EX 7.3 Fault Tracking."

In this project, we will generate a fault exception and look at how it is handled by the NVIC. Once the fault exception has occurred, we can interrogate the stack to find the instruction which caused the error exception.

```
volatile uint32_t op1;
int main (void)
{
int op2 = 0x1234,op3 = 0;
        SCB->CCR = 0x0000010;     //Enable divide by zero usage fault
        op1 = op2/op3;            //perform a divide by zero to generate an usage
exception
while(1);
}
```

The code first enables the "divide by zero usage fault" and then performs a divide by zero to cause the exception.

Build the code and start the debugger.

Set a breakpoint on the line of code that contains the divide instruction (Fig. 7.22).

```
10      SCB->SHCSR = 0x00060000;
11      SCB->CCR = 0x0000010;
● 12    op1 = op2/op3;
```

Figure 7.22
Set a breakpoint on the divide statement.

Run the code until it hits this breakpoint.

Open the Peripherals\Core Peripherals\System Control and Configuration window and check that the Divide by Zero trap has been enabled (Fig. 7.23).

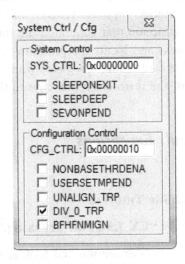

Figure 7.23
Check the Divide by Zero trap is enabled by using the "System Configuration and Control" peripheral view.

Single step the divide instruction (Fig. 7.24).

```
127                     ENDP
128   HardFault_Handler\
129           PROC
130                 EXPORT   HardFault_Handler       [WEAK]
⇨ 131   |           B       .
132           ENDP
```

Figure 7.24
An error exception will cause you to hit the hard fault handler if the other fault exceptions have not been enabled.

A usage fault exception will be raised. We have not enabled the usage fault exception vector so the fault will elevate to a hard fault.

Open the Peripherals\Core Peripherals\Fault Reports window (Fig. 7.25).

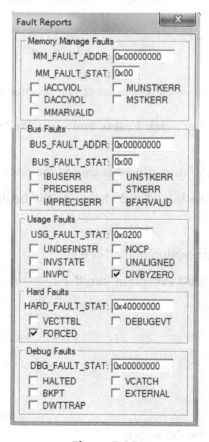

Figure 7.25
If a fault occurs you can view a digest of the fault diagnostic registers in the peripheral fault reports window.

This window shows that the hard fault has been forced by another fault exception. Also, the divide by zero flag has been set in the usage fault status register.

In the Register window, read the contents of R13, the Main Stack Pointer, and open a Memory window at this location (Fig. 7.26).

```
---- R12        0x20000048
---- R13 (SP)   0x20000248
---- R14 (LR)   0xFFFFFFF9
```

```
Address: 0x20000248
```

```
0x20000248:  00000000  00001234  00000000  E000ED14  20000048  0800017B  08000198  21000000
```

Figure 7.26
Use the Register window, read the current address stored in the stack pointer. Then, use the Memory window to read the PC value stored in the stack frame.

```
0x08000198 FB91F2F0    SDIV      r2,r1,r0
0x0800019C 4B08        LDR       r3,[pc,#32]   ; @0x080001C0
```

Figure 7.27
The PC value stored in the stack frame leads us to the SDIV instruction that caused the error exception.

Read the PC value saved in the Stack Frame and open the Disassembly window at this location (Fig. 7.27).

This takes us back to the SDIV instruction that caused the fault.

Exit the debugger and add the line of code below to the beginning of the program.

```
SCB->SHCSR = 0x00060000;
```

This enables the usage fault Exception in the NVIC.

Now add a "C" level usage fault exception handler.

```
void UsageFault_Handler (void)
{
error_address = (uint32_t *)(__get_MSP());    // load the current base address of the
                                                 stack pointer
error_address = error_address + 6;            // Locate the PC value in the last stack
frame
while(1);
}
```

Build the project and start the debugger.

Set a breakpoint on the while loop in the exception function (Fig. 7.28).

```
19    void UsageFault_Handler (void)
20  □ {
21        error_address = (uint32_t *)(__get_MSP());
22        error_address = error_address + 6;
23        while(1);
24    }
```

Figure 7.28
The usage fault exception routine can be used to read the PC value from the stack.

Run the code until the exception is raised and the breakpoint is reached.

When a usage fault occurs, this exception routine will be triggered. It reads the value stored in the stack pointer and extracts the value of the PC stored in the stack frame.

Instruction Trace with the Embedded Trace Macrocell

The Cortex-M3, Cortex-M4 and Cortex-M7 may have an optional debug module fitted by the Silicon Vendor when the microcontroller is designed. The embedded trace macrocell (ETM) is a third trace unit which provides an instruction trace as the application code is executed on the Cortex-M processor. Because the ETM is an additional cost to the Silicon Vendor, it is normally only fitted to higher end Cortex-M3, Cortex-M4 and Cortex-M7 based microcontrollers. When selecting a device, it will be listed as a feature of the microcontroller in the datasheet (Fig. 7.29).

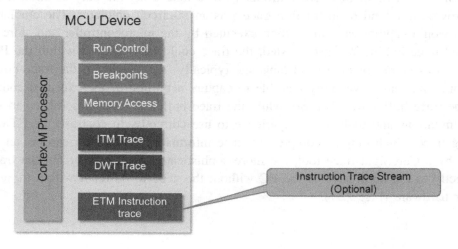

Figure 7.29
The Cortex-M3, Cortex-M4 and Cortex-M7 may optionally be fitted with a third trace unit. The embedded trace macrocell (ETM) supports instruction trace. This allows you to quickly find complex bugs. The ETM also enables code coverage and performance analysis tools which are essential for software validation.

The ETM trace pipe requires four additional pins which are brought out to a larger 20-pin socket which incorporates the SW debug pins and the ETM trace pins. Standard JTAG/ CoreSight debug tools do not support the ETM trace channel so you will more sophisticated debug unit (Fig. 7.30).

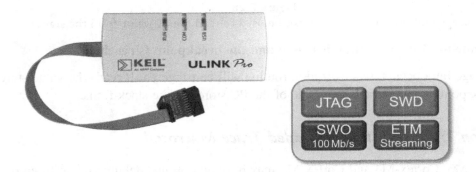

Figure 7.30
An ETM trace unit provides all the features of the standard hardware debugger plus instruction trace.

At the beginning of this chapter, we looked at various debug methods that have been used historically with small microcontrollers. For a long time, the only solution that would provide any kind of instruction trace was an in-circuit emulator. The emulator hardware would capture each instruction executed by the microcontroller and store it in an internal trace buffer. When requested, the trace could be displayed within the PC debugger as assembly or high-level language typically "C." However, the trace buffer had a finite size and it was only possible to capture a portion of the executed code before the trace buffer was full. So, while the trace buffer was very useful it had some serious limitations and took some experience to use correctly. In contrast, the ETM is a streaming trace which outputs compressed trace information. This information can be captured by a CoreSight trace tool, the more sophisticated units will stream the trace data directly to the hard drive of the PC without the need to buffer it within the debugger hardware (Fig. 7.31).

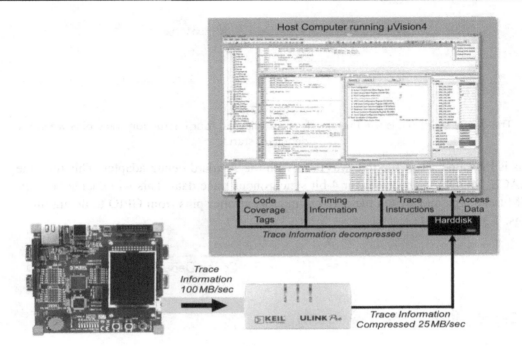

Figure 7.31
A "Streaming" trace unit is capable of recording every instruction directly onto the hard drive of the PC. The size of the trace buffer is only limited by the size of your hard disk.

This streaming trace allows the debugger software to display 100% of the instructions executed along with execution times. The streaming trace buffer is only limited by the size of the PC hard disk, it is also possible to analyze the trace information to provide accurate code coverage and performance analysis information.

Exercise 7.4 Using the ETM Trace

In this exercise, we will look at how to configure the Ulink PRO debug adapter for streaming trace and then review the additional debug features available through the ETM (Fig. 7.32).

Open the Pack Installer.

Select the Boards::Designers Guide Tutorial.

Select the Example tab and Copy "EX 7.4 Instruction Trace."

Open the STM32_TP.ini initialization file.

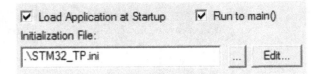

Figure 7.32
The debug script file is used to enable the additional four instruction trace pins when the debugger starts.

This is the same script file that was used with the standard debug adapter. This time the TRACE_MODE has been set for 4-bit synchronous trace data. This will enable both the ETM trace pipe and switch the external microcontroller pins from GPIO to debug pins (Figs. 7.33 and 7.34).

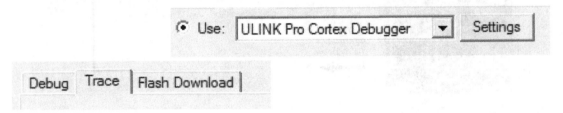

Figure 7.33
Select a debug adapter which is capable of capturing data from the ETM.

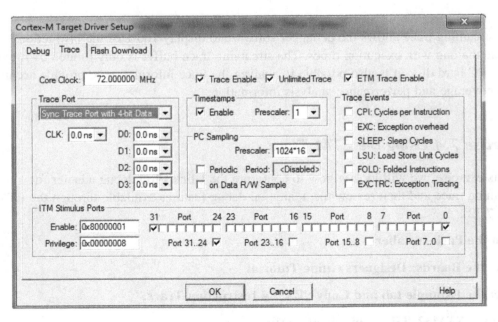

Figure 7.34
The Trace port can now be configured to access the ETM.

Now press the ULINK Pro settings button.

Select the Trace tab.

When the ULINK Pro Trace tool is connected, you have the option to enable the ETM trace. The unlimited trace option allows you to stream every instruction executed to a file on your PC hard disk. The trace buffer is then only limited by the size of the PC hard disk. This makes it possible to trace the executed instructions for days if necessary, yes days.

Click OK to quit back to the μVision editor.

Start the debugger.

Now the debugger has an additional instruction trace window (Fig. 7.35).

Nr.	Time	Address	Opcode	Instruction	Src Code
32,739	0.038 093 547 s	0x00000684	B168	CBZ r0,0x000006A2	
32,740	0.038 093 577 s	0x000006A2	EA950006	EORS r0,r5,r6	if (ad_val ^ ad_val_) { /...
32,741	0.038 093 587 s	0x000006A6	D005	BEQ 0x000006B4	
32,742	0.038 093 617 s	0x000006B4	480D	LDR r0,[pc,#52] ; @0x000006EC	if (clock_1s) {
32,743	0.038 093 637 s	0x000006B6	7800	LDRB r0,[r0,#0x00]	
32,744	0.038 093 657 s	0x000006B8	B130	CBZ r0,0x000006C8	
32,745	0.038 093 687 s	0x000006C8	E7DA	B 0x00000680	while (1) { /* Lo...
32,746	0.038 093 717 s	0x00000680	4815	LDR r0,[pc,#84] ; @0x000006D8	if (AD_done) { /* I...
32,747	0.038 093 737 s	0x00000682	7800	LDRB r0,[r0,#0x00]	
32,748	0.038 093 757 s	0x00000684	B168	CBZ r0,0x000006A2	
32,749	0.038 093 787 s	0x000006A2	EA950006	EORS r0,r5,r6	if (ad_val ^ ad_val_) { /...
32,750	0.038 093 797 s	0x000006A6	D005	BEQ 0x000006B4	
32,751	0.038 093 827 s	0x000006B4	480D	LDR r0,[pc,#52] ; @0x000006EC	if (clock_1s) {
32,752	0.038 093 847 s	0x000006B6	7800	LDRB r0,[r0,#0x00]	
32,753	0.038 093 867 s	0x000006B8	B130	CBZ r0,0x000006C8	

Figure 7.35
The debugger trace window can now display all of the instructions executed by the Cortex-M processor (streaming trace).

In addition to the trace buffer, the ETM also allows us to show the code coverage information that was previously only available in the simulator (Fig. 7.36).

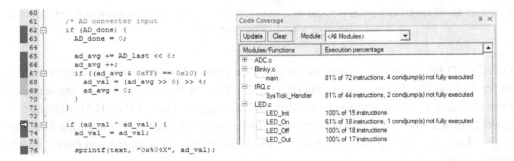

Figure 7.36
A streaming trace allows you to display accurate code coverage information.

Similarly, timing information can be captured and displayed alongside the "C" code or as a performance analysis report (Fig. 7.37).

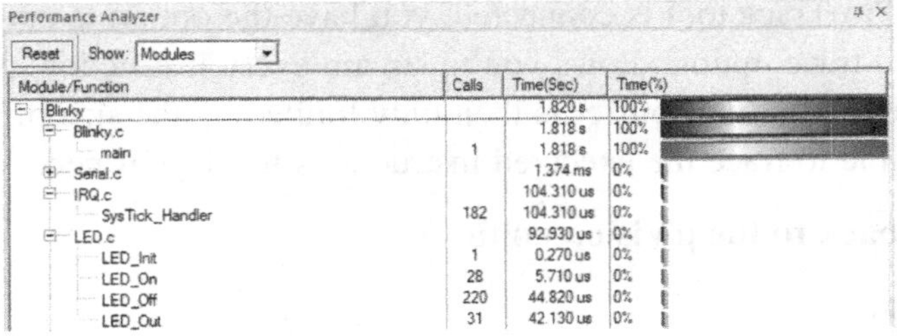

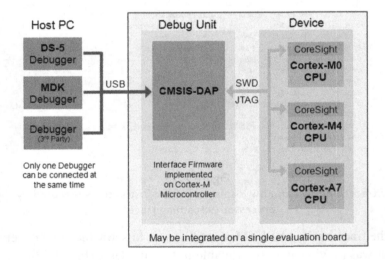

Figure 7.37
The performance analyzer shows the cumulative execution time for each function.

CMSIS-DAP

Figure 7.38
The CMSIS-DAP specification is designed to support interoperability between different debugger hardware and debugger software.

The CMSIS-DAP specification defines the interface protocol between the CoreSight debugger hardware and the PC debugger software (Fig. 7.38). This creates a new level of interoperability between different vendors' software and hardware debuggers. The CMSIS-DAP firmware is designed to turn a low cost Cortex-M microcontroller with a USB peripheral into a Coresight debugger. This allows even the most basic evaluation board to host a common debug interface that can be used with any CMSIS compliant toolchain (Fig. 7.39).

Figure 7.39
The MBED module is the first to support the CMSIS-DAP specification.

The CMSIS-DAP specification is designed to support a USB interface between the target hardware and the PC. This allows many simple modules to be powered directly from the PC USB port (Fig. 7.40).

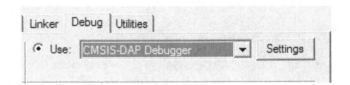

Figure 7.40
The CMSIS-DAP driver must be selected in the debugger menu.

The CMSIS-DAP interface can be selected in the debug menu in place of the proprietary Ulink2. The configuration options are essentially the same as the ulink2 but the options available will depend on the level of firmware implemented by the device manufacturer. The CMSIS-DAP specification supports all of the debug features found in the CoreSight debug architecture including the Cortex-M0+ Micro Trace Buffer (MTB).

Cortex-M0+ MTB

While the ETM is available for the Cortex-M3 and Cortex-M4, no form of instruction trace is currently available for the Cortex-M0. However the Cortex-M0+ has a simple form of instruction trace buffer called the MTB. The MTB uses a region of internal SRAM which is allocated by the developer. When the application code is running, a trace of executed instructions are recorded into this region. When the code is halted, the debugger can read the MTB trace data and display the executed instructions. The MTB trace RAM can be configured as a circular buffer or a one-shot recording. While this is a very limited trace, the circular buffer will allow you to see "what just happened" before the code halted. While the one-shot mode can be triggered by the hardware breakpoints to start and stop allowing you to track down more elusive bugs.

Exercise 7.5 Micro Trace Buffer

This exercise is based on the NXP Freedom board for the MKL25Z microcontroller. This was the first microcontroller available to use the Cortex-M0+.

Connect the freedom board via its USB cable to the PC.

Open the Pack Installer.

Select the Boards::Designers Guide Tutorial.

Select the Example tab and Copy "EX 7.5 Micro Trace Buffer."

Open the Options for Target/debug menu (Fig. 7.41).

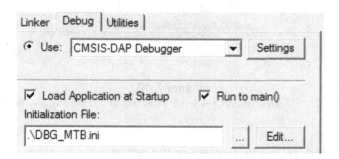

Figure 7.41
Select the CMSIS-DAP debug adapter and the Micro Trace Buffer script file.

Here the CMSIS-DAP interface is selected along with an initializing file for the MTB.

The initializing script file has a wizard that allows you to configure the size and configuration of the MTB (Fig. 7.42).

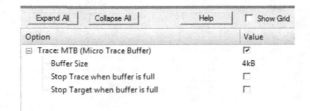

Figure 7.42
The script file used to configure the Micro Trace Buffer.

Here we can select the amount of internal SRAM that is to be used for the trace. It is also possible to configure different debugger actions when the trace is full. Either halt trace recording or halt execution on the target.

By default, the MTB is located at the start of the internal SRAM (0x20000000). So it is necessary to offset the start of user SRAM by the size of memory allocated to the MTB. This is done in the Options For Target\Target dialog (Fig. 7.43).

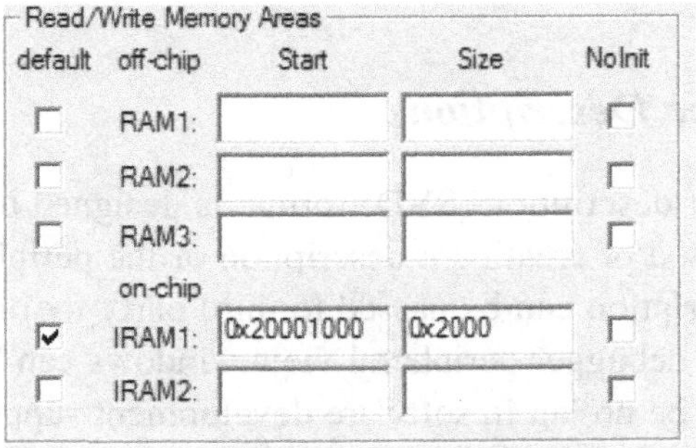

Figure 7.43
You must offset the user RAM from the region used by the Micro Trace Buffer.

Now start the debugger and execute the code.

Halt the debugger and open the Trace window (Fig. 7.44).

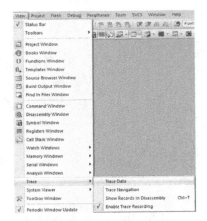

Figure 7.44

The contents of the Micro Trace Buffer are downloaded to the PC and displayed in the debugger Trace window.

While this is a limited trace buffer, it can be used by very low-cost tools and provides a means of tracking down runtime bugs which would be time consuming to find any other way.

CMSIS System Viewer Description

The CMSIS system viewer description (SVD) format is designed to provide silicon manufacturers with a method of creating a description of the peripheral registers in their microcontroller. This description can be passed to third party tool manufactures so that compiler include files and debugger peripheral view windows can be created automatically. This means that there will be no lag in software development support when the next new family of devices are released. As a developer you will not normally need to work with these files but it is useful to understand how the process works so that you can fix any errors that may inevitably occur. It is also possible to create your own additional peripheral debug windows. This would allow you to create a view of an external memory mapped peripheral or provide a debug view of a complex memory object.

When the Silicon Vendor develops a new microcontroller, they also create an XML description of the microcontroller registers. A conversion utility is then used to create a binary version of the file that is used by the debugger to automatically create the peripheral debug windows (Fig. 7.45).

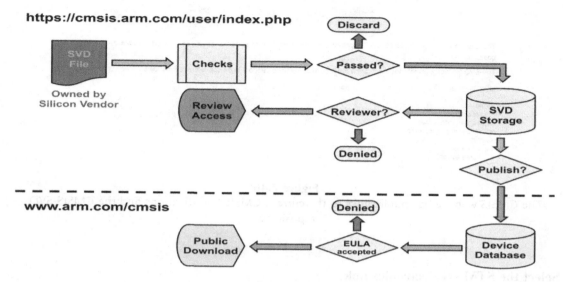

Figure 7.45
ARM have created a repository of "system viewer description" files. This enables tool suppliers to have support for new devices as they are released.

Alongside the XML description definition, ARM has introduced a submission and publishing system for new SVD files. When a Silicon Vendor designs a new microcontroller, its system description file is submitted via the CMSIS website. Once it has been reviewed and validated, it is then published for public download on the main ARM website.

Exercise 7.6 CMSIS-SVD

In this exercise, we will have a look at a typical system viewer file make an addition and rebuild the debugger file, and then check the updated version in the debugger.

For this exercise, you will need an XML editor. If you do not have one, then download a trial or free tool from the internet.

Open your web browser and go to www.arm.com/cmsis.

If you do not have a user account on the ARM website you will need to create one and login.

On the CMSIS page, select the CMSIS-SVD tab (Fig. 7.46).

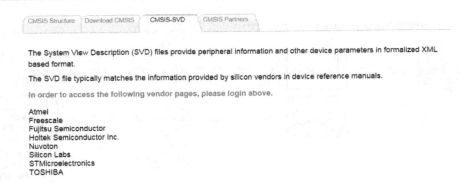

CMSIS Structure   Download CMSIS   CMSIS-SVD   CMSIS Partners

The System View Description (SVD) files provide peripheral information and other device parameters in formalized XML based format.

The SVD file typically matches the information provided by silicon vendors in device reference manuals.

In order to access the following vendor pages, please login above.

Atmel
Freescale
Fujitsu Semiconductor
Holtek Semiconductor Inc.
Nuvoton
Silicon Labs
STMicroelectronics
TOSHIBA

Figure 7.46
The CMSIS website has public links to the current CMSIS specifications and the CMSIS-SVD repository.

Select the STMicroelectronics link.

In the ST window, select the SVD download that supports the STM32F103RB (Fig. 7.47).

☑	STM32F103C8, STM32F103R8, STM32F103RB, STM32F103V8, STM32F103VB, STM32F103CB, STM32F103T8, STM32F103TB,	STM32F103xx.svd	1.0	665.09 KB	21/05/2012

Download   Clear

Figure 7.47
Select the STM32F103 SVD files.

Then click the download button at the bottom of the page.

Unzip the compressed file into a directory.

Open the STM32F103xx.svd file with the XML editor (Fig. 7.48).

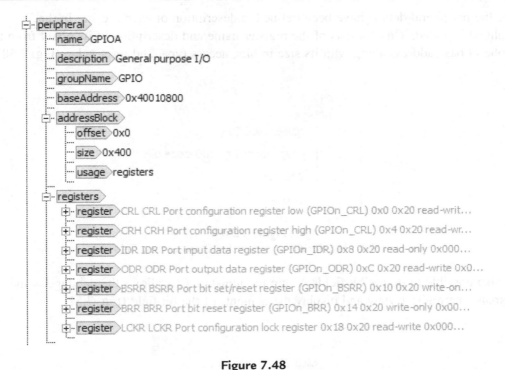

Figure 7.48
The SVD files consist of an XML description of the microcontroller peripheral registers.

Each of the peripheral windows is structured as a series of XML tags which can be edited or a new peripheral pane can be added. This would allow you to display the registers of an external peripheral interfaced onto an external bus. Each peripheral window starts with a name, description, and group name followed by the base address of the peripheral (Fig. 7.49).

Figure 7.49
An SVD description for a GPIO port register.

Once the peripheral details have been defined, a description of each register in the peripheral is added. This consists of the register name and description of its offset from the peripheral base address along with its size in bits, access type, and reset value (Fig. 7.50).

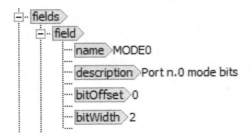

Figure 7.50
An SVD description of a peripheral register bit field.

It is also possible to define bit fields within the register. This allows the debugger window to expand the register view and display the contents of the bit field (Fig. 7.51).

Figure 7.51
The resulting peripheral view.

Make a small change to the XML file and save the results (Fig. 7.52).

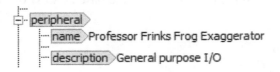

Figure 7.52
Make an edit to the SVD description.

Generate the SFR file by using the SVDConv.exe utility.

This utility is located in c:\keil\arm\cmsis\svd.

The convertor must be invoked with the following command line.

SVDConv STM32F103xx.svd --generate = sfr

This will create the STM32F103xx.SFR.

Now start μVision and open a previous exercise.

Open the "Options for Target" menu and select the STM32F103xx.sfr as the system viewer file (Fig. 7.53).

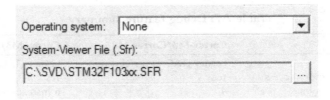

Figure 7.53
Load the SFR file in the project Target menu.

Build the project and start the debugger.

Open the View\System Viewer selection and view the updated peripheral window (Fig. 7.54).

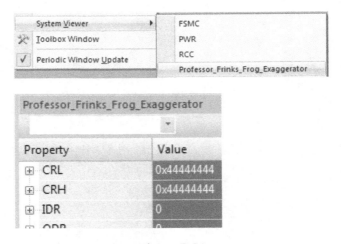

Figure 7.54
The modifications are now visible in the debugger.

Homage to the Simpsons aside, it is useful to know how to add and modify the peripheral windows so you can correct mistakes or add in your own specific debug support.

Conclusion Debug Features Summary

In this chapter, we have had a look through the advanced debug features available within the CoreSight debug architecture for Cortex-M. Table 7.7 summarizes the features available on the different Cortex-M processors.

Table 7.7: Debug feature summary

Feature	Cortex-M0/Cortex-M0+	Cortex-M3/M4/M7
Debug interface	Legacy JTAG or serial wire	Legacy JTAG or serial wire
"On-the-fly" memory access	Yes	Yes
Hardware breakpoint	4	6 Instruction + 2 Literal
Data watch point	2	4
Software breakpoint	Yes	Yes
ETM instruction trace	No	Yes (optional)
Data trace	No	Yes (optional)
Instrumentation trace	No	Yes
Serial wire viewer	No	Yes
Micro Trace Buffer	Yes (Cortex-M0+ only)	No

CHAPTER 8

Practical DSP for Cortex-M4 and Cortex-M7

Introduction

From a developer's perspective, the Cortex-M4 and Cortex-M7 are versions of the Cortex-M processor that have additional features to support Digital Signal Processing (DSP). The key enhancements over the Cortex-M3 are the addition of "Single Instruction Multiple Data" or SIMD instructions, an improved Multiply Accumulate (MAC) unit for integer maths, and the optional addition of a hardware "Floating Point Unit" (FPU). In the case of the Cortex-M4, this is a single precision FPU, while the Cortex-M7 has the option of either a single or double precision FPU. These enhancements give the Cortex-M4 the ability to run DSP algorithms at high enough levels of performance to compete with dedicated 16-bit DSP processors. As we saw in Chapter 6 "Cortex-M7" the Cortex-M7 processor has a more advanced pipeline and the branch cache unit. Both of these features dramatically improve its DSP capability. In this chapter, we will look at using the Cortex-M4/M7 to process real world signals (Fig. 8.1).

Figure 8.1

The Cortex-M4 and Cortex-M7 extends the Cortex-M3 with the addition of DSP instructions and fast maths capabilities. This creates a microcontroller capable of supporting real-time DSP algorithms, a digital signal controller.

Hardware FPU

One of the major features of the Cortex-M4 and Cortex-M7 processors is the addition of a hardware FPU. The FPU supports floating point arithmetic operations to the IEEE 754 standard.

The Designer's Guide to the Cortex-M Processor Family.
DOI: http://dx.doi.org/10.1016/B978-0-08-100629-0.00018-9

Initially, the FPU can be thought of as a coprocessor that is accessed by dedicated instructions to perform most floating point arithmetic operations in a few cycles (Table 8.1).

Table 8.1: Floating point unit performance

Operation	Cycle Count
Add/Subtract	1
Divide	14
Multiply	1
Multiply accumulate (MAC)	3
Fused MAC	3
Square root	14

The FPU consists of a group of control and status registers and 31 single precision scalar registers. The scalar registers can also be viewed as 16 double word registers (Fig. 8.2).

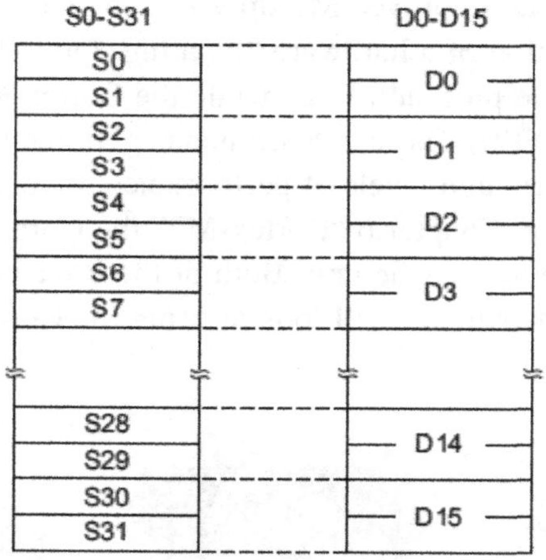

Figure 8.2
The FPU 32-bit scalar registers may also be viewed as 64-bit double word registers. This supports very efficient casting between C types.

While the FPU is designed for floating point operations, it is possible to load and store fixed point and integer values. It is also possible to convert between floating point to fixed point and integer values. This means "C" casting between floating point and integer values can be done in a single cycle.

FPU Integration

While it is possible to consider the FPU as a coprocessor adjacent to the Cortex-M4 and Cortex-M7 processors this is not really true. The FPU is an integral part of the Cortex-M processor, the floating point instructions are executed within the FPU in a parallel pipeline to the Cortex-M processor instructions. Whilst this increases the FPU performance it is "invisible" to the application code and does not introduce any strange side effects (Fig. 8.3).

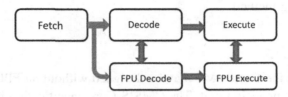

Figure 8.3
The FPU is described as a coprocessor in the register documentation. In reality, it is very tightly coupled to the main instruction pipeline.

FPU Registers

In addition to the scalar registers, the FPU has a block of control and status registers (Table 8.2).

Table 8.2: FPU control registers

Register	Description
Coprocessor access control	Controls the privilege access level to the FPU
Floating point context control	Configures stacking and lazy stacking options
Floating point context address	Holds the address of the unpopulated FPU stack space
Floating point default status control	Holds the FPU condition codes and FPU configuration options
Floating point status control	Holds the default status control values

All of the FPU registers are memory mapped except the Floating Point Status Control (FPSC) Register which is a CPU register accessed by the Move Register to Special Register (MRS) and Move Special Registers to Register (MSR) instructions. Access to this function is supported by a CMSIS-Core function.

```
uint32_t   __get_FPSCR(void );
void       __set_FPSCR(uint32_t (fpscr)
```

The FPSCR register contains three groups of bits. The top four bits contain condition code flags N, Z, C, and V that match the condition code flags in the xPSR. These flags are set and cleared in a similar manner by results of floating point operations. The next groups of bits contain configuration options for the FPU. These bits allow you to change the operation of the FPU from the IEEE 754 standard. Unless you have a strong reason to do this, it is recommended to leave them alone. The final group of bits are status flags for the FPU exceptions. If the FPU encounters an error during execution, an exception will be raised and the matching status flag will be set. The exception line is permanently enabled in the FPU and just needs to be enabled in the NVIC to become active. When the exception is raised, you will need to interrogate these flags to work out the cause of the error. Before returning from the FPU exception, the status flags must be cleared. How this is done depends on the FPU exception stacking method.

Cortex-M7 FPU

Microcontrollers using the Cortex-M7 may be designed without an FPU or may be fitted with a single or double precision unit. The CMSIS-Core specification has been extended with a function to report the Cortex-M7 processor configuration.

```
uint32_t SBC_GetFPUType(void)

0 = No FPU
1 = Single Precision
2 = Dual Precision
```

This function reads a configuration register in the System Control Block, the "Media and VFP feature register" then returns the configuration of the Cortex-M7 processor. This function can only be used with a Cortex-M7.

Enabling the FPU

When the Cortex-M4 or Cortex-M7 leaves the reset vector, the FPU is disabled. The FPU is enabled by setting the coprocessor 10 and 11 bits in the CPARC register. It is necessary to use the data barrier instruction to ensure that the write is made before the code continues. The instruction barrier command is also used to ensure the pipeline is flushed before the code continues.

```
SCB->CPACR |= ((3UL ≪ 10*2) | (3UL ≪ 11*2 ));   // Set CP10 & CP11 Full Access
__DSB();                                          //Data barrier
__ISB();                                          //Instruction barrier
```

In order to write to the CPARC register, the processor must be in privileged mode. Once enabled, the FPU may be used in privileged and unprivileged modes.

Exceptions and the FPU

When the FPU is enabled, an extended stack frame will be pushed when an exception is raised. In addition to the standard stack frame, the Cortex-M processor also pushes the first 16 FPU scalar registers and the FPSCR. This extends the stack frame from 32 to 100 bytes. Clearly pushing this amount of data onto the stack, every interrupt would increase the interrupt latency significantly. To keep the 12 cycle interrupt latency, the Cortex-M processor uses a technique called "lazy stacking." When an interrupt is raised, the normal stack frame is pushed onto the stack and the stack pointer is incremented to leave space for the FPU registers, but their values are not pushed onto the stack. This leaves a void space in the stack. The start address of this void space is automatically stored in the "Floating Point Context Address Register" (FPCAR). If the interrupt routine uses floating point calculations, the FPU registers will be pushed into this space using the address stored in the FPCAR as a base address. The "Floating Point Context Control Register" is used to select the stacking method used. Lazy stacking is enabled by default when the FPU is first enabled. The stacking method is controlled by the most significant 2 bits in the "Floating Point Context Control Register"; these are the Automatic State Preservation Enable (ASPEN) and Lazy State Preservation Enable (LSPEN) (Table 8.3).

Table 8.3: Lazy stacking options

LSPEN	ASPEN	Configuration
0	0	No automatic state presevation. Only use when the interrupts do not use floating point
0	1	Lazy stacking disabled
1	0	Lazy stacking enabled
1	1	Invalid configuration

Using the FPU

Once you have enabled the FPU, the compiler will start to use hardware floating point calculations in place of software libraries. The exception is the square root instruction sqrt (), which is part of the math.h library. If you have enabled the FPU, the ARM compiler provides an intrinsic instruction to use the FPU square root instruction.

```
float __sqrtf( float x);
```

Note: The intrinsic square root function differs from the ANSI sqrt() library function in that it takes and returns a float rather than a double.

Exercise 8.1 Floating Point Unit

This exercise performs a few simple floating point calculations using the Cortex-M4 processor so that we can compare the performance of the software and hardware floating point execution times.

Open the Pack Installer.

Select the Boards::Designers Guide Tutorial.

Select the Example tab and Copy "EX 8.1 Floating Point Unit."

The code in the main loop is a mixture of maths operations to exercise the FPU.

```
#include<math.h>
float a,b,c,d,e;
int f,g = 100;

while(1){
        a = 10.1234;
        b = 100.2222;
        c = a*b;
        d = c-a;
        e = d+b;
        f = (int)a;
        f = f*g;
        a1 = (unsigned int) a;
        a = __sqrtf(e);
        //a = sqrt(e);
        a = c/f;
        e = a/0;
        }
}
```

Before we can build this project, we need to make sure that the compiler will build code to use the FPU.

Open the Options for Target and select the Target menu (Fig. 8.4).

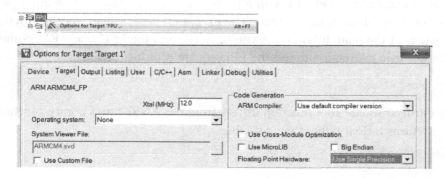

Figure 8.4
Hardware floating point support is enabled in the Project Target options.

We can enable floating point support by selecting "Use FPU" in the floating point hardware box. This will enable the necessary compiler options and load the correct simulator model.

Close the Options for Target menu and return to the editor (Fig. 8.5).

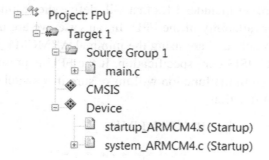

Figure 8.5
This example project is using a simulation of the Corex-M4 processor.

In addition to our source code, the project includes the CMSIS startup and system files for the Cortex-M4.

Now build the project and note down the build size (Fig. 8.6).

```
Build target 'FPU'
compiling system_ARMCM4.c...
assembling startup_ARMCM4.s...
compiling main.c...
linking...
Program Size: Code=224 RO-data=224 RW-data=4 ZI-data=1028
"cortexM4.axf" - 0 Error(s), 0 Warning(s).
```

Figure 8.6
Note the code size for a project using the hardware FPU.

Start the debugger.

When the simulator runs the code to main, it will hit a breakpoint that has been preset in the system_ARMCM4.c file (Fig. 8.7).

```
61   void SystemInit (void)
62   {
63      #if (__FPU_USED == 1)
64         SCB->CPACR |= ((3UL << 10*2) | (3UL << 11*2)  );
65      #endif
66
67      SystemCoreClock = __SYSTEM_CLOCK;
68
69   }
```

Figure 8.7
The CMSIS-Core SystemInit() function will enable the FPU on startup.

The standard microcontroller includes files that will define the feature set of the Cortex-M processor including the availability of the FPU. In our case, we are using a simulation model of the Cortex-M4 only and are using the minimal ARMCM4_FP.h include file provided as part of the CMSIS core specification. If the FPU is present on the microcontroller, the SystemInit() function will make sure it is switched on before you reach your applications main() function.

The processor definitions are in ARMCM4_FP.H

```
/* = = = = = = = = = = = = = = = = = = = = = = = = = = = = = = = = = = = = = = = = = */
/* = = = = = = = = = =    Processor and Core Peripheral Section   = = = = = = = = = = */
/* = = = = = = = = = = = = = = = = = = = = = = = = = = = = = = = = = = = = = = = = = */
/* ——————— Configuration of the Cortex-M4 Processor and Core Peripherals ———————    */
#define __CM4_REV           0x0001  /*!< Core revision r0p1                     */
#define __MPU_PRESENT        1      /*!< MPU present or not                     */
#define __NVIC_PRIO_BITS      3     /*!< Number of Bits used for Priority Levels   */
#define __Vendor_SysTickConfig  0   /*!< Set to 1 if different SysTick Config is used */
#define __FPU_PRESENT         1     /*!< FPU present or not       */

Then __FPU_USED is defined in core_cm4.h

#if (__FPU_PRESENT = = 1U)
    #define __FPU_USED    1U
```

Open the main.c module and run the code to the main while() loop (Fig. 8.8).

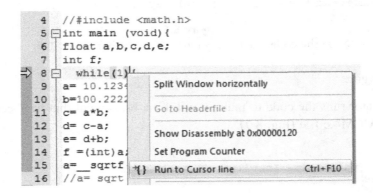

Figure 8.8
Use the local debugger options to run to the main() while(1) loop.

Now open the Disassembly window (Fig. 8.9).

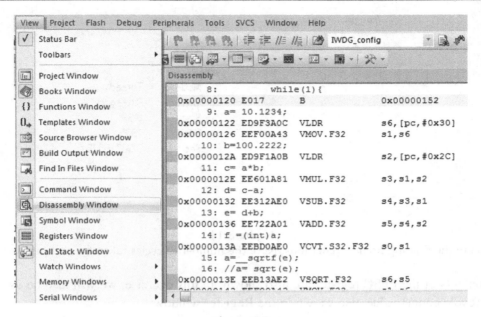

Figure 8.9
The Disassembler window will show you the calls to the hardware FPU.

This will show the "C" source code interleaved with the Cortex-M4 assembly instructions.

In the Project window select the Register window (Fig. 8.10).

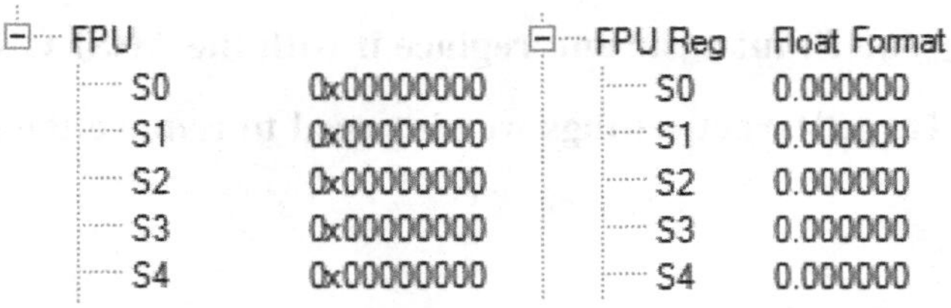

FPU		FPU Reg	Float Format
S0	0x00000000	S0	0.000000
S1	0x00000000	S1	0.000000
S2	0x00000000	S2	0.000000
S3	0x00000000	S3	0.000000
S4	0x00000000	S4	0.000000

Figure 8.10
The Register window allows you to view the FPU registers.

The Register window now shows the 31 scalar registers in their raw format and the IEE754 format (Fig. 8.11).

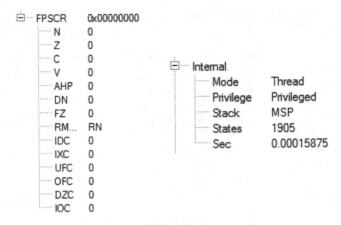

Figure 8.11
Execute the floating point calculations. Note the number of cycles taken for each operation.

The contents of the FPSC register are also shown. In this exercise, we will also be using the states (cycle count) value also shown in the Register window.

Highlight the assembly window and step through each operation noting the cycle time for each calculation.

Now quit the debugger and change the code to use software floating point libraries.

```
 a1 = (unsigned int) a;
//a = __sqrtf(e);
 a = sqrt(e);
```

Comment out the __sqrtf() intrinsic and replace it with the ANSI C sqrt() function.

In the Options for Target\target settings we also need to remove the FPU support (Fig. 8.12).

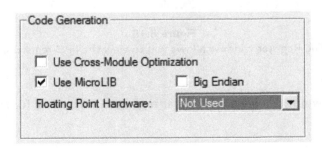

Figure 8.12
Disable the FPU support in the project target menu.

Also change the processor header file for the Cortex-M4 without the FPU.

```
#include <ARMCM4.h>
```

Rebuild the code and compare the build size to the original version (Fig. 8.13).

```
Build target 'FPU'
compiling main.c...
linking...
Program Size: Code=1440 RO-data=224 RW-data=8 ZI-data=1024
"cortexM4.axf" - 0 Error(s), 0 Warning(s).
```

Figure 8.13
Rebuild the project and compare the code size in the original project.

Now restart the debugger run and run the code to main.

In the disassembly window, step through the code and compare the number of cycles used for each operation to the number of cycles used by the FPU.

By the end of this exercise, you can clearly see not only the vast performance improvement provided by the FPU but also its impact on project code size. The only downside is the additional cost to use a microcontroller fitted with the FPU and the additional power consumption when it is running.

Cortex-M4/M7 DSP and SIMD Instructions

The Thumb-2 instruction set has a number of additional instructions that are useful in DSP algorithms (Table 8.4).

Table 8.4: Thumb-2 DSP instructions

Instruction	Description
CLZ	Count leading zeros
REV, REV16, REVSH, and RBIT	Reverse instructions
BFI	Fit field insert
BFC	Bit field clear
UDIV and SDIV	Hardware divide
SXT and UXT	Sign and zero extend

The Cortex-M4 and M7 instruction set includes a new group of instructions that can perform multiple arithmetic calculations in a single cycle. The SIMD instructions allow 16 bit or 8 bit data packed into two 32-bit words to be operated on in parallel. So, for example, you can perform two 16-bit multiplies and a 32-bit or 64-bit accumulate or a quad 8-bit addition in one processor cycle. Since many DSP algorithms work on a pipeline of data, the SIMD instructions can be used to dramatically boost performance (Fig. 8.14).

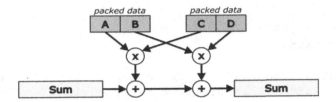

Figure 8.14
The SIMD instructions support multiple arithmetic operations in a single cycle. The operand data must be packed into 32-bit words.

The SIMD instructions have an additional field in the xPSR register. The "Greater than or Equal" (GE) field contains 4 bits, which correspond to the 4 bytes in the SIMD instruction result operand. If the result operand byte is GE to zero then the matching GE flag will be set (Fig. 8.15).

Figure 8.15
The Cortex-M4 xPSR register has an additional greater than or equal field. Each of the four GE bits are updated when a SIMD instruction is executed.

The SIMD instructions can be considered as three distinct groups: Add and subtract operations, multiply operations, and supporting instructions. The add and subtract operations can be performed on 8- or 16-bit signed and unsigned quantities. A signed and unsigned halving instruction is also provided; this instruction adds or subtracts the 8 or 16-bit quantities and then halves the result as shown in Table 8.5.

Table 8.5: SIMD add halving and subtract halving instructions

Instruction	Description	Operation
UHSUB16	Unsigned halving 16-bit subtract	Res[15:0] = (Op1[15:0] − Op2[15:0])/2 Res[31:16] = (Op1[31:16] − Op2[31:16])/2
UHADD16	Unsigned halving 16-bit add	Res[15:0] = (Op1[15:0]+Op2[15:0])/2 Res[31:16] = (Op1[31:16]+Op2[31:16])/2

The SIMD instructions also include an Add and Subtract with Exchange (ASX) and a Subtract and Add with Exchange (SAX). These instructions perform and add and subtract on the two halfwords and store the results in the upper and lower halfwords of the destination register (Table 8.6).

Table 8.6: SIMD add exchange and subtract exchange instructions

Instruction	Description	Operation
USAX	Unsigned 16-bit subtract and add with exchange	Res[15:0]=Op1[15:0]+Op2[31:16] Res[31:16] = Op1[31:16] − Op2[15:0]
UASX	Unsigned 16-bit add and subtract with exchange	Res[15:0] = Op1[15:0]+Op2[31:16] Res[31:16] = Op1[31:16] − Op2[15:0]

A further group of instructions combine these two operations in a subtract and add (or add and subtract) with exchange halving instruction. This gives quite few possible permutations. A summary of the add and subtract SIMD instructions is shown in Table 8.7.

Table 8.7: Permutations of the SIMD add, subtract, halving, and saturating instructions

Instruction	Signed	Signed Saturating	Signed Halving	Unsigned	Unsigned Saturating	Unsigned Halving
ADD8	SADD8	QADD8	SHADD8	UADD8	UQADD8	UHADD8
SUB8	SSUB8	QSUB8	SHSUB8	USUB8	UQSUB8	UHSUB8
ADD16	SADD16	QADD16	SHADD16	UADD16	UQADD16	UHADD16
SUB16	SSUB16	QSUB16	SHSUB16	USUB16	UQSUB16	UHSUB16
ASX	SASX	QASX	SHASX	UASX	UQASX	UHASX
SAX	SSAX	QSAX	SHSAX	USAX	UQSAX	UHSAX

The SIMD instructions also include a group of multiply instructions that operate on packed 16-bit signed values. Like the add and subtract instructions, the multiply instructions also support saturated values. As well as multiply and multiply accumulate the SIMD multiply instructions supports multiply subtract and multiply add as shown in Table 8.8.

Table 8.8: SIMD multiply instructions

Instruction	Description	Operation
SMLAD	Q setting dual 16-bit signed multiply with single 32-bit accumulator	$X = X + (A \times B) + (C \times D)$
SMLALD	Dual 16-bit signed multiply with single 64-bit accumulator	$X = X + (A \times B) + (C \times D)$
SMLSD	Q setting dual 16-bit signed multiply subtract with 32-bit accumulate	$X = X + (A \times B) - (B \times C)$
SMLSLD	Q setting dual 16-bit signed multiply subtract with 64-bit accumulate	$X = X + (A \times B) - (B \times C)$
SMUAD	Q setting sum of dual 16-bit signed multiply	$X = (A \times B) + (C \times D)$
SMUSD	Dual 16-bit signed multiply returning difference.	$X = (A \times B) - (C \times D)$

To make the SIMD instructions more efficient, a group of supporting pack and unpack instructions have also been added to the instruction set. The pack/unpack instructions can be used to extract 8- and 16-bit values from a register and move them to a destination register. The unused bits in the 32-bit word can be set to zero (unsigned) or one (signed). The pack instructions can also take two 16-bit quantities and load them into the upper and lower halfwords of a destination register (Fig. 8.16) (Table 8.9).

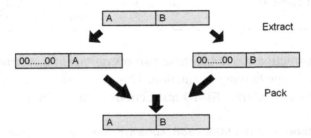

Figure 8.16
The SIMD instruction group includes support instructions to pack 32-bit words with 8- and 16-bit quantities.

Table 8.9: SIMD supporting instructions

Mnemonic	Description
PKH	Pack halfword
SXTAB	Extend 8-bit signed value to 32 bits and add
SXTAB16	Dual extend 8-bit signed value to 16 bits and add
SXTAH	Extend 16-bit signed value to 32 bits and add
SXTB	Sign extend a byte
SXTB16	Dual extend 8-bit signed values to 16 bits and add
SXTH	Sign extend a halfword
UXTAB	Extend 8-bit signed value to 32 bits and add
UXTAB16	Dual extend 8 to 16 bits and add
UXTAH	Extend a 16-bit value and add
UXTB	Zero extend a byte
UXTB16	Dual zero extend 8 to 16 bits and add
UXTH	Zero extend a halfword

When a SIMD instruction is executed, it will set or clear the xPSR GE bits depending on the values in the resulting bytes or halfwords. An addition select (SEL) instruction is provided to access these bits. The SEL instruction is used to select bytes or halfwords from two input operands depending on the condition of the GE flags (Table 8.10).

Table 8.10: xPSR "Greater than or equal" bit field results

GE Bit[3:0]	GE Bit = 1	GE Bit = 0
0	Res[7:0] = OP1[7:0]	Res[7:0] = OP2[7:0]
1	Res[15:8] = OP1[15:8]	Res[15:8] = OP2[15:8]
2	Res[23:16] = OP1[23:16]	Res[23:16] = OP2[23:16]
3	Res[31:24] = OP1[31:24]	Res[31:24] = OP2[31:24]

Exercise 8.2 SIMD Instructions

In this exercise, we will have a first look at using the Cortex-M4/M7 SIMD instructions. We will simply multiply and accumulate two 16-bit arrays first using a SIMD instruction and then using the standard "C" statements.

First open the CMSIS-Core documentation and the SIMD signed MAC intrinsic __SMLAD (Fig. 8.17).

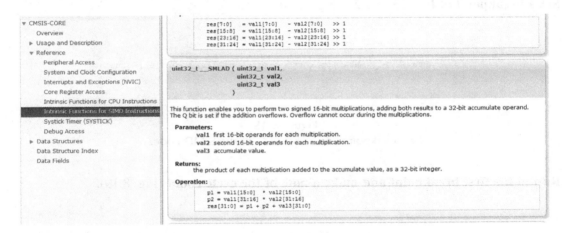

Figure 8.17
CMSIS-DSP documentation for the __SMLAD() intrinsic.

Open the Pack Installer.

Select the Boards::Designers Guide Tutorial.

Select the Example tab and Copy "EX 8.2 SIMD.".

The application code defines two sets of arrays as a union of 16-bit and 32-bit quantities.

```
union _test{
int16_t    Arry_halfword[100];
int32_t    Arry_word[50];
};
```

The code first initializes the arrays with the values 0–100.

```
for(n = 0;n<100;n++){
op1.Arry_halfword[n] = op2.Arry_halfword[n] = n;   }
```

We then do a multiply accumulate first using the SIMD instruction, then the standard MAC Instruction.

```
for(n = 0;n<50;n++){
Result = __SMLAD(op1.Arry_word[n],op2.Arry_word[n],Result);}
```

Result is then reset and the calculation is repeated without using the SIMD instruction.

```
Result = 0;
for(n = 0;n<100;n++){
Result = Result + (op1.Arry_halfword[n] * op2.Arry_halfword[n]);}
```

Build the code and start the debugger.

Set a breakpoint at line 23 and 28 (Fig. 8.18).

```
●23     for(n=0;n<50;n++){
 24
 25        Result = __SMLAD(op1.Arry_word[n],op2.Arry_word[n],Result);
 26
 27     }
●28     Result = 0;
```

Figure 8.18
Set breakpoints on either side of the SIMD code.

Run to the first breakpoint and make a note of the cycle count (Fig. 8.19).

Internal	
Mode	Thread
Privilege	Privileged
Stack	MSP
States	4066
Sec	0.00033883

Figure 8.19
Note the start cycle count.

Run the code until it hits the second breakpoint. See how many cycles have been used to execute the SIMD instruction (Fig. 8.20).

Figure 8.20
Note the final cycle count.

Cycles used $= 5172 - 4066 = 1106$

Set a breakpoint at the final while loop (Fig. 8.21).

```
30   for(n=0;n<100;n++){
31      Result = Result + (op1.Arry_halfword[n] * op2.Arry_halfword[n]);
32   }
33   while(1){
```

Figure 8.21
Now set a breakpoint after the array copy routine.

Run the code and see how many cycles are used to perform the calculation without using the SIMD instruction (Fig. 8.22).

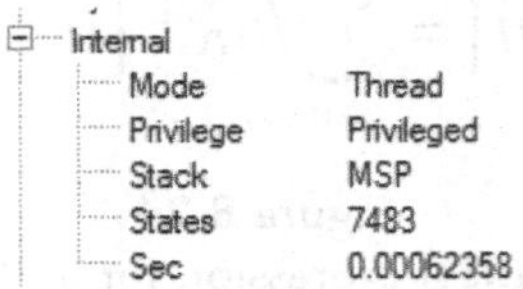

Internal	
Mode	Thread
Privilege	Privileged
Stack	MSP
States	7483
Sec	0.00062358

Figure 8.22
Note the final cycle count.

Cycles used $= 7483 - 5172 = 2311$

Compare the number of cycles used to perform the same calculation without using the SIMD instructions.

As expected, the SIMD instructions are much more efficient when performing calculations on large data sets.

The primary use for the SIMD instructions is to optimize the performance of DSP algorithms. In the next exercise, we will look at various techniques in addition to the SIMD instructions that can be used to boost increase the efficiency of a given algorithm.

Exercise 8.3 Optimizing DSP Algorithms

In this exercise, we will look at optimizing an Finite Impulse Response filter (FIR). This is a classic algorithm that is widely used in DSP applications (Fig. 8.23).

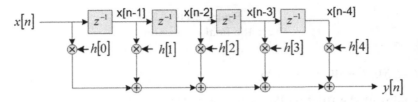

Figure 8.23
An FIR filter is an averaging filter with its characteristics defined by a series of coefficients applied to each sample in a series of "taps."

The FIR filter is an averaging filter that consists of a number of "taps." Each tap has a coefficient and as the filter runs each sample is multiplied against coefficient in the first tap and then shifted to the next tap to be multiplied against its coefficient when the next sample arrives (Fig. 8.24). The output of each tap is summed to give the filter output. Or to put it mathematically,

$$y[n] = \sum_{k=0}^{N-1} h[k]x[n-k]$$

Figure 8.24
Mathematical expression for an FIR filter.

In this exercise, we will use the Cortex-M4 simulation model to look at several techniques that can be used to optimize a DSP algorithm running on a Cortex-M processor.

Open the Pack Installer.

Select the Boards::Designers Guide Tutorial.

Select the Example tab and Copy "EX 8. FIR Optimization."

Build the project and start the debugger.

The main function consists of four FIR functions that introduce different optimizations to the standard FIR algorithm.

```
int main (void){
fir                 (data_in,data_out,  coeff,&index,  FILTERLEN, BLOCKSIZE);
fir_block           (data_in,data_out,  coeff,&index,  FILTERLEN, BLOCKSIZE);
fir_unrolling       (data_in,data_out,  coeff,&index,  FILTERLEN, BLOCKSIZE);
fir_SIMD            (data_in,data_out,  coeff,&index,  FILTERLEN, BLOCKSIZE);
fir_SuperUnrolling  (data_in,data_out,  coeff,&index,  FILTERLEN, BLOCKSIZE);

while(1);
}
```

Step into the first function and examine the code.

The filter function is implemented in "C" as shown below. This is a standard implementation of an FIR filter written purely in "C".

```
void fir(q31_t *in, q31_t *out, q31_t *coeffs, int *stateIndexPtr,
                   int filtLen, int blockSize)
{
   int sample;
   int k;
   q31_t sum;
   int stateIndex = *stateIndexPtr;
   for(sample = 0; sample < blockSize; sample++)
      {
        state[stateIndex++] = in[sample];
        sum = 0;
        for(k = 0;k<filtLen;k++)
               {
                 sum + = coeffs[k] * state[stateIndex];
                 stateIndex--;
                 if (stateIndex < 0)
                    {
                       stateIndex = filtLen-1;
                    }
               }
        out[sample] = sum;
      }
      *stateIndexPtr = stateIndex;
}
```

While this compiles and runs fine, it does not take full advantage of the Cortex-M4 DSP enhancements. To get the best out of the Cortex-M4, we need to optimize this algorithm, particularly the inner loop. The inner loop performs the FIR multiply and accumulate for each tap.

```
for(k = 0;k<filtLen;k++)
               {
                 sum + = coeffs[k] * state[stateIndex];
                 stateIndex--;
                 if (stateIndex < 0)
```

```
      {
         stateIndex = filtLen-1;
      }
   }
```

The inner loop processes the samples by implementing a circular buffer in software. While this works fine, we have to perform a test for each loop to wrap the pointer when it reaches the end of the buffer (Fig. 8.25).

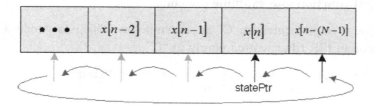

Figure 8.25
Processing data in a circular buffer requires the Cortex-M4 to check for the end of the buffer on each iteration. This increases the execution time.

Run to the start of the inner loop and set a breakpoint (Fig. 8.26).

```
16          for(k=0;k<filtLen;k++)
17  {
18      sum += coeffs[k] * state[stateIndex];
19      stateIndex--;
20      if (stateIndex < 0)
21          {
```

Figure 8.26
Set a breakpoint at the start of the inner loop.

Run the code so it does one iteration of the inner loop and note the number of cycles used.

Circular addressing requires us to perform an end of buffer test on each iteration. A dedicated DSP device can support circular buffers in hardware without any such overhead, so this is one area we need to improve. By passing our FIR filter function, a block of data rather than individual samples allows us to use block processing as an alternative to circular addressing. This improves the efficiency on the critical inner loop (Fig. 8.27).

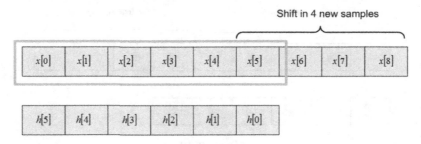

Figure 8.27
Block processing increases the size of the buffer but increases the efficiency of the inner processing loop.

By increasing the size of the state buffer to number of filter taps + processing block size, we can eliminate the need for circular addressing. In the outer loop, the block of samples is loaded into the top of the state buffer (Fig. 8.28).

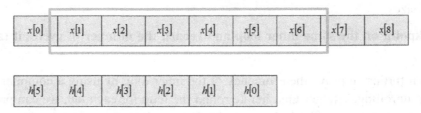

Figure 8.28
With block processing, the fixed size buffer is processed without the need to check for the end of the buffer.

The inner loop then performs the filter calculations for each sample of the block by sliding the filter window one element to the right for each pass through the loop. So the inner loop now becomes

```
for(k = 0; k<filtLen; k++)

{
  sum + = coeffs[k] * state[stateIndex];
  stateIndex++;
}
```

Once the inner loop has finished processing, the current block of sample data held in the state buffer must be shifted to the right and a new block of data loaded (Fig. 8.29).

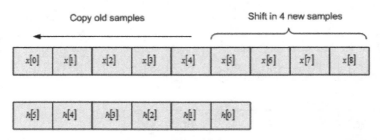

Figure 8.29
Once the block of data has been processed, the outer loop shifts the samples one block to the left and adds a new block of data.

Now step into the second FIR function.

Examine how the code has been modified to process blocks of data.

There is some extra code in the outer loop, but this is only executed one per tap and becomes insignificant compared to the savings made within the inner loop particularly for large block sizes.

Set a breakpoint on the same inner loop and record the number of cycles it takes to run.

Next, we can further improve the efficiency of the inner loop by using a compiler trick called "loop unrolling." Rather than iterate round the loop for each tap, we can process several taps in each iteration by in-lining multiple tap calculations per loop.

```
I = filtLen≫2
for(k = 0;k<filtLen;k++)
{
  sum + = coeffs[k] * state[stateIndex];
  stateIndex++;
  sum + = coeffs[k] * state[stateIndex];
  stateIndex++;
  sum + = coeffs[k] * state[stateIndex];
  stateIndex++;
  sum + = coeffs[k] * state[stateIndex];
  stateIndex++;
}
```

Now step into the third FIR function.

Set a breakpoint on the same inner loop and record the number of cycles it takes to run. Divide this by four and compare it to the previous implementations.

The next step is to make use of the SIMD instructions. By packing the coefficient and sample data into 32-bit words, the single MACs can be replaced by dual signed MAC which allows us to extend the loop unrolling from four summations to eight for the same number of cycles.

```
for(k = 0;k<filtLen;k++)
{
   sum + = coeffs[k] * state[stateIndex];
   stateIndex++;
   sum + = coeffs[k] * state[stateIndex];
   stateIndex++;
   sum + = coeffs[k] * state[stateIndex];
   stateIndex++;
   sum + = coeffs[k] * state[stateIndex];
   stateIndex++;
}
```

Step into the fourth FIR function and again calculate the number of cycles used per tap for the inner loop.

Remember we are now calculating eight summations, so divide the raw loop cycle count by eight.

To reduce the cycle count of the inner loop even further, we can extend the loop unrolling to calculate several results simultaneously (Fig. 8.30).

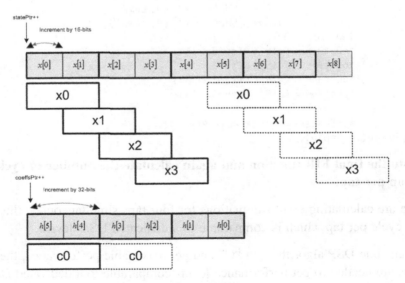

Figure 8.30
Super loop unrolling extends loop unrolling to process multiple output samples simultaneously.

This is kind of a "super loop unrolling" where we perform each of the inner loop calculations for a block of data in one pass.

```
sample = blockSize/4;
        do
        {
                sum0 = sum1 = sum2 = sum3 = 0;
                statePtr = stateBasePtr;
                coeffPtr = (q31_t *)(S->coeffs);
                x0 = *(q31_t *)(statePtr++);
                x1 = *(q31_t *)(statePtr++);
                i = numTaps>>2;
                do
                {
                        c0 = *(coeffPtr++);
                        x2 = *(q31_t *)(statePtr++);
                        x3 = *(q31_t *)(statePtr++);
                        sum0 = __SMLALD(x0, c0, sum0);
                        sum1 = __SMLALD(x1, c0, sum1);
                        sum2 = __SMLALD(x2, c0, sum2);
                        sum3 = __SMLALD(x3, c0, sum3);
                        c0 = *(coeffPtr++);
                        x0 = *(q31_t *)(statePtr++);
                        x1 = *(q31_t *)(statePtr++);
                        sum0 = __SMLALD(x0, c0, sum0);
                        sum1 = __SMLALD(x1, c0, sum1);
                        sum2 = __SMLALD (x2, c0, sum2);
                        sum3 = __SMLALD (x3, c0, sum3);
                } while(--i);
                *pDst++ = (q15_t) (sum0>>15);
                *pDst++ = (q15_t) (sum1>>15);
                *pDst++ = (q15_t) (sum2>>15);
                *pDst++ = (q15_t) (sum3>>15);

                stateBasePtr = stateBasePtr + 4;
        } while(--sample);
```

Now step into the final FIR function and again calculate the number of cycles used by the inner loop per tap.

This time we are calculating eight summations for four taps simultaneously; this brings us close to one cycle per tap which is comparable to a dedicated DSP device.

While you can code DSP algorithms in "C" and get reasonable performance, these kinds of optimizations are needed to get performance levels comparable to a dedicated DSP device. This kind of code development needs experience of the Cortex-M4/M7 and the DSP algorithms you wish to implement. Fortunately, ARM provides a free DSP library already optimized for the Cortex-M processors.

The CMSIS-DSP Library

While it is possible to code all of your own DSP functions, this can be time consuming and requires a lot of domain specific knowledge. To make it easier to add common DSP functions to your application, ARM have published a library of 61 common DSP functions which make up the CMSIS-DSP (Cortex Microcontroller Software Interface Standard-Digital Signal Processing) specification. Each of these functions are optimized for the Cortex-M4/M7 but can also be compiled to run on the Cortex-M3 and even on the Cortex-M0. The CMSIS-DSP library is a free download and is licensed for use in any commercial or noncommercial project. The CMSIS-DSP library is also included as part of the MDK-ARM installation and just needs to be added to your project by selecting the CMSIS::DSP option within the Run-Time Environment (RTE) manager. The installation includes a prebuilt library for each of the Cortex-M processors and all of the source code.

Documentation for the library is included as part of the CMSIS help which can opened from within the "Manage Run-Time Environment" window (Figure 8.31).

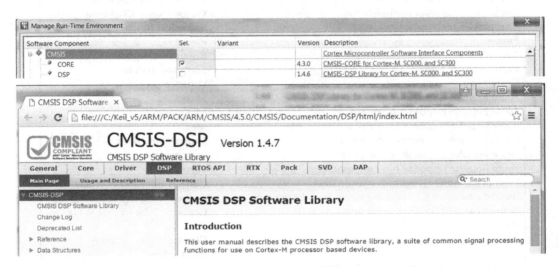

Figure 8.31
The CMSIS-DSP documentation accessed by clicking on the description link available in the Manage Run-Time Environment window.

CMSIS-DSP Library Functions

The CMSIS-DSP library provides easy to use functions for the most commonly used signal processing algorithms. The functions included in the library are shown in Table 8.11.

Table 8.11: CMSIS-DSP library functions

Basic Maths Functions	Matrix Functions
Vector multiplication	Matrix initialization
Vector subtraction	Complex matrix multiplication
Vector addition	Matrix addition
Vector scale	Matrix subtraction
Vector shift	Matrix multiplication
Vector offset	Matrix inverse
Vector negate	Matrix transpose
Vector absolute	Matrix scale
Vector dot product	**Transforms**
Fast maths functions	Complex FFT functions
Cosine	Real FFT functions
Sine	DCT type IV functions
Square root	**Controller functions**
Complex maths functions	Sine cosine
Complex conjugate	PID motor control
Complex dot product	Vector park transform
Complex magnitude	Vector inverse park transform
Complex magnitude squared	Vector Clarke transform
Complex by complex multiplication	Vector inverse Clarke transform
Complex by real multiplication	**Statistical functions**
Filters	Power
Convolution	Root mean square
Partial convolution	Standard deviation
Correlation	Variance
Finite impulse response filter	Maximum
Finite impulse response decemator	Minimum
Finite impulse response lattice filter	Mean
Infinite impulse response lattice filter	**Support functions**
Finite impulse response sparse filter	Vector copy
Finite impulse response filter interpolation	Vector fill
High precision Q31 Biquad cascade filter	Convert 8-bit integer value
Biquad cascade IIR filter using direct form I structure	Convert 16-bit integer value
Biquad cascade IIR filter using direct form II transposed structure	Convert 32-bit integer value
Least mean squares FIR filter	Convert 32-bit floating point value
Least mean squares normalized FIR filter	**Interpolation functions**
	Linear interpolate function
	Bilinear interpolate function

Exercise 8.3 Using the CMSIS-DSP Library

In this project, we will have a first look at using the CMSIS-DSP library by setting up a project to experiment with the Proportional Integral Derivative (PID) control algorithm (Fig. 8.32).

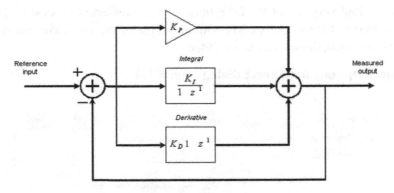

Figure 8.32
A PID control loop consists of proportional, integral, and derivative control blocks.

Open the Pack Installer.

Select the Boards::Designers Guide Tutorial.

Select the Example tab and Copy "EX 8.3 CMSIS-DSP PID" (Fig. 8.33).

Figure 8.33
The PID example project is configured with the DSP added as a CMSIS software component.

The project is targeted at a Cortex-M4 with FPU. It includes the CMSIS startup and system files for the Cortex-M4 and the CMSIS-DSP functions as a precompiled library. The CMSIS-DSP library is located in c:\CMSIS\lib with subdirectories for the ARM compiler and GCC versions of the library (Fig. 8.34).

Software Component	Sel.	Variant	Version	Description
◈ CMSIS				Cortex Microcontroller Software Interface Components
● CORE	☑		4.3.0	CMSIS-CORE for Cortex-M, SC000, and SC300
● DSP	☑		1.4.6	CMSIS-DSP Library for Cortex-M, SC000, and SC300

Figure 8.34
The CMSIS-DSP library is added to the project as a software component in the RTE.

There are precompiled versions of the DSP library for each Cortex processor. There is also a version of the library for the Cortex-M4 with and without the FPU. For our project, the Cortex-M4 floating point library has been added.

Open the project\Options for Target dialog (Fig. 8.35).

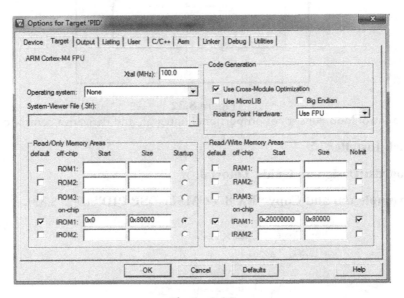

Figure 8.35
The FPU is enabled in the Project Target settings.

In this example, we are using a simulation model for the Cortex-M4 processor only. A 512K memory region has been defined for code and data. The FPU is enabled and the CPU clock has been set to 100 MHz (Fig. 8.36).

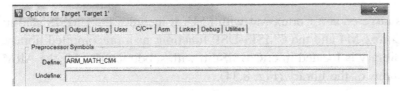

Figure 8.36
The FPU must be enabled by the startup code.

In the compiler Options tab, the _FPU_PRESENT define is set to one. Normally you will not need to configure this option as this will be done in the microcontroller include file.

We also need to add a #define to configure the DSP header file for the processor we are using. The library defines are:

```
ARM_MATH_CM4
ARM_MATH_CM3
ARM_MATH_CM0
```

The source code for the library is located in C:\keil\arm\CMSIS\DSP_Lib\Source.

Open the file pid_example_f32.c which contains the application code.

To access the library, we need to add its header file to our code.

```
#include "arm_math.h"
```

This header file is located in c:\keil\arm\cmsis\include.

In this project, we are going to use the PID algorithm. All of the main functions in the DSP library have two function calls: an initializing function and a process function.

```
void arm_pid_init_f32 ( arm_pid_instance_f32 *S,int32_t resetStateFlag)
__STATIC_INLINE void arm_pid_f32 (arm_pid_instance_f32 *s, float32_t in)
```

The initializing function is passed a configuration structure which is unique to the algorithm. The configuration structure holds constants for the algorithm, derived values, and arrays for state memory. This allows multiple instances of each function to be created.

```
typedef struct
  {
    float32_t A0;        /**< The derived gain, A0 = Kp + Ki + Kd . */
    float32_t A1;        /**< The derived gain, A1 = -Kp - 2Kd. */
    float32_t A2;        /**< The derived gain, A2 = Kd . */
    float32_t state[3]; /**< The state array of length 3. */
    float32_t Kp;        /**< The proportional gain. */
    float32_t Ki;        /**< The integral gain. */
    float32_t Kd;        /**< The derivative gain. */
  } arm_pid_instance_f32;
```

The PID configuration structure allows you to define values for the proportional, integral, and derivative gains. The structure also includes variables for the derived gains A0, A1, and A2 as well as a small array to hold the local state variables.

```
int32_t main(void){
int i; S.Kp = 1; S.Ki = 1; S.Kd = 1;
setPoint = 10;
arm_pid_init_f32(&S,0);
while(1){
        error = setPoint-motorOut;
        motorIn = arm_pid_f32        (&S,error );
        motorOut = transferFunction(motorIn,time);
```

```
        time + = 1;
        for(i = 0;i<100000;i++);
}}
```

The application code sets the PID gain values and initializes the PID function. The main loop calculates the error value before calling the PID process function. The PID output is fed into a simulated hardware transfer function. The output of the transfer function is fed back into the error calculation to close the feedback loop. The time variable provides a pseudotime reference.

Build the project and start the debugger.

Add the key variables to the Logic Analyzer and start the code running (Fig. 8.37).

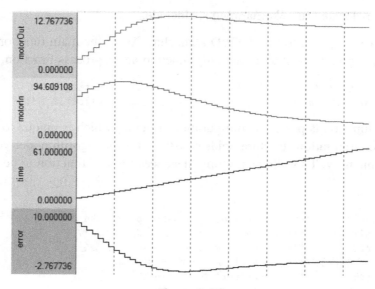

Figure 8.37
The Logic Analyzer is invaluable for visualizing real-time signals.

The Logic Analyzer is invaluable for visualizing data in a real-time algorithm and can be used in the simulator or can capture data from the CoreSight data watch trace unit as we saw in Chapter 7 "Debugging with CoreSight."

Experiment with the gain values to tune the PID function.

The performance of the PID control algorithm is tuned by adjusting the gain values. As a rough guide, each of the gain values has the following effect:

Kp   Effects the rise time of the control signal.

Ki   Effects the steady state error of the control signal.

Kd   Effects the overshoot of the control signal.

DSP Data Processing Techniques

One of the major challenges of a DSP application is managing the flow of data from the sensors and Analogue to Digital Converter (ADC) through the DSP algorithm and back out to the real world via the Digital to Analogue Converter (DAC) (Fig. 8.38).

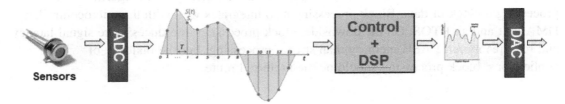

Figure 8.38
A typical DSP system consists of an analog sample stage, microcontroller with DSP algorithm, and an output DAC. In this chapter, we microcontroller software in isolation from the hardware design.

In a typical system, each sampled value is a discrete value at a point in time. The sample rate must be at least twice the signal bandwidth or up to four times the bandwidth for a high-quality oversampled audio system. Clearly the volume of data is going to ramp up very quickly and it becomes a major challenge to process the data in real time. In terms of processing the sampled data, there are two basic approaches, stream processing or block processing (Fig. 8.39).

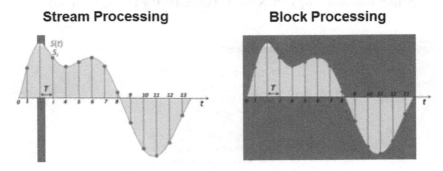

Figure 8.39
Analogue data can be processed as single samples with minimum latency or as a block of samples for maximum processing efficiency.

In stream processing, each sampled value is processed individually. This gives the lowest signal latency and also the minimum memory requirements. However, it has the disadvantage of making the DSP algorithm more complex. The DSP algorithm has to be run every time an ADC conversion is made, which can cause problems with other high-priority interrupt routines. The alternative to stream processing is block processing. Here, a number of ADC results are stored in a buffer, typically about 32 samples, and then this buffer is processed by the DSP algorithm as a block of data. This lowers the number of times that the DSP algorithm has to run. As we have seen in the optimization exercise, there are a number of techniques that can improve the efficiency of an algorithm when processing a block of data. Block processing also integrates well with the microcontroller DMA unit and an RTOS. On the downside, block processing introduces more signal latency and requires more FLASH memory than stream processing. For the majority of applications, block processing should be the preferred route.

Exercise 8.4 FIR Filter with Block Processing

In this exercise, we will implement an FIR filter, this time by using the CMSIS-DSP functions. This exercise uses the same project template as the PID program. The characteristics of the filter are defined by the filter coefficients. It is possible to calculate the coefficient values manually or by using a design tool. Calculating the coefficients is outside the scope of this book, but the appendices list some excellent design tools and DSP books for further reading.

Open the Pack Installer.

Select the Boards::Designers Guide Tutorial.

Select the Example tab and Copy "EX 8.4 CMSIS-DSP FIR" (Fig. 8.40).

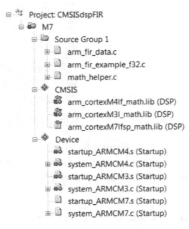

Figure 8.40
The FIR project has builds for the Cortex-M7, M4, and M3.

There is an additional data file that holds a sampled data set and an additional math_helper. "C" file which contains some ancillary functions.

```
int32_t main(void)
{
    uint32_t i;
    arm_fir_instance_f32 S;
    arm_status status;
    float32_t *inputF32, *outputF32;
    /* Initialize input and output buffer pointers */
    inputF32 = &testInput_f32_1kHz_15kHz[0];
    outputF32 = &testOutput[0];
    /* Call FIR init function to initialize the instance structure. */
    arm_fir_init_f32(&S, NUM_TAPS, (float32_t *)&firCoeffs32[0], &firStateF32[0],
blockSize);
    for(i = 0; i < numBlocks; i++)
    {
        arm_fir_f32(&S, inputF32 + (i * blockSize), outputF32 + (i * blockSize), blockSize);
    }
    snr = arm_snr_f32(&refOutput[0], &testOutput[0], TEST_LENGTH_SAMPLES);
    if (snr < SNR_THRESHOLD_F32)
    {
        status = ARM_MATH_TEST_FAILURE;
    } else  {
status = ARM_MATH_SUCCESS;
}
```

The code first creates an instance of a 29-tap FIR filter. The processing block size is 32 bytes. An FIR state array is also created to hold the working state values for each tap. The size of this array is calculated as the "block size + number of taps − 1". Once the filter has been initialized, we can pass it to the sample data in 32-byte blocks and store the resulting processed data in the output array. The final filtered result is then compared to a precalculated result.

Build the project and start the debugger.

Step through the project to examine the code.

Lookup the CMSIS-DSP functions used in the help documentation.

Set a breakpoint at the end of the project (Fig. 8.41).

```
209      if( status != ARM_MATH_SUCCESS)
210  □   {
211          while(1);
212      }
213
● 214      while(1);
215  }
```

Figure 8.41
Set a breakpoint on the final while() loop.

Reset the project and run the code until it hits the breakpoint.

Now look at the cycle count in the Registers window (Fig. 8.42).

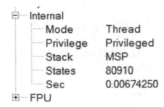

Figure 8.42
Use the register windows states counter to monitor the number of cycles executed.

Open the project in c:\exercises\CMSIS _FIR\CM3.

This is the same project built for the Cortex-M3.

Build the project and start the debugger.

Set a breakpoint in the same place as the previous Cortex-M4 example.

Reset the project and run the code until it hits the breakpoint.

Now compare the cycle count used by the Cortex-M3 to the Cortex-M4 version (Fig. 8.43).

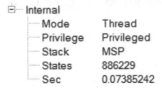

Figure 8.43
Compare the Cortex-M3 cycle count to the Cortex-M4.

When using floating point numbers, the Cortex-M4 is nearly and order faster than the Cortex-M3 using software floating point libraries. When using fixed point maths, the Cortex-M4 still has a considerable advantage over the Cortex-M3 (Fig. 8.44).

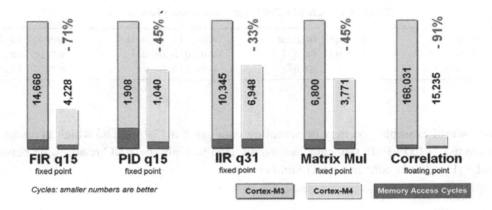

Figure 8.44
The bar graphs show a comparison between the Cortex-M3 and Cortex-M4 for common DSP algorithms.

Fixed Point DSP with Q Numbers

The functions in the DSP library support floating point and fixed point data. The fixed point data is held in a Q number format. Q numbers are fixed point fractional numbers held in integer variables. A Q number has a sign bit followed by a fixed number of bits to represent the integer value. The remaining bits represent the fractional part of the number.

Signed = S IIIIIIIIII.FFFFF

A Q number is defined as the number of bits in the integer portion of the variable and the number of bits in the fractional portion of the variable. Signed Q numbers are stored as twos compliment values. A Q number is typically referred to by the number of fractional bits it uses so Q10 has 10 fractional places. The CMSIS-DSP library functions are designed to take input values between +1 and −1. The supported integer values are shown in Table 8.12.

Table 8.12: Floating point unit performance

CMSIS-DSP Type Def	Q Number
Q31_t	Q31
Q15_t	Q15
Q7_t	Q7

The library includes a group of conversion functions to change between floating point numbers and the integer Q numbers. Support functions are also provided to convert between different Q number resolutions (Table 8.13).

Table 8.13: CMSIS-DSP type conversion functions

arm_float_to_q31	arm_q31_to_float	arm_q15_to_float	arm_q7_to_float
arm_float_to_q15	arm_q31_to_q15	arm_q15_to_q31	arm_q7_to_q31
arm_float_to_q7	arm_q31_to_q7	arm_q15_to_q7	arm_q7_to_q15

As a real world example you may be sampling data using a 12-bit ADC which gives an output from $1 - 0xFFF$. In this case, we would need to scale the ADC result to be between $+1$ and -1 and then convert to a Q number.

```
Q31_t  ADC_FixedPoint;
float  temp;
```

Read the ADC register to a float variable. Scale between $+1$ and -1

```
temp = ((float32_t)((ADC_DATA_REGISTER) & 0xFFF) / (0xFFF / 2)) - 1;
```

Convert the float value to a fixed point Q31 value.

```
arm_float_to_q31(&ADC_FixedPoint, &temp, 1);
```

Similarly, after the DSP function has run, it is necessary to convert back to a floating point value before using the result. Here, we are converting from a Q31 result to a 10-bit integer value prior to outputting the value to a DAC peripheral.

```
arm_q31_to_float(&temp, &DAC_Float, 1);
DAC_DATA_REGISTER = (((uint32_t)((DAC_Float))) & 0x03FF) ;
```

Exercise 8.5 Fixed Point FFT Transform

In this project, we will use the Fast Fourier transform to analyze a signal.

Open the Pack Installer.

Select the Boards::Designers Guide Tutorial.

Select the Example tab and Copy "EX 8.5 CMSIS-DSP FFT" (Fig. 8.45).

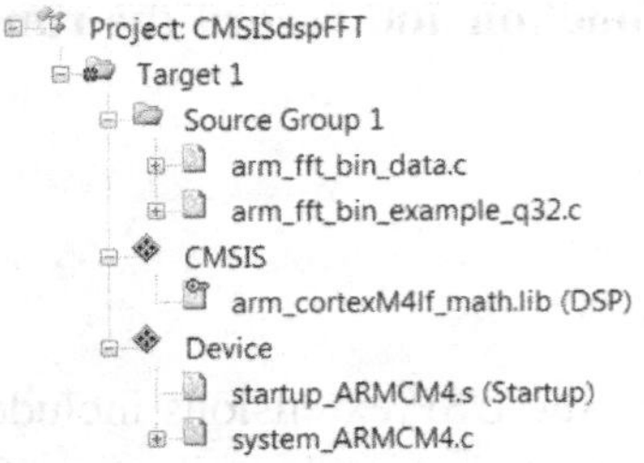

Figure 8.45
The project is configured for the Cortex-M4. Signal data is included in the arm_fft_bin_data.c file.

The FFT project uses the same template as the PID example with the addition of a data file that holds an array of sample data. This data is a 10 kHz signal plus random "white noise."

```
int32_t main(void) {
          arm_status status;
          arm_cfft_radix4_instance_q31 S;
          q31_t maxValue;
          status = ARM_MATH_SUCCESS;
          /* Convert the floating point values the Q31 fixed point format */
          arm_float_to_q31 (testInput_f32_10khz,testInput_q31_10khz ,2048);
          /* Initialize the CFFT/CIFFT module */
          status = arm_cfft_radix4_init_q31(&S, fftSize, ifftFlag, doBitReverse);
          /* Process the data through the CFFT/CIFFT module */
          arm_cfft_radix4_q31(&S, testInput_q31_10khz);
          /* Process the data through the Complex Magnitude Module for
          calculating the magnitude at each bin */
          arm_cmplx_mag_q31(testInput_q31_10khz, testOutput, fftSize);
          /* Calculates maxValue and returns corresponding BIN value */
          arm_max_q31(testOutput, fftSize, &maxValue, &testIndex);
          if(testIndex != refIndex)
          {
                    status = ARM_MATH_TEST_FAILURE;
          }
```

The code initializes the complex FFT transform with a block size of 1024 output bins. When the FFT function is called, the sample data set is first converted to fixed point Q31 format and is then passed to the transform as one block. Next the complex output is converted to a scalar magnitude and then scanned to find the maximum value. Finally, we compare this value to an expected result.

Work through the project code and look up the CMSIS-DSP functions used.

Note: In an FFT transform, the number of output bins will be half the size of the sample data size.

Build the project and start the debugger.

Execute the code function by function and record the run time required to perform the DSP functions.

Conclusion

In this chapter, we have looked at the DSP extensions included in the Cortex-M4/M7 and how to make best use of these features in a real application. The CMSIS-DSP library provides many common DSP functions that have been optimized for the Cortex-M4/M7. In Chapter 10 "RTOS Techniques," we will look at how to integrate continuous real-time DSP processing and event driven microcontroller code into the same project.

Cortex Microcontroller Software Interface Standard-Real-Time Operating System

Introduction

This chapter is an introduction to using a small footprint real-time operating system (RTOS) on a Cortex-M microcontroller. As we saw in Chapter 4 "Cortex Microcontroller Software Interface Standard (CMSIS)" defines a standard API for Cortex-M RTOS. If you are used to writing procedural-based "C" code on small 8-/16-bit microcontrollers, you may be doubtful about the need for such an operating system. If you are not familiar with using an RTOS in real-time embedded systems, you should read this chapter before dismissing the idea. The use of an RTOS represents a more sophisticated design approach, inherently fostering structured code development which is enforced by the RTOS application programming interface (API).

The RTOS structure allows you to take a more object-orientated design approach, while still programming in "C". The RTOS also provides you with multithreaded support on a small microcontroller. These two features actually create quite a shift in design philosophy, moving us away from thinking about procedural "C" code and flow charts. Instead we consider the fundamental program threads and the flow of data between them. The use of an RTOS also has several additional benefits which may not be immediately obvious. Since an RTOS-based project is composed of well-defined threads it helps to improve project management, code reuse, and software testing.

The tradeoff for this is that an RTOS has additional memory requirements and increased interrupt latency. Typically, the Keil RTX RTOS will require 500 bytes of RAM and 5k bytes of code, but remember that some of the RTOS code would be replicated in your program anyway. We now have a generation of small low-cost microcontrollers that have enough on-chip memory and processing power to support the use of an RTOS. Developing using this approach is therefore much more accessible.

In this chapter we will first look at setting up an introductory RTOS project for a Cortex-M based microcontroller. Next, we will go through each of the RTOS primitives and how they influence the design of our application code. Finally, when we have a clear understanding of the RTOS features, we will take a closer look at the RTOS configuration options. If you are used to programming a microcontroller without using an RTOS, that is bare metal, there

The Designer's Guide to the Cortex-M Processor Family.
DOI: http://dx.doi.org/10.1016/B978-0-08-100629-0.00009-8

are two key things to understand as you work through this tutorial Concurrency and synchronization. In the following sections we will focus on creating and managing threads. The key concept here is to consider them running as parallel concurrent objects. In the section "Inter-Thread Communication" we will look at how to communicate between threads. In this section the key concept is synchronization of the concurrent threads.

First Steps with CMSIS-RTOS

The RTOS itself consists of a scheduler (Fig. 9.1) which supports round-robin, preemptive, and cooperative multitasking of program threads, as well as time and memory management services. Inter-thread communication is supported by additional RTOS objects, including signal triggering, semaphores, mutex, and a mailbox system. As we will see in Chapter 10 "RTOS Techniques", interrupt handling can also be accomplished by prioritized threads which are scheduled by the RTOS kernel.

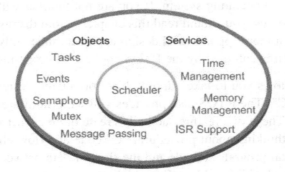

Figure 9.1

The RTOS kernel contains a scheduler that runs program code as tasks. Communication between tasks is accomplished by RTOS objects such as events, semaphores, mutexes, and mailboxes. Additional RTOS services include time and memory management and interrupt support.

Accessing the CMSIS-RTOS API

To access any of the CMSIS-RTOS features in our application code it is necessary to include the following header file:

```
#include <cmsis_os.h>
```

This header file is maintained by ARM as part of the CMSIS-RTOS standard. For the Keil RTX RTOS this is the default API. Other RTOS will have their own proprietary API but may provide a wrapper layer to implement the CMSIS-RTOS API so they can be used where compatibility with the CMSIS standard is required.

Threads

The building blocks of a typical "C" program are functions which we call to perform a specific procedure and which then return to the calling function. In CMSIS-RTOS the basic unit of execution is a "thread." A thread is very similar to a "C" procedure but has some very fundamental differences.

```
unsigned int procedure (void)   void thread (void)
{                               {
                                    while(1)
                                    {
    ......                              ......
    return(ch);                     }
}                               }
```

While we always return from our "C" function, once started an RTOS thread must contain a loop so that it never terminates and thus runs forever. You can think of a thread as a mini self-contained program that runs within the RTOS.

An RTOS program is made up of a number of threads, which are controlled by the RTOS scheduler. This scheduler uses the SysTick timer to generate a periodic interrupt as a time base. The scheduler will allot a certain amount of execution time to each thread. So thread1 will run for 5 milliseconds then be de-scheduled to allow thread2 to run for a similar period; thread2 will give way to thread3 and finally control passes back to thread1. By allocating these slices of run-time to each thread in a round-robin fashion, we get the appearance of all three threads running in parallel to each other.

Conceptually, we can think of each thread as performing a specific functional unit of our program with all threads running simultaneously. This leads us to a more object-orientated design, where each functional block can be coded and tested in isolation and then integrated into a fully running program. This not only imposes a structure on the design of our final application but also aids debugging, as a particular bug can be easily isolated to a specific thread. It also aids code reuse in later projects. When a thread is created, it is also allocated its own thread ID. This is a variable which acts as a handle for each thread and is used when we want to manage the activity of the thread.

```
osThreadId id1,id2,id3;
```

In order to make the thread-switching process happen, we have the code overhead of the RTOS and we have to dedicate a CPU hardware timer to provide the RTOS time reference. In addition, each time we switch running threads, we have to save the state of all the thread variables to a thread stack. Also, all the run-time information about a thread is stored in a thread control block, which is managed by the RTOS kernel. Thus the "context switch time," that is, the time to save the current thread state and load up and start the next thread, is a crucial figure and will depend on both the RTOS kernel and the design of the underlying hardware.

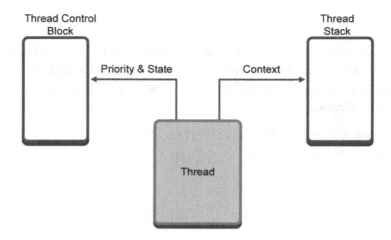

Figure 9.2
Each thread has its own stack for saving its data during a context switch. The thread control block is used by the kernel to manage the active thread.

The Thread Control Block (Fig. 9.2) contains information about the status of a thread. Part of this information is its run state. In a given system only one thread can be running and all the others will be suspended but ready to run. The RTOS has various methods of inter-thread communication (signals, semaphores, messages). Here, a thread may be suspended to wait to be signaled by another thread or interrupt before it resumes its ready state, whereupon it can be placed into running state by the RTOS scheduler (Table 9.1).

Table 9.1: Threads may be in one of three states

Running	The currently running thread
Ready	Threads ready to run
Wait	Blocked threads waiting for an OS event

The remaining threads will be ready to run and will be scheduled by the kernel. Threads may also be waiting pending an OS event. When this occurs they will return to the ready state and be scheduled by the kernel.

Starting the RTOS

To build a simple RTOS program we declare each thread as a standard "C" function and also declare a thread ID variable for each function.

```
void thread1 (void);
void thread2 (void);
osThreadId thrdID1, thrdID2;
```

By default the CMSIS-RTOS scheduler will be running when main() is entered and the main() function becomes the first active thread. Once in main(), we can stop the scheduler thread switching by calling osKernelInitialize (). While the RTOS is halted we can create further threads and other RTOS objects. Once the system is in a defined state we can restart the RTOS scheduler with osKernelStart().

You can run any initializing code you want before starting the RTOS.

```
void main (void)
{
  osKernelInitialize ();
  IODIR1 = 0x00FF0000;      // Do any C code you want
  Init_Thread();            //Create a Thread
  osKernelStart();          //Start the RTOS
}
```

When threads are created they are also assigned a priority. If there are a number of threads ready to run and they all have the same priority, they will be allotted run-time in a round-robin fashion. However, if a thread with a higher priority becomes ready to run, the RTOS scheduler will de-schedule the currently running thread and start the high priority thread running. This is called preemptive priority-based scheduling. When assigning priorities you have to be careful because the high priority thread will continue to run until it enters a waiting state or until a thread of equal or higher priority is ready to run (Fig. 9.3).

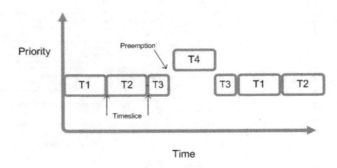

Figure 9.3
Threads of equal priority will be scheduled in a round-robin fashion. High priority tasks will preempt low priority tasks and enter the running state "on demand."

Exercise 9.1 A First CMSIS-RTOS Project

This project will take you through the steps necessary to create and debug a CMSIS-RTOS-based project. We will create this project from scratch—a reference example is included as example 9.1 in the examples pack.

Start μVision and select Project → New μVision Project (Fig. 9.4).

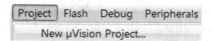

Figure 9.4
Create a new project.

In the new project dialog enter a suitable project name and directory and click Save.

Next the device database will open. Navigate through to the STMicroelectronics:: STM32F103:STM32F103RB (Fig. 9.5).

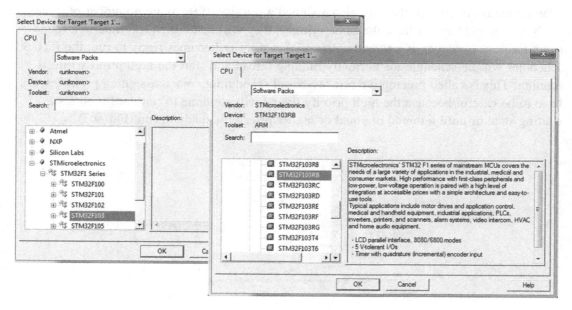

Figure 9.5
Select the microcontroller.

Once you have selected this device click OK.

Once the microcontroller variant has been selected the Run-Time Environment Manager will open (Fig. 9.6).

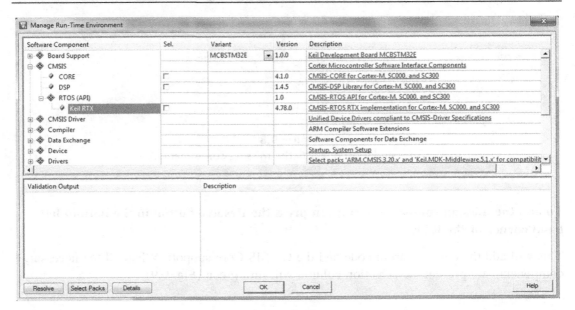

Figure 9.6
Add the RTOS.

This allows you to configure the platform of software components you are going to use in a given project. As well as displaying the available components the RTE understands their dependencies on other components.

To configure the project for use with the CMSIS-RTOS Keil RTX, simply tick the CMSIS::RTOS (API):Keil RTX box (Fig. 9.7).

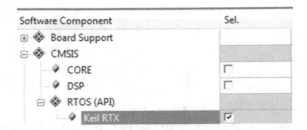

Figure 9.7
The Sel. column elements turn orange (light gray in print version). The RTOS requires other components to be added.

This will cause the selection box to turn orange meaning that additional components are required. The required component will be displayed in the Validation Output window (Fig. 9.8).

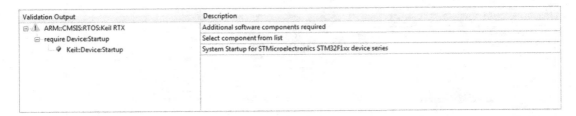

Figure 9.8
The validation box lists the missing components.

To add the missing components you can press the Resolve button in the bottom left hand corner of the RTE.

This will add the device startup code and the CMSIS-Core support. When all the necessary components are present the selection column will turn green (Fig. 9.9).

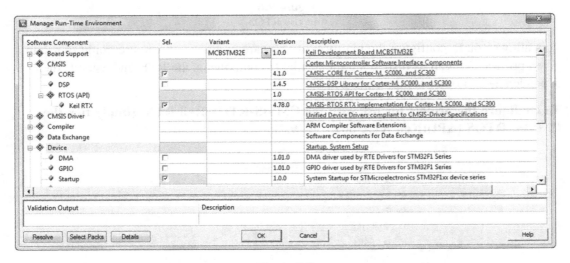

Figure 9.9
Pressing the resolve button adds the missing components and the Sel. column turns green (light gray in print version).

It is also possible to access a component's help files by clicking on the blue hyperlink in the Description column.

Now press the OK button and all the selected components will be added to the new project (Fig. 9.10).

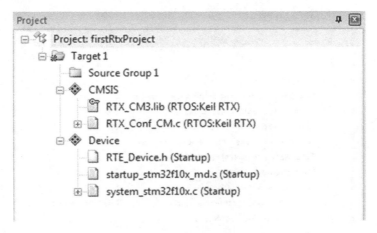

Figure 9.10
The configured project platform.

The CMSIS components are added to folders displayed as a green diamond. There are two types of files here. The first type is a library file which is held within the toolchain and is not editable. This file is shown with a yellow key to show that it is "locked" (read-only). The second type of file is a configuration file. These files are copied to your project directory and can be edited as necessary. Each of these files can be displayed as a text files but it is also possible to view the configuration options as a set of pick lists and drop down menus.

To see this open the RTX_Conf_CM.c file and at the bottom of the editor window select the "Configuration Wizard" tab (Fig. 9.11).

Figure 9.11
Selecting the configuration wizard.

Click on "Expand All" to see all of the configuration options as a graphical pick list (Fig. 9.12).

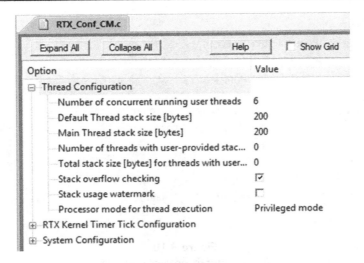

Figure 9.12
The RTX configuration options.

For now it is not necessary to make any changes here and these options will be examined towards the end of this chapter.

Our project contains four configuration files, three of which are standard CMSIS files (Table 9.2).

Table 9.2: Project configuration files

File Name	Description
Startup_STM32F10x_md.s	Assembler vector table
System_STM32F10x.c	C code to initialize key system peripherals, such as clock tree, PLL external memory interface.
RTE_Device.h	Configures the pin multiplex
RTX_Conf_CM.c	Configures Keil RTX

Now that we have the basic platform for our project in place we can add some user source code which will start the RTOS and create a running thread (Fig. 9.13).

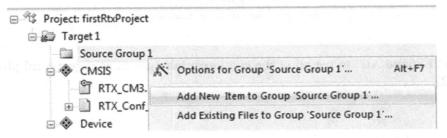

Figure 9.13
Adding a source module.

To do this right-click the "Source Group 1" folder and select "Add new item to Source Group 1".

In the Add new Item dialog select the "User code template" Icon and in the CMSIS section select the "CMSIS-RTOS 'main' function" and click Add (Fig. 9.14).

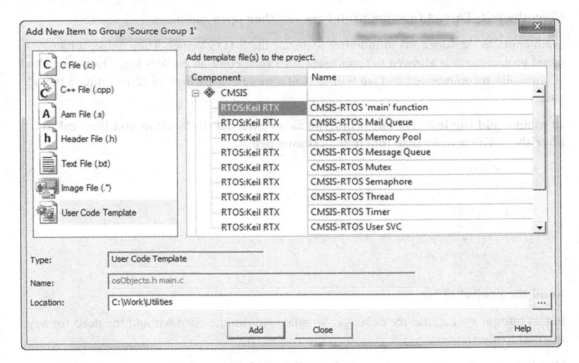

Figure 9.14
Selecting a CMSIS-RTOS template.

Repeat this but this time select "CMSIS-RTOS Thread".

This will now add two source files to our project main.c and thread.c (Fig. 9.15).

Figure 9.15
The project with main and thread code.

Open thread.c in the editor.

We will look at the RTOS definitions in this project in the next section. For now this file contains two functions Init_Thread() which is used to start the thread running and the actual thread function.

Copy the Init_Thread function prototype and then open main.c.

Main contains the functions to initialize and start the RTOS kernel. Then unlike a bare metal project main is allowed to terminate rather than enter an endless loop. However, this is not really recommended and we will look at a more elegant way of terminating a thread later.

In main.c add the Init_Thread prototype as an external declaration and then call it after the osKernelInitialize function as shown below.

```
#define osObjectsPublic
#include "osObjects.h"
extern int Init_Thread (void);     //Add this line
int main (void) {
        osKernelInitialize ();
        Init_Thread ();            //Add this line
        osKernelStart ();
}
```

Build the project (F7).

In this tutorial we can use the debugger simulator to run the code without the need for any external hardware.

Highlight the Target 1 root folder, right click, and select "Options for target 1."

Select the debugger tab.

This menu is in two halves: the left side configures the simulator and the right half configures the hardware debugger.

Select the Simulator radio button and check that "Dialog DLL" is set to DARMSTM. DLL with parameter -pSTM32F103RB (Fig. 9.16).

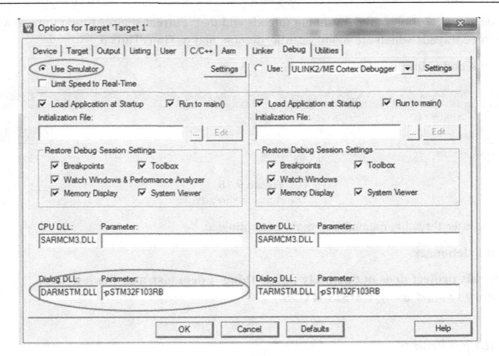

Figure 9.16
Configuring the Simulator.

Click ok to close the options for target menu.

Start the debugger (Ctrl + F5).

This will run the code up to main.

Open the Debug → OS Support → System and Thread Viewer (Fig. 9.17).

Item	Value					
Tick Timer:	1.000 mSec					
Round Robin Timeout:	5.000 mSec					
Default Thread Stack Size:	200					
Thread Stack Overflow Check:	Yes					
Thread Usage:	Available: 7, Used: 3 + os_idle_demon					

ID	Name	Priority	State	Delay	Event Value	Event Mask	Stack Usage
1	osTimerThread	High	Wait_MBX				32%
2	main	Normal	Running				0%
255	os_idle_demon	None	Ready				

Figure 9.17
The RTOS System and Thread viewer.

This debug view shows all the running threads and their current state. At the moment we have three threads which are main, os_idle_demon, and osTimerThread.

Start the code running (F5) (Fig. 9.18).

ID	Name	Priority	State	Delay	Event Value	Event Mask	Stack Usage
1	osTimerThread	High	Wait_MBX				32%
3	Thread	Normal	Running				16%
255	os_idle_demon	None	Ready				32%

Figure 9.18
The running threads.

Now the user thread is created and main is terminated.

Exit the debugger.

While this project does not actually do anything it demonstrates the few steps necessary to start using CMSIS-RTOS.

Creating Threads

Once the RTOS is running, there are a number of system calls that are used to manage and control the active threads. By default, the main() function is automatically created as the first thread running. In the first example we used it to create an additional thread then let it terminate by running through the closing brace. However, if we want to we can continue to use main as a thread in its own right. If we want to control main as a thread we must get its thread ID. The first RTOS function we must therefore call is osThreadGetId() which returns the thread ID number of the currently running thread. This is then stored in its ID handle. When we want to refer to this thread in future OS calls, we use this handle rather than the function name of the thread.

```
osThreadId main_id;   //create the thread handle
void main (void)
{
  /* Read the Thread-ID of the main thread */
  main_id = osThreadGetId ();
  while(1)
  {
    .........
  }
}
```

Now that we have an ID handle for main we could create the application threads and then call osTerminate(main_id) to end the main thread. This is the best way to end a thread rather than let it run off the end of the closing brace. Alternatively we can add a while(1) loop as shown above and continue to use main in our application.

As we saw in the first example the main thread is used as a launcher thread to create the application threads. This is done in two stages. First a thread structure is defined; this allows us to define the thread operating parameters.

```
osThreadId thread1_id;                      //thread handle
void thread1 (void const *argument);        //function prototype for thread1
osThreadDef(thread1, osPriorityNormal, 1, 0);   //thread definition structure
```

The thread structure requires us to define the name of the thread function, its thread priority, the number of instances of the thread that will be created, and its stack size. We will look at these parameters in more detail later. Once the thread structure has been defined the thread can be created using the osThreadCreate() API call. Then the thread is created from within the application code; this is often done within the main thread but can be at any point in the code.

```
thread1_id = osThreadCreate(osThread(thread1), NULL);
```

This creates the thread and starts it running. It is also possible to pass a parameter to the thread when it starts.

```
uint32_t startupParameter = 0x23;
thread1_id = osThreadCreate(osThread(thread1), startupParameter);
```

When each thread is created, it is also assigned its own stack for storing data during the context switch. This should not be confused with the native Cortex processor stack; it is really a block of memory that is allocated to the thread. A default stack size is defined in the RTOS configuration file (we will see this later) and this amount of memory will be allocated to each thread unless we override it to allocate a custom size. The default stack size will be assigned to a thread if the stack size value in the thread definition structure is set to zero. If necessary a thread can be given additional memory resources by defining a bigger stack size in the thread structure.

```
osThreadDef(thread1, osPriorityNormal, 1, 0);  //assign default stack size to this
                                                  thread
osThreadDef(thread2, osPriorityNormal, 1, 1024); //assign 1KB of stack to this thread
```

However, if you allocate a larger stack size to a thread then the additional memory must be allocated in the RTOS configuration file; again we will see this later.

Exercise 9.2 Creating and Managing Threads

In this project we will create and manage some additional threads. Each of the threads created will toggle a GPIO pin on GPIO port B to simulate flashing an LED. We can then view this activity in the simulator.

Open the Pack Installer.

Select the Boards::Designers Guide Tutorial.

Select the example tab and Copy "EX 9.2 and 9.3 CMSIS-RTOS Threads".

A reference copy of the first exercise is included as Exercise 9.1.

This will install the project to a directory of your choice and open the project in µ**Vision**.

Open the Run-Time Environment Manager.

In the board support section the MCBSTM32E:LED box is ticked. This adds support functions to control the state of a bank of LED's on the Microcontroller's GPIO port B (Fig. 9.19).

Software Component	Sel.	Variant	Version	Description
⊟ ◆ Board Support		MCBSTM32E ▼	1.0.0	Keil Development Board MCBSTM32E
⊟ ◆ MCBSTM32E				
◆ A/D Converter	□		1.0.0	A/D Converter driver for Keil MCBSTM32E Development Board
◆ Graphic LCD	□		1.0.0	Graphic LCD driver for Keil MCBSTM32E Development Board
◆ Joystick	□		1.0.0	Joystick driver for Keil MCBSTM32E Development Board
◆ Keyboard	□		1.0.0	Keyboard driver for Keil MCBSTM32E Development Board
◆ LED	☑		1.0.0	LED driver for Keil MCBSTM32E Development Board
◆ emWin LCD	□	16-bit IF	1.0.0	emWin LCD driver (16-bit Interface) for Keil MCBSTM32E Development Board

Figure 9.19
Selecting the board support components.

When the RTOS starts main() runs as a thread and we will create two additional threads. First we create handles for each of the threads and then define the parameters of each thread. These include the priority the thread will run at, the number of instances of each thread we will create and its stack size (the amount of memory allocated to it) zero indicates it will have the default stack size.

```
osThreadId main_ID,led_ID1,led_ID2;
osThreadDef(led_thread2, osPriorityNormal, 1, 0);
osThreadDef(led_thread1, osPriorityNormal, 1, 0);
```

Then in the main() function the two threads are created.

```
led_ID2 = osThreadCreate(osThread(led_thread2), NULL);
led_ID1 = osThreadCreate(osThread(led_thread1), NULL);
```

When the thread is created we can pass it as a parameter in place of the NULL define.

Build the project and start the debugger.

Start the code running and open the Debug → OS Support → System and Thread Viewer (Fig. 9.20).

ID	Name	Priority	State	Delay	Event Value	Event Mask	Stack Usage
1	osTimerThread	High	Wait_MBX				32%
3	led_thread2	Normal	Running				0%
4	led_thread1	Normal	Ready				32%
255	os_idle_demon	None	Ready				

Figure 9.20
The running Threads.

Now we have four active threads with one running and the others ready.

Open the Debug → OS Support → Event Viewer (Fig. 9.21).

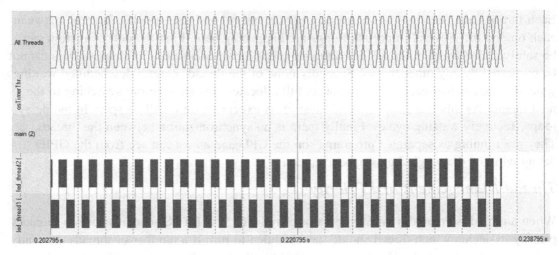

Figure 9.21
The event viewer shows the thread switching history.

The event viewer shows the execution of each thread as a trace against time. This allows you to visualize the activity of each thread and get a feel for amount of CPU time consumed by each thread.

Now open the Peripherals → General Purpose IO → GPIOB window (Fig. 9.22).

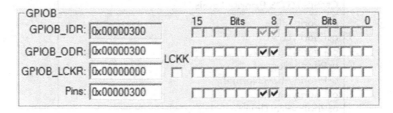

Figure 9.22
The peripheral window shows the LED pin activity.

Our two LED threads are each toggling a GPIO port pin. Leave the code running and watch the pins toggle for a few seconds.

If you do not see the debug windows updating, check the view/periodic window update option is ticked.

```
void led_thread2 (void const *argument) {
for (;;) {
  LED_On(1);
  delay(500);
  LED_Off(1);
  delay(500);
}}
```

Each thread calls functions to switch an LED on and off and uses a delay function between each on and off. Several important things are happening here. First the delay function can be safely called by each thread. Each thread keeps local variables in its stack so they cannot be corrupted by any other thread. Secondly none of the threads enter a descheduled waiting state, this means that each one runs for its full allocated timeslice before switching to the next thread. As this is a simple thread most of its execution time will be spent in the delay loop effectively wasting cycles. Finally there is no synchronization between the threads. They are running as separate "programs" on the CPU and as we can see from the GPIO debug window the toggled pins appear random.

Thread Management and Priority

When a thread is created it is assigned a priority level. The RTOS scheduler uses a thread's priority to decide which thread should be scheduled to run. If a number of threads are ready to run, the thread with the highest priority will be placed in the run state. If a high priority thread becomes ready to run it will preempt a running thread of lower priority. Importantly, a high priority thread running on the CPU will not stop running unless it blocks on an RTOS API call or is preempted by a higher priority thread. A thread's priority is defined in the thread structure and the following priority definitions are available. The default priority is osPriorityNormal (Table 9.3).

Table 9.3: CMSIS-RTOS priority levels

CMSIS-RTOS Priority Levels
osPriorityIdle
osPriorityLow
osPriorityBelowNormal
osPriorityNormal
osPriorityAboveNormal
osPriorityHigh
osPriorityRealTime
osPriorityError

Once the threads are running, there are a small number of OS system calls which are used to manage the running threads. It is also then possible to elevate or lower a thread's priority either from another function or from within its own code.

```
osStatus osThreadSetPriority(threadID, priority);
osPriority osThreadGetPriority(threadID);
```

As well as creating threads, it is also possible for a thread to delete itself or another active thread from the RTOS. Again we use the thread ID rather than the function name of the thread.

```
osStatus = osThreadTerminate (threadID1);
```

Finally, there is a special case of thread switching where the running thread passes control to the next ready thread of the same priority. This is used to implement a third form of scheduling called cooperative thread switching.

```
osStatus osThreadYield();   //switch to next ready to run thread
```

Exercise 9.3 Creating and Managing Threads II

In this exercise we will look at assigning different priorities to threads and also how to create and terminate threads dynamically.

Go back to the project "Ex 9.2 and 9.3 Threads".

Change the priority of LED Thread2 to Above Normal.

```
osThreadDef(led_thread2, osPriorityAboveNormal, 1, 0);
osThreadDef(led_thread1, osPriorityNormal, 1, 0);
```

Build the project and start the debugger.

Start the code running.

Open the Debug → OS Support → Event Viewer window (Fig. 9.23).

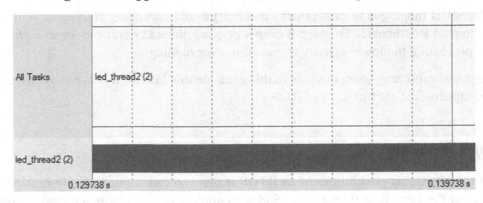

Figure 9.23
Only led_thread2 is running.

Here we can see thread2 running but no sign of thread1. Looking at the coverage monitor for the two threads shows us that led_thread1 has not run (Fig. 9.24).

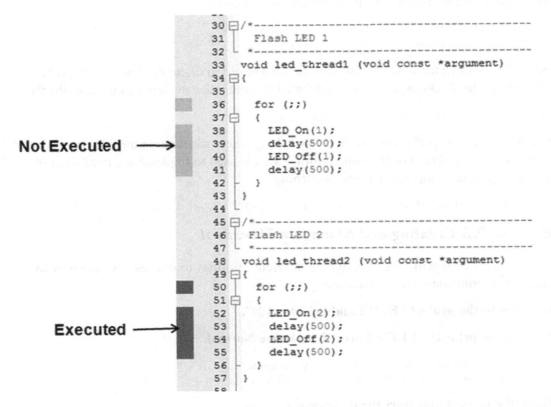

Figure 9.24
The coverage monitor shows what code has executed by coloring the margin green (dark gray in print version).

Led_thread1 is running at normal priority and led_thread2 is running at a higher priority so has preempted led_thread1. To make it even worse led_thread2 never blocks so it will run forever preventing the lower priority thread from ever running.

Although this error may seem obvious in this example this kind of mistake is very common when designers first start to use an RTOS.

Multiple Instances

One of the interesting possibilities of an RTOS is that you can create multiple running instances of the same base thread code. So for example you could write a thread to control a UART and then create two running instances of the same thread code. Here each instance of the UART code could manage a different UART.

First we create the thread structure and set the number of thread instances to two:

```
osThreadDef(thread1, osPriorityNormal, 2, 0);
```

Then we can create two instances of the thread assigned to different thread handles. A parameter is also passed to allow each instance to identify which UART it is responsible for.

```
ThreadID_1_0 = osThreadCreate(osThread(thread1), UART1);
ThreadID_1_1 = osThreadCreate(osThread(thread1), UART2);
```

Exercise 9.4 Multiple Thread Instances

In this project we will look at creating one thread and then create multiple run-time instances of the same thread.

In the Pack Installer select "Ex 9.4 Multiple Instances" and copy it to your tutorial directory.

This project performs the same function as the previous LED flasher program. However, we now have one led switcher function that uses an argument passed as a parameter to decide which LED to flash.

```
void ledSwitcher (void const *argument) {
  for (;;) {
    LED_On((uint32_t)argument);
    delay(500);
    LED_Off((uint32_t)argument);
    delay(500);
  }
}
```

When we define the thread we adjust the instances parameter to two.

```
osThreadDef(ledSwitcher, osPriorityNormal, 2, 0);
```

Then in the main thread we create two threads which are different instances of the same base code. We pass a different parameter which corresponds to the led that will be toggled by the instance of the thread.

```
led_ID1 = osThreadCreate(osThread(ledSwitcher),(void *) 1UL);
led_ID2 = osThreadCreate(osThread(ledSwitcher),(void *) 2UL);
```

Build the code and start the debugger.

Start the code running and open the RTX tasks and system window (Fig. 9.25).

ID	Name	Priority	State
255	os_idle_demon	0	Ready
3	ledSwitcher	4	Ready
2	ledSwitcher	4	Running

Figure 9.25
Multiple instances of thread running

Here we can see both instances of the ledSwitcher task each with a different ID.

Examine the Call stack + locals window (Fig. 9.26).

ledSwitcher : 2	0x08000318
⊞ ● delay	0x08000310
⊟ ● ledSwitcher	0x0800032A
⊞ ●● argument	0x00000001
⊟ ● ledSwitcher : 3	0x08000318
⊞ ● delay	0x08000314
⊟ ● ledSwitcher	0x08000338
⊞ ●● argument	0x00000002

Figure 9.26
The watch window is thread aware.

Here we can see both instances of the ledSwitcher threads and the state of their variables. A different argument has been passed to each instance of the thread.

Time Management

As well as running your application code as threads, the RTOS also provides some timing services which can be accessed through RTOS system calls.

Time Delay

The most basic of these timing services is a simple timer delay function. This is an easy way of providing timing delays within your application. Although the RTOS kernel size is quoted as 5k bytes, features such as delay loops and simple scheduling loops are often part of a non-RTOS application and would consume code bytes anyway, so the overhead of the RTOS can be less than it immediately appears.

```
void osDelay ( uint32_t millisec )
```

This call will place the calling thread into the WAIT_DELAY state for the specified number of milliseconds. The scheduler will pass execution to the next thread in the READY state (Fig. 9.27).

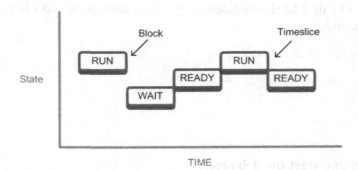

Figure 9.27

During their lifetime threads move through many states. Here a running thread is blocked by an osDelay call so it enters a wait state. When the delay expires, it moves to ready. The scheduler will place it in the run state. If its timeslice expires, it will move back to ready.

When the timer expires, the thread will leave the wait_delay state and move to the READY state. The thread will resume running when the scheduler moves it to the RUNNING state. If the thread then continues executing without any further blocking OS calls, it will be descheduled at the end of its timeslice and be placed in the ready state, assuming another thread of the same priority is ready to run.

Waiting for an Event

In addition to a pure time delay it is possible to make a thread halt and enter the waiting state until the thread is triggered by another RTOS event. RTOS events can be a signal, message, or mail event. The osWait() API call also has a timeout period defined in milliseconds that allows the thread to wake up and continue execution if no event occurs.

```
osStatus osWait ( uint32_t millisec )
```

When the interval expires, the thread moves from the wait to the READY state and will be placed into the running state by the scheduler. osWait is an optional API call within the CMSIS-RTOS specification. If you intend to use this function you must first check it is supported by the RTOS you are using. The osWait API call is not supported by the Keil RTX RTOS.

Exercise 9.5 Time Management

In this exercise we will look at using the basic time delay function.

In the Pack Installer select "Ex 9.5 Time Management" and copy it to your tutorial directory.

This is our original LED flasher program but the simple delay function has been replaced by the osDelay API call. LED2 is toggled every 100 milliseconds and LED1 is toggled every 500 milliseconds.

```
void ledOn (void const *argument) {
for (;;) {
  LED_On(1);
  osDelay(500);
  LED_Off(1);
  osDelay(500);
}}
```

Build the project and start the debugger.

Start the code running and open the event viewer window (Fig. 9.28).

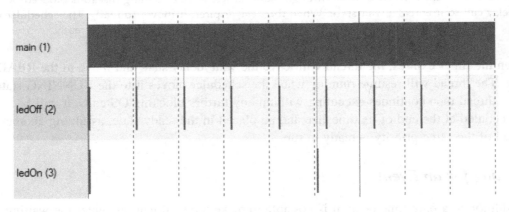

Figure 9.28
Using osDelay() allows the thread to block when it is idle.

Now we can see that the activity of the code is very different. When each of the LED tasks reaches the osDelay API call it "blocks" and moves to a waiting state. The main task will be in a ready state so the scheduler will start it running. When the delay period has timed out the LED tasks will move to the ready state and will be placed into the running state by the scheduler. This gives us a multithreaded program where CPU run-time is efficiently shared between tasks. This sharing of CPU runtime allows us to think of our threads running concurrently when we are designing the system.

Virtual Timers

The CMSIS-RTOS API can be used to define any number of virtual timers which act as count down timers. When they expire, they will run a user call-back function to perform a specific action. Each timer can be configured as a one shot or repeat timer. A virtual timer is created by first defining a timer structure.

```
osTimerDef(timer0,led_function);
```

This defines a name for the timer and the name of the call back function. The timer must then be instantiated in an RTOS thread.

```
osTimerId timer0_handle = osTimerCreate (timer(timer0), osTimerPeriodic, (void *)0);
```

This creates the timer and defines it as a periodic timer or a single shot timer (osTimerOnce). The final parameter passes an argument to the call back function when the timer expires.

```
osTimerStart ( timer0_handle,0x100);
```

The timer can then be started at any point in a thread the timer start function invokes the timer by its handle and defines a count period in milliseconds.

Exercise 9.6 Virtual Timer

In this exercise we will configure a number of virtual timers to trigger a callback function at various frequencies.

In the Pack Installer select "Ex 9.6 Virtual Timers" and copy it to your tutorial directory.

This is our original LED flasher program and code has been added to create four virtual timers to trigger a callback function. Depending on which timer has expired, this function will toggle an additional LED.

The timers are defined at the start of the code:

```
osTimerDef(timer0_handle, callback);
osTimerDef(timer1_handle, callback);
osTimerDef(timer2_handle, callback);
osTimerDef(timer3_handle, callback);
```

They are then initialized in the main function.

```
osTimerId timer0 = osTimerCreate(osTimer(timer0_handle),osTimerPeriodic, (void *)0);
osTimerId timer1 = osTimerCreate(osTimer(timer1_handle),osTimerPeriodic, (void *)1);
osTimerId timer2 = osTimerCreate(osTimer(timer2_handle),osTimerPeriodic, (void *)2);
osTimerId timer3 = osTimerCreate(osTimer(timer3_handle),osTimerPeriodic, (void *)3);
```

Each timer has a different handle and ID and passed as different parameter to the common callback function.

```
void callback(void const *param){
switch( (uint32_t) param){
  case 0:
        GPIOB->ODR ^= 0x8;
  break;
  case 1:
        GPIOB->ODR ^= 0x4;
  break;
  case 2:
        GPIOB->ODR ^= 0x2;
  break;
}}
```

When triggered, the callback function uses the passed parameter as an index to toggle the desired LED.

In addition to the configuring the virtual timers in the source code, the timer thread must be enabled in the RTX configuration file.

Open the RTX_Conf_CM.c file and press the configuration wizard tab (Fig. 9.29).

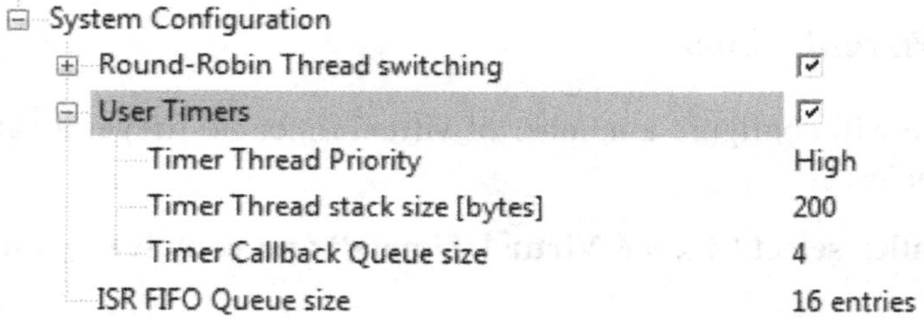

⊟ System Configuration		
⊞ Round-Robin Thread switching	☑	
⊟ User Timers	☑	
Timer Thread Priority	High	
Timer Thread stack size [bytes]	200	
Timer Callback Queue size	4	
ISR FIFO Queue size	16 entries	

Figure 9.29
Configuring the virtual timers.

In the system configuration section make sure the User Timers box is ticked. If this thread is not created the timers will not work.

Build the project and start the debugger.

Run the code and observe the activity of the GPIOB pins in the peripheral window (Fig. 9.30).

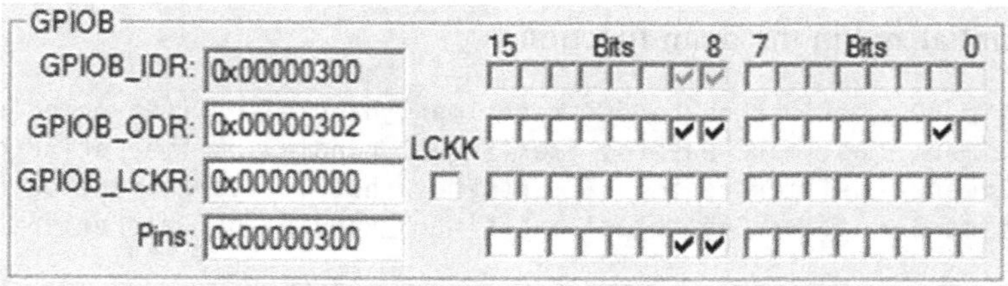

Figure 9.30
The user timers toggle additional LED pins.

There will also be an additional thread running in the System and Thread Viewer window (Fig. 9.31).

ID	Name	Priority	State	Delay	Event Value	Event Mask	Stack Usage
1	osTimerThread	High	Wait_MBX				32%
3	ledThread1	Normal	Wait_AND		0x0000	0x0001	40%
4	ledThread2	Normal	Wait_DLY	328			32%
255	os_idle_demon	None	Running				

Figure 9.31
The user timers create an additional osTimerThread.

The osDelay() function provides a relative delay from the point at which the delay is started. The virtual timers provide an absolute delay which allows you to schedule code to run at fixed intervals.

Sub-Millisecond Delays

While the various CMSIS-RTOS time functions have a resolution of 1 millisecond, it is possible to create delays with a resolution in microseconds using the raw SysTick count. This form of delay does not deschedule the task, but it simply halts its execution for the desired period. To create a delay we can first get the SysTick count.

```
int32_t tick,delayPeriod;
tick = osKernelSysTick();    // get start value of the Kernel system tick
```

Then we can scale a period in microseconds to a SysTick count value.

```
delayPeriod = osKernelTickMicroSec(100));
```

This then allows us to create a delay for the required period.

```
do {              // Delay for 100 microseconds
} while ((osKernelSysTick() - tick) < delayPeriod);
```

Idle Demon

The final timer service provided by the RTOS is not really a timer, but this is probably the best place to discuss it. If during our RTOS program we have no thread running and no thread ready to run (eg, they are all waiting on delay functions) then the RTOS will use the spare run-time to call an "Idle Demon" that is again located in the RTX_Conf_CM.c file. This idle code is in effect a low priority thread within the RTOS which only runs when nothing else is ready.

```
void os_idle_demon (void)
{
    for (;;) {
    /* HERE: include here optional user code to be executed when no thread runs.*/
    }
}
```

You can add any code to this thread, but it has to obey the same rules as user threads. The simplest use of the idle demon is to place the microcontroller into a low-power mode when it is not doing anything.

```
void os_idle_demon (void) {
    for (;;) {
                __wfe();
}}
```

What happens next depends on the power mode selected in the microcontroller. At a minimum the CPU will halt until an interrupt is generated by the SysTick timer and execution of the scheduler will resume. If there is a thread ready to run then execution of the application code will resume. Otherwise, the idle demon will be reentered and the system will go back to sleep.

Exercise 9.7 Idle Thread

In the Pack Installer select "Ex 9.7 Idle" and copy it to your tutorial directory.

This is a copy of the virtual timer project.

Open the RTX_Conf_CM.c file and click the text editor tab.

Locate the os_idle_demon thread.

```
void os_idle_demon (void) {
int32_t i;
for (;;) {
  //wfe();
}}
```

Build the code and start the debugger.

Run the code and observe the activity of the threads in the event Viewer.

This is a simple program that spends most of its time in the idle demon so this code will run almost continuously (Fig. 9.32).

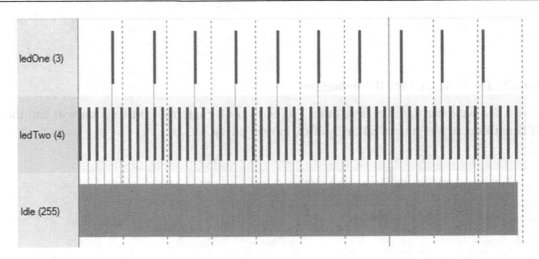

Figure 9.32
The idle thread is a good guide to the CPU loading.

You can also see the activity of the idle demon in the event viewer. In a real project, the amount of time spent in the idle demon is an indication of spare CPU cycles.

Open the View → Analysis Windows → Performance Analyzer (Fig. 9.33).

Module/Function	Calls	Time(Sec)	Time(%)	
CMSISrtxIdle		151.249 ms	100%	
RTE/CMSIS/RTX_Conf_CM.c		148.247 ms	98%	
os_idle_demon	1	148.247 ms	98%	
os_error	0	0us	0%	
__user_perthread_libspace	0	0us	0%	
_mutex_initialize	0	0us	0%	
_mutex_acquire	0	0us	0%	
_mutex_release	0	0us	0%	

Figure 9.33
The performance analyzer shows that most of the run-time is being spent in the idle loop.

This window shows the cumulative run-time for each function in the project. In this simple project the os_idle_demon is using most of the run-time because there is very little application code.

Exit the debugger.

Remove the delay loop and the toggle instruction and add a __wfe() instruction in the for loop, so the code now looks like this.

```
void os_idle_demon (void) {
    for (;;) {
            __wfe();
}}
```

Rebuild the code, restart the debugger.

Now when we enter the idle thread the __wfe() (wait for interrupt) instruction will halt the CPU until there is a peripheral or SysTick interrupt (Fig. 9.34).

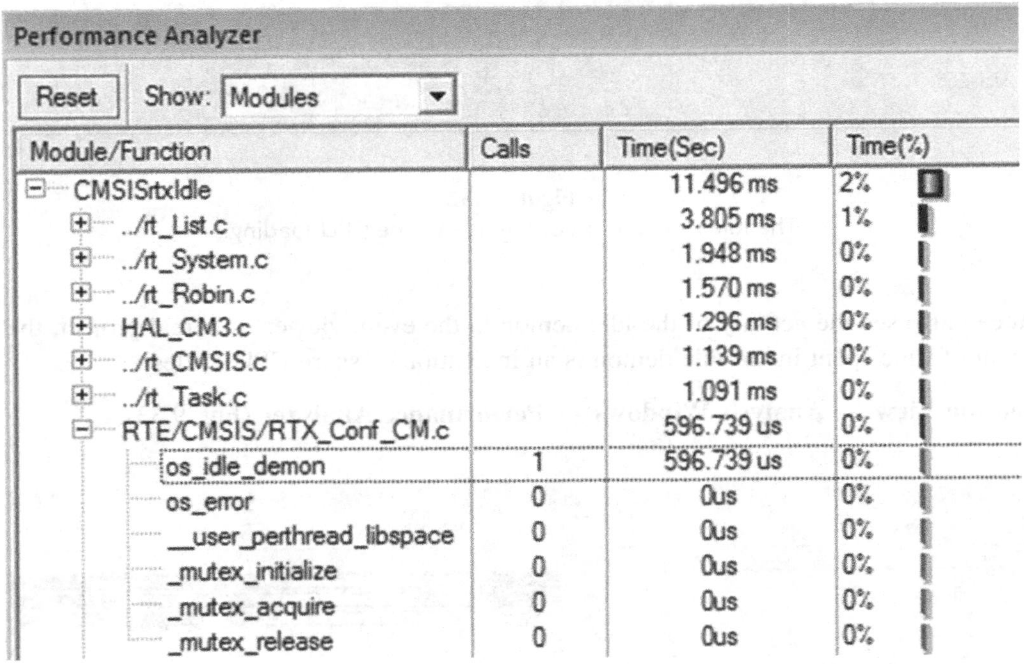

Performance Analyzer

Module/Function	Calls	Time(Sec)	Time(%)
⊟ CMSISrtxIdle		11.496 ms	2%
⊞ ../rt_List.c		3.805 ms	1%
⊞ ../rt_System.c		1.948 ms	0%
⊞ ../rt_Robin.c		1.570 ms	0%
⊞ HAL_CM3.c		1.296 ms	0%
⊞ ../rt_CMSIS.c		1.139 ms	0%
⊞ ../rt_Task.c		1.091 ms	0%
⊟ RTE/CMSIS/RTX_Conf_CM.c		596.739 us	0%
os_idle_demon	1	596.739 us	0%
os_error	0	0us	0%
__user_perthread_libspace	0	0us	0%
_mutex_initialize	0	0us	0%
_mutex_acquire	0	0us	0%
_mutex_release	0	0us	0%

Figure 9.34

The __wfe() intrinsic halts the CPU when it enters the idle loop. Saving cycles and run-time energy.

Performance analysis during hardware debugging.

The code coverage and performance analysis tools are available when you are debugging on real hardware rather than simulation. However, to use these features you need two things: First, you need a microcontroller that has been fitted with the optional Embedded Trace Macrocell (ETM). Second, you need to use Keil ULINK*pro* debug adapter that supports instruction trace via the ETM.

Inter-Thread Communication

So far we have seen how application code can be defined as independent threads and how we can access the timing services provided by the RTOS. In a real application we need to be able to communicate between threads in order to make an application useful. To this end, a typical RTOS supports several different communication objects which can be used to link the threads together to form a meaningful program. The CMSIS-RTOS API supports inter-thread communication with signals, semaphores, mutexes, mailboxes, and message queues. In the first section the key concept was concurrency. In this section the key concept is synchronizing the activity of multiple threads.

Signals

CMSIS-RTOS Keil RTX supports up to 16 signal flags for each thread. These signals are stored in the thread control block. It is possible to halt the execution of a thread until a particular signal flag or group of signal flags are set by another thread in the system (Fig. 9.35).

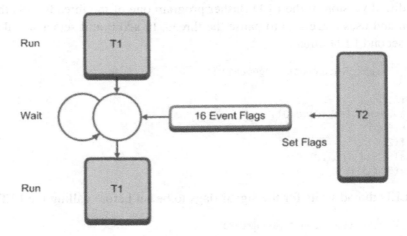

Figure 9.35
Each thread has 16 signal flags. A thread may be placed into a waiting state until a pattern of flags is set by another thread. When this happens, it will return to the ready state and wait to be scheduled by the kernel.

The signal wait system calls will suspend execution of the thread and place it into the wait_evnt state. Execution of the thread will not start until all the flags set in the signal wait API call have been set. It is also possible to define a periodic timeout after which the waiting thread will move back to the ready state, so that it can resume execution when selected by the scheduler. A value of 0xFFFF defines an infinite timeout period.

```
osEvent osSignalWait ( int32_t signals,uint32_t millisec);
```

If the signals variable is set to zero when osSignalWait is called then setting any flag will cause the thread to resume execution. You can see which flag was set by reading the osEvent.value.signals return value.

Any thread can set or clear a signal on any other thread.

```
int32_t osSignalSet ( osThreadId thread_id, int32_t signals);
int32_t osSignalClear ( osThreadId thread_id, int32_t signals);
```

Exercise 9.8 Signals

In this exercise we will look at using signals to trigger activity between two threads. While this is a simple program it introduces the concept of synchronizing the activity of threads together.

In the Pack Installer select "Ex 9.8 Signals" and copy it to your tutorial directory.

This is a modified version of the LED flasher program one of the threads calls the same LED function and uses osDelay() to pause the thread. In addition it sets a signal flag to wake up the second LED thread.

```
void led_Thread2 (void const *argument) {
for (;;) {
  LED_On(2);
  osSignalSet (T_led_ID1,0x01);
  osDelay(500);
  LED_Off(2);
  osSignalSet (T_led_ID1,0x01);
  osDelay(500);}}
```

The second LED thread waits for the signal flags to be set before calling the LED functions.

```
void led_Thread1 (void const *argument) {
for (;;) {
  osSignalWait (0x01,osWaitForever);
  LED_On(1);
  osSignalWait (0x01,osWaitForever);
  LED_Off(1);
}}
```

Build the project and start the debugger.

Open the GPIOB peripheral window and start the code running.

Now the port pins will appear to be switching on and off together. Synchronizing the threads gives the illusion that both threads are running in parallel.

This is a simple exercise but it illustrates the key concept of synchronizing activity between threads in an RTOS-based application.

Semaphores

Like signals, semaphores are a method of synchronizing activity between two or more threads. Put simply, a semaphore is a container that holds a number of tokens. As a thread executes, it will reach an RTOS call to acquire a semaphore token. If the semaphore contains one or more tokens, the thread will continue executing and the number of tokens in the semaphore will be decremented by one. If there are currently no tokens in the semaphore, the thread will be placed in a waiting state until a token becomes available. At any point in its execution, a thread may add a token to the semaphore causing its token count to increment by one (Fig. 9.36).

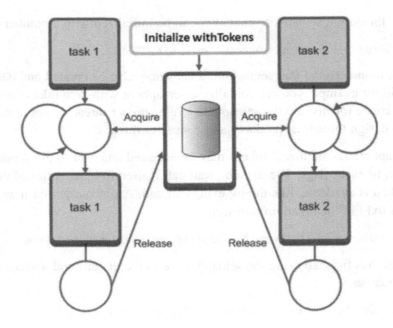

Figure 9.36
Semaphores help to control access to program resources. Before a thread can access a resource, it must acquire a token. If none is available, it waits. When it is finished with the resource, it must return the token.

The diagram above illustrates the use of a semaphore to synchronize two threads. First, the semaphore must be created and initialized with an initial token count. In this case the semaphore is initialized with a single token. Both threads will run and reach a point in their code where they will attempt to acquire a token from the semaphore. The first thread to reach this point will acquire the token from the semaphore and continue execution. The second thread will also attempt to acquire a token, but as the semaphore is empty it will halt execution and be placed into a waiting state until a semaphore token is available.

Meanwhile, the executing thread can release a token back to the semaphore. When this happens, the waiting thread will acquire the token and leave the waiting state for the ready state. Once in the ready state the scheduler will place the thread into the run state so that thread execution can continue. While semaphores have a simple set of OS calls they can be one of the more difficult OS objects to fully understand. In this section we will first look at how to add semaphores to an RTOS program and then go on to look at the most useful semaphore applications.

To use a semaphore in the CMSIS-RTOS you must first declare a semaphore container:

```
osSemaphoreId sem1;
osSemaphoreDef(sem1);
```

Then within a thread the semaphore container can be initialized with a number of tokens.

```
sem1 = osSemaphoreCreate(osSemaphore(sem1), SIX_TOKENS);
```

It is important to understand that semaphore tokens may also be created and destroyed as threads run. So for example you can initialize a semaphore with zero tokens and then use one thread to create tokens into the semaphore while another thread removes them. This allows you to design threads as producer and consumer threads.

Once the semaphore is initialized, tokens may be acquired and sent to the semaphore in a similar fashion to event flags. The os_sem_wait call is used to block a thread until a semaphore token is available, like the os_evnt_wait call. A timeout period may also be specified with 0xFFFF being an infinite wait.

```
osStatus osSemaphoreWait(osSemaphoreId semaphore_id, uint32_t millisec);
```

Once the thread has finished using the semaphore resource, it can send a token to the semaphore container.

```
osStatus osSemaphoreRelease(osSemaphoreId semaphore_id);
```

Exercise 9.9 Semaphore Signaling

In this exercise we will look at the configuration of a semaphore and use it to signal between two tasks.

In the Pack Installer select "Ex 9.9 Interrupt Signals" and copy it to your tutorial directory.

First, the code creates a semaphore called sem1 and initializes it with zero tokens.

```
osSemaphoreId sem1;
osSemaphoreDef(sem1);
int main (void) {
sem1 = osSemaphoreCreate(osSemaphore(sem1), 0);
```

The first task waits for a token to be sent to the semaphore.

```
void led_Thread1 (void const *argument) {
for (;;) {
    osSemaphoreWait(sem1, osWaitForever);
    LED_On(1);
    osDelay(500);
    LED_Off(1);
  }
}
```

While the second task periodically sends a token to the semaphore.

```
void led_Thread2 (void const *argument) {
  for (;;) {
            LED_On(2);
            osSemaphoreRelease(sem1);
            osDelay(500);
            LED_Off(2);
            osDelay(500);
}}
```

Build the project and start the debugger.

Set a breakpoint in the led_Thread2 task (Fig. 9.37).

```
● 44          osSemaphoreRelease(sem1);
```

Figure 9.37
Breakpoint on the semaphore release call in led_Thread2.

Run the code and observe the state of the threads when the breakpoint is reached (Fig. 9.38).

ID	Name	Priority	State
255	os_idle_demon	0	Ready
3	led_Thread1	5	Wait_SEM
2	led_Thread2	4	Running
1	main	4	Ready

Figure 9.38
Led_Thread1 is waiting to acquire a semaphore.

Now led_thread1 is blocked waiting to acquire a token from the semaphore. A led_Thread1 has been created with a higher priority than led_thread2 so as soon as a token is placed in the semaphore it will move to the ready state and preempt the lower priority thread and start running. When it reaches the osSemaphoreWait() call it will again block.

Now block step the code (F10) and observe the action of the threads and the semaphore.

Using Semaphores

Although semaphores have a simple set of OS calls, they have a wide range of synchronizing applications. This makes them perhaps the most challenging RTOS object to understand. In this section we will look at the most common uses of semaphores. These are taken from "The Little Book of Semaphores" by Allen B. Downey. This book may be freely downloaded from the URL given in the bibliography at the end of this book.

Signaling

Synchronizing the execution of two threads is the simplest use of a semaphore:

```
        osSemaphoreId sem1;
        osSemaphoreDef(sem1);
        void thread1 (void)
{
    sem1 = osSemaphoreCreate(osSemaphore(sem1), 0);
    while(1)
    {
            FuncA();
            osSemaphoreRelease(sem1)
    }
}
void task2 (void)
{
    while(1)
    {
            osSemaphoreWait(sem1,osWaitForever)
            FuncB();
    }
}
```

In this case the semaphore is used to ensure that the code in FuncA() is executed before the code in FuncB().

Multiplex

A multiplex is used to limit the number of threads that can access a critical section of code. For example, this could be a routine that accesses memory resources and can only support a limited number of calls.

```
osSemaphoreId multiplex;
osSemaphoreDef(multiplex);
void thread1 (void)
{
multiplex = osSemaphoreCreate(osSemaphore(multiplex), FIVE_TOKENS);
while(1){
            osSemaphoreWait(multiplex,osWaitForever)
            ProcessBuffer();
            osSemaphoreRelease(multiplex);
}}
```

In this example we initialize the multiplex semaphore with five tokens. Before a thread can call the ProcessBuffer() function, it must acquire a semaphore token. Once the function has completed, the token is sent back to the semaphore. If more than five threads are attempting to call ProcessBuffer(), the sixth must wait until a thread has finished with ProcessBuffer() and returns its token. Thus the multiplex semaphore ensures that a maximum of five threads can call the ProcessBuffer() function "simultaneously."

Exercise 9.10 Multiplex

In this exercise we will look at using a semaphore to control access to a function by creating a multiplex.

In the Pack Installer select "Ex 9.10 Multiplex" and copy it to your tutorial directory.

The project creates a semaphore called semMultiplex which contains one token. Next, six instances of a thread containing a semaphore multiplex are created.

Build the code and start the debugger.

Open the Peripherals → General Purpose IO → GPIOB window.

Run the code and observe how the tasks set the port pins.

As the code runs only one thread at a time can access the LED functions so only one port pin is set.

Exit the debugger and increase the number of tokens allocated to the semaphore when it is created.

```
semMultiplex = osSemaphoreCreate(osSemaphore(semMultiplex), 3);
```

Build the code and start the debugger.

Run the code and observe the GPIOB pins.

Now three threads can access the LED functions "concurrently."

Rendezvous

A more generalized form of semaphore signaling is a rendezvous. A rendezvous ensures that two threads reach a certain point of execution. Neither may continue until both have reached the rendezvous point.

```
        osSemaphoreId arrived1,arrived2;
        osSemaphoreDef(arrived1);
        osSemaphoreDef(arrived2);
    void thread1 (void){
      Arrived1 = osSemaphoreCreate(osSemaphore(arrived1),ZERO_TOKENS);
      Arrived2 = osSemaphoreCreate(osSemaphore(arrived2),ZERO_TOKENS);
      while(1){
        FuncA1();
        osSemaphoreRelease(Arrived1);
        osSemaphoreWait(Arrived2,osWaitForever);
        FuncA2();
    }}
    void thread2 (void) {
    while(1){
        FuncB1();
        os_sem_send(Arrived2);
        os_sem_wait(Arrived1,osWaitForever);
        FuncB2();
    }}
```

In the above case the two semaphores will ensure that both threads will rendezvous and then proceed to execute FuncA2() and FuncB2().

Exercise 9.11 Rendezvous

In this project we will create two threads and make sure that they have reached a semaphore rendezvous before running the LED functions.

In the Pack Installer select "Ex 9.11 Rendezvous" and copy it to your tutorial directory.

Build the project and start the debugger.

Open the Peripherals\General Purpose IO\GPIOB window.

Run the code.

Initially the semaphore code in each of the LED threads is commented out. Since the threads are not synchronized the GPIO pins will toggle randomly.

Exit the debugger.

Uncomment the semaphore code in the LED tasks.

Built the project and start the debugger.

Run the code and observe the activity of the pins in the GPIOB window.

Now the threads are synchronized by the semaphore and run the LED functions "concurrently."

Barrier Turnstile

Although a rendezvous is very useful for synchronizing the execution of code, it only works for two threads. A barrier is a more generalized form of rendezvous which works to synchronize multiple threads.

```
      osSemaphoreId count,barrier;
      osSemaphoreDef(counter);
      osSemaphoreDef(barrier);
      unsigned int count;
  void thread1 (void)
  {
      count = osSemaphoreCreate(osSemaphore(count),ONE_TOKEN);
     barrier = osSemaphoreCreate(osSemaphore(barrier),ZERO_TOKENS);
     while(1){
              //Allow only one task at a time to run this code
              osSemaphoreWait(counter);
              count = count+1;
              if count == 5 os_sem_send(barrier, osWaitForever);
              osSemaphoreRelease(counter);
              //when all five tasks have arrived the barrier is opened
              os_sem_wait(barrier, osWaitForever);
              os_sem_send(barrier);
         critical_Function();
  }}
```

In this code we use a global variable to count the number of threads that have arrived at the barrier. As each function arrives at the barrier it will wait until it can acquire a token from the counter semaphore. Once acquired, the count variable will be incremented by one. Once we have incremented the count variable, a token is sent to the counter semaphore so that other waiting threads can proceed. Next, the barrier code reads the count variable. If this is equal to the number of threads which are waiting to arrive at the barrier, we send a token to the barrier semaphore.

In the example above we are synchronizing five threads. The first four threads will increment the count variable and then wait at the barrier semaphore. The fifth and last thread to arrive will increment the count variable and send a token to the barrier semaphore. This will allow it to immediately acquire a barrier semaphore token and continue execution. After passing through the barrier it immediately sends another token to the barrier semaphore. This allows one of the other waiting threads to resume execution. This thread places another token in the barrier semaphore which triggers another waiting thread and so on. This final section of the barrier code is called a turnstile because it allows one thread at a time to pass the barrier. In our model of concurrent execution this means that each thread waits at the barrier until the last arrives then all resume simultaneously. In the following exercise we create five instances of one thread containing barrier code. However the barrier could be used to synchronize five unique threads.

Exercise 9.12 Semaphore Barrier

In this exercise we will use semaphores to create a barrier to synchronize multiple tasks.

In the Pack Installer select "Ex 9.12 Barrier" and copy it to your tutorial directory.

Build the project and start the debugger.

Open the Peripherals\General Purpose IO\GPIOB window.k.

Run the code.

Initially, the semaphore code in each of the threads is commented out. Since the threads are not synchronized the GPIO pins will toggle randomly like in the rendezvous example.

Exit the debugger.

Remove the comments on lines 34, 45, 53, and 64 to enable the barrier code.

Build the project and start the debugger.

Run the code and observe the activity of the pins in the GPIOB window.

Now the threads are synchronized by the semaphore and run the LED functions "concurrently."

Semaphore Caveats

Semaphores are an extremely useful feature of any RTOS. However semaphores can be misused. You must always remember that the number of tokens in a semaphore is not fixed. During the run-time of a program semaphore tokens may be created and destroyed. Sometimes this is useful, but if your code depends on having a fixed number of tokens available to a semaphore you must be very careful to always return tokens back to it. You should also rule out the possibility of accidently creating additional new tokens.

Mutex

Mutex stands for "Mutual Exclusion." In reality, a mutex is a specialized version of semaphore. Like a semaphore, a mutex is a container for tokens. The difference is that a mutex can only contain one token which cannot be created or destroyed. The principle use of a mutex is to control access to a chip resource such as a peripheral. For this reason a mutex token is binary and bounded. Apart from this it really works in the same way as a semaphore. First of all we must declare the mutex container and initialize the mutex:

```
osMutexId uart_mutex;
osMutexDef (uart_mutex);
```

Once declared the mutex must be created in a thread.

```
uart_mutex = osMutexCreate(osMutex(uart_mutex));
```

Then any thread needing to access the peripheral must first acquire the mutex token:

```
osMutexWait(osMutexId mutex_id,uint32_t millisec;
```

Finally, when we are finished with the peripheral the mutex must be released:

```
osMutexRelease(osMutexId mutex_id);
```

Mutex use is much more rigid than semaphore use, but is a much safer mechanism when controlling absolute access to underlying chip registers.

Exercise 9.13 Mutex

In this exercise our program writes streams of characters to the microcontroller UART from different threads. We will declare and use a mutex to guarantee that each thread has exclusive access to the UART until it has finished writing its block of characters.

In the Pack Installer select "Ex 9.13 Mutex" and copy it to your tutorial directory.

This project declares two threads which both write blocks of characters to the UART. Initially, the mutex is commented out.

```
void uart_Thread1 (void const *argument) {
uint32_t i;
for (;;) {
  //osMutexWait(uart_mutex, osWaitForever);
  for( i = 0;i<10;i++)  SendChar('1');
  SendChar('\n');
```

```
    SendChar('\r');
    //osMutexRelease(uart_mutex);
}}
```

In each thread the code prints out the thread number. At the end of each block of characters it then prints the carriage return and new line characters.

Build the code and start the debugger.

Open the UART1 console window with View\Serial Windows\UART #1 (Fig. 9.39).

Figure 9.39
Open the UART console window.

Start the code running and observe the output in the console window (Fig. 9.40).

Figure 9.40
The miss ordered serial output.

Here we can see that the output data stream is corrupted by each thread writing to the UART without any access control.

Exit the debugger.

Uncomment the mutex calls in each thread.

Build the code and start the debugger.

Observe the output of each task in the console window (Fig. 9.41).

Figure 9.41
Order restored by using a mutex.

Now the mutex guarantees each task exclusive access to the UART while it writes each block of characters.

Mutex Caveats

Clearly you must take care to return the mutex token when you are finished with the chip resource, or you will have effectively prevented any other thread from accessing it. You must also be extremely careful about using the osThreadTerminate() call on functions which control a mutex token. Keil RTX is designed to be a small footprint RTOS so that it can run on even the very small Cortex-M microcontrollers. Consequently there is no thread deletion safety. This means that if you delete a thread which is controlling a mutex token, you will destroy the mutex token and prevent any further access to the guarded peripheral.

Data Exchange

So far all of the inter-thread communication methods have only been used to trigger execution of threads; they do not support the exchange of program data between threads. Clearly, in a real program we will need to move data between threads. This could be done by reading and writing to globally declared variables. In anything but a very simple program, trying to guarantee data coherence would be extremely difficult and prone to unforeseen errors. The exchange of data between threads needs a more formal asynchronous method of communication.

CMSIS-RTOS provides two methods of data transfer between threads. The first method is a message queue which creates a buffered data "pipe" between two threads. The message queue is designed to transfer integer values (Fig. 9.42).

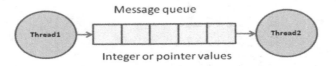

Figure 9.42
Message queue acts as a FIFO buffer between threads.

The second form of data transfer is a mail queue. This is very similar to a message queue except that it transfers blocks of data rather than a single integer (Fig. 9.43).

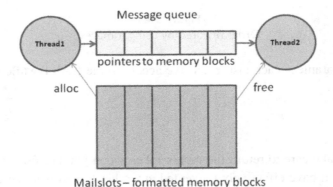

Figure 9.43
A mail queue can transfer blocks of structured data between threads.

Message and mail queues both provide a method for transferring data between threads. This allows you to view your design as a collection of objects (threads) interconnected by data flows. The data flow is implemented by message and mail queues. This provides both a buffered transfer of data and a well-defined communication interface between threads. Starting with a system level design based on threads connected by mail and message queues allows you to code different subsystems of your project, especially useful if you are working in a team. Also as each thread has well-defined inputs and outputs it is easy to isolate for testing and code reuse (Fig. 9.44).

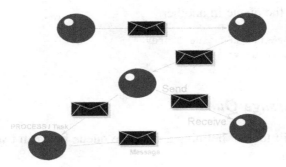

Figure 9.44
The system level view of an RTOS-based project consists of thread objects connected by data flows in the form of message and mail queues.

Message Queue

To setup a message queue we first need to allocate the memory resources.

```
osMessageQId Q_LED;
osMessageQDef (Q_LED,16_Message_Slots,unsigned int);
```

This defines a message queue with sixteen storage elements. In this particular queue each element is defined as an unsigned int. While we can post data directly into the message queue, it is also possible to post a pointer to a data object.

```
osEvent result;
```

We also need to define an osEvent variable which will be used to retrieve the queue data. The osEvent variable is a union that allows you to retrieve data from the message queue in a number of formats.

```
union{
  uint32_t v
  void *p
  int32_t signals
}value
```

The osEvent union allows you to read the data posted to the message queue as an unsigned int or a void pointer. Once the memory resources are created we can declare the message queue in a thread.

```
Q_LED = osMessageCreate(osMessageQ(Q_LED),NULL);
```

Once the message queue has been created we can put data into the queue from one thread.

```
osMessagePut(Q_LED,0x0,osWaitForever);
```

and then read it from the queue in another.

```
result = osMessageGet(Q_LED,osWaitForever);
LED_data = result.value.v;
```

Exercise 9.14 Message Queue

In this exercise we will look at defining a message queue between two threads and then use it to send process data.

In the Pack Installer select "Ex 9.14 Message Queue" and copy it to your tutorial directory.

Open Main.c and view the message queue initialization code.

```
osMessageQId Q_LED;
osMessageQDef (Q_LED,0x16,unsigned char);
osEvent result;
int main (void) {
  LED_Init ();
  Q_LED = osMessageCreate(osMessageQ(Q_LED),NULL);
```

We define and create the message queue in the main thread along with the event structure.

```
osMessagePut(Q_LED,0x1,osWaitForever);
osDelay(100);
```

Then in one of the threads we can post data and receive it in the second.

```
result = osMessageGet(Q_LED,osWaitForever);
LED_On(result.value.v);
```

Build the project and start the debugger.

Set a breakpoint in led_thread1 (Fig. 9.45).

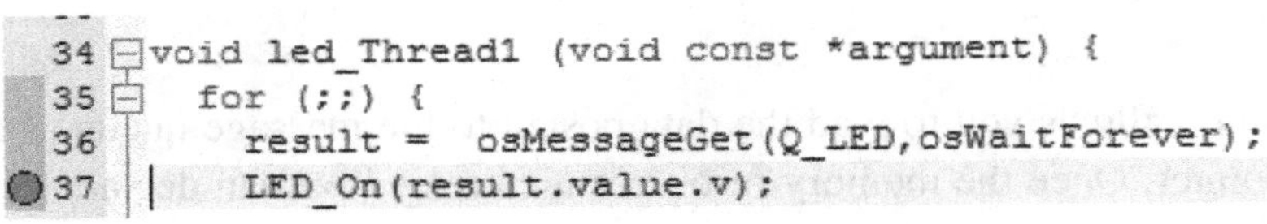

```
34 void led_Thread1 (void const *argument) {
35   for (;;) {
36     result =  osMessageGet(Q_LED,osWaitForever);
37     LED_On(result.value.v);
```

Figure 9.45
Set a breakpoint on the receiving thread.

Now run the code and observe the data as it arrives.

Memory Pool

While it is possible to post simple data values into the message queue it is also possible to post a pointer to a more complex object. CMSIS-RTOS supports the dynamic allocation of memory in the form of a memory pool. Here we can declare a structure which combines a number of data elements.

```
typedef struct {
  uint8_t LED0;
  uint8_t LED1;
  uint8_t LED2;
  uint8_t LED3;
} memory_block_t;
```

Then we can create a pool of these objects as blocks of memory.

```
osPoolDef(led_pool,ten_blocks,memory_block_t);
osPoolId( led_pool );
```

Then we can create the memory pool by declaring it in a thread.

```
led_pool = osPoolCreate(osPool(led_pool));
```

Now we can allocate a memory pool within a thread.

```
memory_block_t *led_data;
*led_data = (memory_block_t *) osPoolAlloc(led_pool);
```

and then populate it with data;

```
led_data->LED0 = 0;
led_data->LED1 = 1;
led_data->LED2 = 2;
led_data->LED3 = 3;
```

It is then possible to place the pointer to the memory block in a message queue.

```
osMessagePut(Q_LED,(uint32_t)led_data,osWaitForever);
```

so the data can be accessed by another task.

```
osEvent event; memory_block_t * received;
event = osMessageGet(Q_LED,osWatiForever);
*received = (memory_block *)event.value.p;
led_on(received->LED0);
```

Once the data in the memory block has been used the block must be released back to the memory pool for reuse.

```
osPoolFree(led_pool,received);
```

Exercise 9.15 Memory Pool

This exercise demonstrates the configuration of a memory pool and message queue to transfer complex data between threads.

In the Pack Installer select "Ex 9.15 Memory Pool" and copy it to your tutorial directory.

This exercise creates a memory pool and a message queue. A producer thread acquires a buffer from the memory pool and fills it with data. A pointer to the memory pool buffer is then placed in the message queue. A second thread reads the pointer from the message queue and then accesses the data stored in the memory pool buffer before freeing the buffer back to the memory pool. This allows large amounts of data to be moved from one thread to another in a safe synchronized way. This is called a "zero copy" memory queue as only the pointer is moved through the message queue, but the actual data does not move memory locations.

At the beginning of main.c the memory pool and message queue are defined.

```
typedef struct {
  uint8_t canData[8];
} message_t;
osPoolDef(mpool, 16, message_t);
  osPoolId mpool;
osMessageQDef(queue, 16, message_t);
osMessageQId queue;
```

In the producer thread acquires a message buffer, and fills it with the testData variable.;

```
message = (message_t*)osPoolAlloc(mpool);
for(index =0;index<8;index++){
  message->canData[index] = testData + index;}
  osMessagePut(queue, (uint32_t)message, osWaitForever);
```

Then in the consumer thread we can read the message queue using the event.value.p pointer object and then access data stored in the memory pool buffer. Once we have used the data in the buffer it must be released back to the memory pool.

```
for(index = 0;index<8;index++){
  message_t *message = (message_t*)evt.value.p;
  LED_On((uint32_t)message->canData[index]);}
  osPoolFree(mpool, message);
```

Build the code and start the debugger.

Place breakpoints on the osMessagePut and osmessageGet functions (Fig. 9.46).

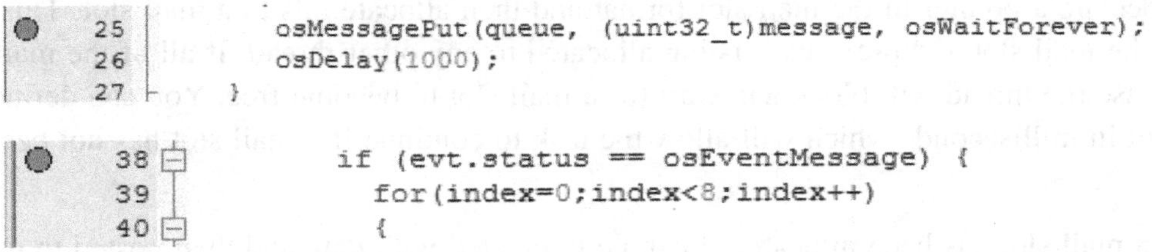

```
25        osMessagePut(queue, (uint32_t)message, osWaitForever);
26        osDelay(1000);
27   }

38   if (evt.status == osEventMessage) {
39       for(index=0;index<8;index++)
40       {
```

Figure 9.46
Set breakpoints on the sending and receiving threads.

Run the code and observe the data being transferred between the threads.

Mail Queue

While memory pools can be used as data buffers within a thread, CMSIS-RTOS also implements a mail queue which is a combination of memory pool and message queue. The Mail queue uses a memory pool to create formatted memory blocks and passes pointers to these blocks in a message queue. This allows the data to stay in an allocated memory block while we only move a pointer between the different threads. A simple mail queue API makes this easy to setup and use. First we need to declare a structure for the mail slot similar to the one we used for the memory pool.

```
typedef struct {
  uint8_t LED0;
  uint8_t LED1;
  uint8_t LED2;
  uint8_t LED3;
} mail_format;
```

This message structure is the format of the memory block that is allocated in the mail queue. Now we can create the mail queue and define the number of memory block "slots" in the mail queue.

```
osMailQDef(mail_box, sixteen_mail_slots, mail_format);
osMailQId mail_box;
```

Once the memory requirements have been allocated we can create the mail queue in a thread.

```
mail_box = osMailCreate(osMailQ(mail_box), NULL);
```

Once the mail queue has been instantiated we can post a message. This is different from the message queue in that we must first allocate a mail slot and populate it with data.

```
mail_format *LEDtx;
LEDtx = (mail_format*)osMailAlloc(mail_box, osWaitForever);
```

First declare a pointer in the mail slot format and then allocate this to a mail slot. This locks the mail slot and prevents it being allocated to any other thread. If all of the mail slots are in use the thread will block and wait for a mail slot to become free. You can define a timeout in milliseconds which will allow the task to continue if a mail slot has not become free.

Once a mail slot has been allocated it can be populated with data and then posted to the mail queue.

```
LEDtx->LED0 = led0[index];
LEDtx->LED1 = led1[index];
LEDtx->LED2 = led2[index];
LEDtx->LED3 = led3[index];
osMailPut(mail_box, LEDtx);
```

The receiving thread must declare a pointer in the mail slot format and an osEvent structure.

```
osEvent evt;
mail_format *LEDrx;
```

Then in the thread loop we can wait for a mail message to arrive.

```
evt = osMailGet(mail_box, osWaitForever);
```

We can then check the event structure to see if it is indeed a mail message and extract the data.

```
if(evt.status == osEventMail) {
  LEDrx = (mail_format*)evt.value.p;
```

Once the data in the mail message has been used the mail slot must be released so it can be reused.

```
osMailFree(mail_box, LEDrx);
```

Exercise 9.16 Mailbox

This exercise demonstrates configuration of a mailbox and using it to post messages between tasks.

In the Pack Installer select "Ex 9.16 Mailbox" and copy it to your tutorial directory.

The project creates a sixteen slot mailbox to send LED data between threads.

```
typedef struct {
  uint8_t LED0;
  uint8_t LED1;
```

```
  uint8_t LED2;
  uint8_t LED3;
} mail_format;
osMailQDef(mail_box, 16, mail_format);
osMailQId mail_box;
  int main (void) {
    LED_Init();
    mail_box = osMailCreate(osMailQ(mail_box), NULL);
```

A producer task allocates a mail slot, then fills it with data and posts it to the mail queue.

```
LEDtx = (mail_format*)osMailAlloc(mail_box, osWaitForever);
LEDtx->LED0 = led0[index];
LEDtx->LED1 = led1[index];
LEDtx->LED2 = led2[index];
LEDtx->LED3 = led3[index];
osMailPut(mail_box, LEDtx);
```

The receiving task waits for a mail message to arrive then reads the data. Once the data has been used the mail slot is released.

```
evt = osMailGet(mail_box, osWaitForever);
if(evt.status == osEventMail){
  LEDrx = (mail_format*)evt.value.p;
  LED_Out((LEDrx->LED0|LEDrx->LED1|LEDrx->LED2|LEDrx);
  osMailFree(mail_box, LEDrx);
}
```

Build the code and start the debugger.

Set a breakpoint in the consumer and producer threads and run the code (Fig. 9.47).

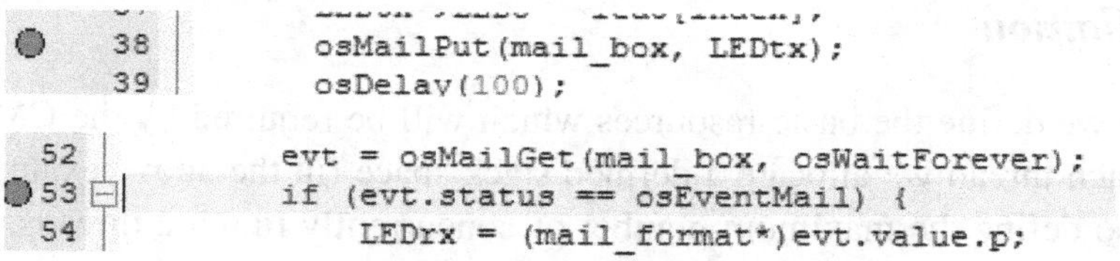

```
38        osMailPut(mail_box, LEDtx);
39        osDelay(100);

52        evt = osMailGet(mail_box, osWaitForever);
53        if (evt.status == osEventMail) {
54            LEDrx = (mail_format*)evt.value.p;
```

Figure 9.47
Set breakpoints on the sending and receiving threads.

Observe the mailbox messages arriving at the consumer thread.

Configuration

So far we have looked at the CMSIS-RTOS API. This includes thread management functions, time management, and inter-thread communication. Now that we have a clear idea of exactly what the RTOS kernel is capable of, we can take a more detailed look at the configuration file. There is one configuration file for all of the Cortex-M-based microcontrollers called RTX_Conf_CM.c (Fig. 9.48).

⊟ Thread Configuration	
Number of concurrent running user threads	6
Default Thread stack size [bytes]	200
Main Thread stack size [bytes]	200
Number of threads with user-provided stack ...	0
Total stack size [bytes] for threads with user-...	0
Stack overflow checking	☑
Stack usage watermark	☑
Processor mode for thread execution	Privileged mode
⊟ RTX Kernel Timer Tick Configuration	
Use Cortex-M SysTick timer as RTX Kernel Ti...	☑
RTOS Kernel Timer input clock frequency [Hz]	12000000
RTX Timer tick interval value [us]	1000
⊟ System Configuration	
⊞ Round-Robin Thread switching	☑
⊞ User Timers	☑
ISR FIFO Queue size	16 entries

Figure 9.48
RTX is configured using one central configuration file.

Like the other configuration files, the RTX_Conf_CM.c file is a template file which presents all the necessary configurations as a set of menu options.

Thread Definition

In this section we define the basic resources which will be required by the CMSIS-RTOS threads. For each thread we allocate a defined stack space (in the above example this is 200 bytes.) We also define the maximum number of concurrently running threads. Thus, the amount of RAM required for the above example can easily be computed as 200×6 or 1200 bytes. If some of our threads need a larger stack space, then a larger stack can be allocated when the task is created. In addition the total custom stack size must be allocated in the configuration file along with the total number of threads with custom stack size. Again, the RAM requirement is easily calculated.

Kernel Debug Support

During development, CMSIS-RTOS can trap stack overflows. When this option is enabled, an overflow of a thread stack space will cause the RTOS kernel to call the os_error function which is located in the RTX_Conf_CM.c file. This function is passed an error code and then sits in an infinite loop. The stack checking option is intended for use during debugging and should be disabled on the final application to minimize the kernel overhead. However, it is possible to modify the os_error() function if enhanced error protection is required in the final release.

```
#define OS_ERROR_STACK_OVF    1
#define OS_ERROR_FIFO_OVF     2
#define OS_ERROR_MBX_OVF      3
extern osThreadId svcThreadGetId (void);
void os_error (uint32_t error_code) {
  switch (error_code) {
      case OS_ERROR_STACK_OVF:
      /* Stack overflow detected for the currently running task. */
      /* Thread can be identified by calling svcThreadGetId(). */
      break;
      case OS_ERROR_FIFO_OVF:
          /* ISR FIFO Queue buffer overflow detected. */
      break;
      case OS_ERROR_MBX_OVF:
          /* Mailbox overflow detected. */
      break;
  }
  for (;;);
}
```

It is also possible to monitor the maximum stack memory usage during run-time. If you check the "Stack Usage Watermark" option, a pattern (0xCC) is written into each stack space. During run-time this watermark is used to calculate the maximum memory usage. This figure is reported in the threads section of the "System and Event Viewer" window (Fig. 9.49).

ID	Name	Priority	State	Delay	Event Value	Event Mask	Stack Usage
1	osTimerThread	High	Wait_MBX				cur: 32%, max: 32% [64/200]
3	Thread	Normal	Running				cur: 32%, max: 32% [64/200]
255	os_idle_demon	None	Ready				

Figure 9.49
Thread stack watermarking allows the debugger to calculate maximum memory usage for each thread.

The thread definition section also allows us to select whether the threads are running in privileged or unprivileged mode. This defaults to privileged, for most applications this is OK. If you are writing a safety critical or high-security application then selecting unprivileged will protect the Cortex-M processor registers from accidental or malicious access.

System Timer Configuration

The default timer for use with CMSIS-RTOS is the Cortex-M SysTick timer which is present on nearly all Cortex-M processors. The input to the SysTick timer will generally be the CPU clock. It is possible to use a different timer by unchecking the "Use SysTick" option. If you do this there are two function stubs in the RTX_Conf_CM.c file that allow you to initialize the alternative timer and acknowledge its interrupt.

```
int os_tick_init (void) {
  return (-1); /* Return IRQ number of timer (0..239) */
}
void os_tick_irqack (void) {
/* ... */
}
```

Whichever timer you use you must next setup its input clock value. Next we must define our timer tick rate. This is the rate at which timer interrupts are generated. On each timer tick the RTOS kernel will run the scheduler to determine if it is necessary to perform a context switch and replace the running thread. The timer tick value will depend on your application, but the default starting value is set to 1 millisecond.

Timeslice Configuration

The final configuration setting allows you to enable round robin scheduling and define the timeslice period. This is a multiple of the timer tick rate so in the above example each thread will run for five ticks or 5 milliseconds before it will pass execution to another thread of the same priority that is ready to run. If no thread of the same priority is ready to run, it will continue execution. The system configuration options also allow you to enable and configure the virtual timer thread. If you are going to use the virtual timers this option must be configured or the timers will not work. Then lastly if you are going to trigger a thread from an interrupt routine using event flags then it is possible to define a FIFO queue for triggered signals. This buffers signal triggers in the event of bursts of interrupt activity.

Scheduling Options

CMSIS-RTOS allows you to build an application with three different kernel scheduling options. These are round-robin scheduling, preemptive scheduling, and cooperative multitasking. A summary of these options are as follows:

Preemptive Scheduling

If the round-robin option is disabled in the RTX_Config_CM.c file, each thread must be declared with a different priority. When the RTOS is started and the threads are created, the thread with the highest priority will run (Fig. 9.50).

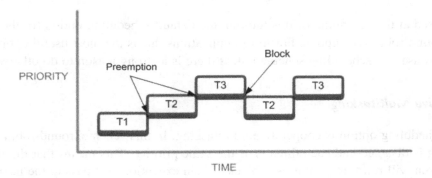

Figure 9.50
In a preemptive RTOS each thread has a different priority level and will run until it is preempted or has reached a blocking OS call.

This thread will run until it blocks, that is, it is forced to wait for an event flag, semaphore, or other object. When it blocks, the next ready thread with the highest priority will be scheduled and will run until it blocks, or a higher priority thread becomes ready to run. So with preemptive scheduling we build a hierarchy of thread execution, with each thread consuming variable amounts of run-time.

Round-Robin Scheduling

A round-robin-based scheduling scheme can be created by enabling the round-robin option in the RTX_Conf_CM.c file and declaring each thread with the same priority (Fig. 9.51).

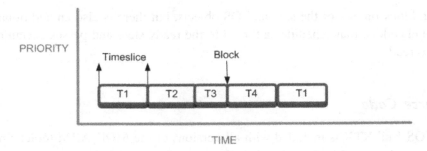

Figure 9.51
In a round robin RTOS threads will run for a fixed period or timeslice or until they reach a blocking OS call.

In this scheme, each thread will be allotted a fixed amount of run-time before execution is passed to the next ready thread. If a thread blocks before its timeslice has expired, execution will be passed to the next ready thread.

Round-Robin Preemptive Scheduling

As discussed at the beginning of this tutorial, the default scheduling option for the Keil RTX is round-robin preemptive. For most applications this is the most useful option and you should use this scheduling scheme unless there is a strong reason to do otherwise.

Cooperative Multitasking

A final scheduling option is cooperative multitasking. In this scheme, round-robin scheduling is disabled and each thread has the same priority. This means that the first thread to run will run forever unless it blocks. Then execution will pass to the next ready thread (Fig. 9.52).

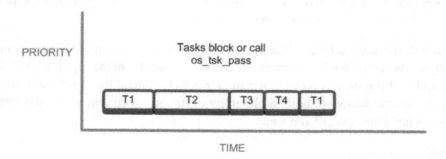

Figure 9.52
In a cooperative RTOS each thread will run until it reaches a blocking OS call or uses the osThreadYield() call.

Threads can block on any of the standard OS objects, but there is also an additional OS call, osThreadYeild(), that schedules a thread to the ready state and passes execution to the next ready thread.

RTX Source Code

CMSIS-RTOS Keil RTX is included with all versions of the MDK-ARM toolchain. The source code can be found in the following directory of the toolchain:

C:\Keil\ARM\Pack\ARM\CMSIS\ < version > \CMSIS\RTOS\RTX

If you want to perform source level debugging of the RTOS code create a text file containing the following command line where the path is the RTX source directory.

SET SRC = < path >

Now add this file to the initialization box in the debugger menu (Fig. 9.53).

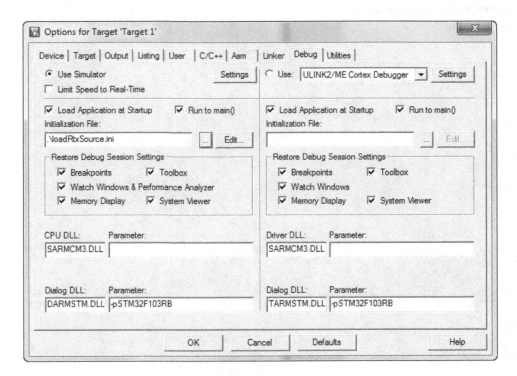

Figure 9.53
Add the script to the debugger. This will load the RTX symbols when the debugger starts.

Now when you start the debug session the RTX source will be loaded.

RTX License

CMSIS-RTOS Keil RTX is provided under a three clause BSD license and may be used freely without cost for commercial and noncommercial projects. RTX will also compile using the IAR and GCC tools. For more information use the URL below:

https://www.keil.com/demo/eval/rtx.htm

Conclusion

In this chapter we have worked our way through the CMSIS-RTOS API and introduced some of the key concepts associated with using an RTOS. The only real way to learn how to develop with an RTOS is to actually use one in a real project. In the next chapter we will look at some proven techniques that can be used when developing using an RTOS running on a Cortex-M microcontroller.

CHAPTER 10

RTOS Techniques

Introduction

In this chapter, we will look through some techniques that can be used when developing with a real-time operating system (RTOS). Each of these techniques has been used in a real project so they are tried and tested. First, we will look at how to design a system that integrates RTOS threads and peripheral interrupt routines. We will also look at how to add power management and watchdog support to a multithreaded RTOS project. Next, we will see how to design a system that maintains real-time processing of continuous data but is also able to respond to event-driven tasks such as a user interface. Finally, we will see how we can use the RTOS application programming interface (API) to extend the debug capabilities of your application by adding a range of diagnostic messages.

RTOS and Interrupts

In the last chapter, we saw that our application code can execute in RTOS threads. However, in a real system, we will also have a number of Interrupt Service Routines (ISRs) which will be triggered by events in the microcontroller peripherals. The RTOS does not affect the raw interrupt latency and you can service an interrupt in exactly the same way you would on a non-RTOS-based system. However, as we have seen the scheduler and any RTOS API calls are also generating Cortex-M processor exceptions (Fig. 10.1).

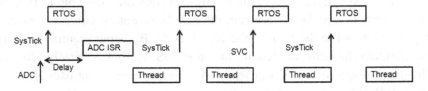

Figure 10.1

The RTOS will generate SysTick and SVC exceptions. These have to be integrated with the user peripheral interrupts to make a successful system.

The RTOS scheduler uses the SysTick timer exception and the RTOS API calls use the supervisor call (SVC) instruction. These exceptions also contend with the peripheral interrupts through the NVIC priority scheme. If the SysTick and SVC exceptions are

The Designer's Guide to the Cortex-M Processor Family.
DOI: http://dx.doi.org/10.1016/B978-0-08-100629-0.00010-4

configured with a high priority, they will preempt the user peripherals and cause a delay to the servicing of the peripheral ISR. If we increase the priority of the user peripheral interrupts then the peripheral ISR will be served but the RTOS exceptions will be delayed and this will destroy the real-time features of the RTOS (Fig. 10.2).

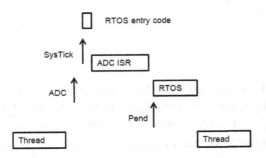

Figure 10.2

The SysTick exception runs the minimal amount of RTOS code to maintain its real-time features. It then sets the PEND exception and quits. This allows any peripheral interrupts to execute followed by the PEND exception. The PEND exception will execute the remainder of the RTOS code.

Fortunately, the Cortex-M processors have an additional exception called the PEND exception which is used to mitigate against this problem. The PEND interrupt is simply an interrupt channel that is connected to a software register. In the Cortex-M system control block (USERSETMPEND bit in the CFG_CTRL register) rather than a device peripheral. When the software register is written to a, PEND exception is raised and the NVIC will trigger execution on the PEND exception in the same fashion as any other interrupt source. The PEND exception is also configured to have the lowest interrupt priority available. So in an RTOS, the SysTick timer will generate periodic exceptions at the highest priority. In the SysTick exception, we run the absolute minimum amount of RTOS code to maintain its real-time features, then set the PEND exception and then quit the SysTick exception. Once we quit the SysTick exception, any pending peripheral IRS are now able to run. When all the peripheral interrupts have been serviced, the PEND exception will be serviced. In the PEND ISR, we can now run the bulk of the RTOS code. By using this mechanism, we can guarantee to maintain the real-time feature of the RTOS without adversely affecting the peripheral interrupt response. So it is possible to design a system that uses RTOS threads combined with ISR running at the native processor interrupt latency. In the next section, we will see how to more fully integrate peripheral interrupt handling with the RTOS threads.

RTOS Interrupt Handling

When working with an RTOS, it is desirable to keep ISR functions as short as possible. Ideally an ISR will just be a few lines of code taking a minimal amount of run time. One way to achieve this is to move the ISR code and place it in a dedicated high-priority thread, a servicing thread. The servicing thread will be designed to sit in a blocked state as soon as it starts allowing the application threads to run. When the peripheral interrupt occurs, we will enter the ISR as normal but rather than process the interrupt the ISR will use an RTOS object (signal, message queue semaphore etc.) to wake up the blocked servicing thread. Any of the RTOS interprocessor communication objects can be used to link an ISR to a thread which can, in turn, service the peripheral interrupt. The servicing thread code will then process the interrupt and then go back to the blocked state. This method keeps the ISR code to a minimum and allows the RTOS scheduler to decide which thread should be running. The downside is the increased context switch time required to schedule the servicing thread. However, for many applications this is not a real problem. If you do need the absolute minimum interrupt latency for key peripherals, it is still possible to service this interrupt with a dedicated ISR.

In Fig. 10.3, we will look at integrating an ISR and servicing thread execution using the RTOS signal functions. The first line of code in the servicing thread should make it wait for a signal flag. When an interrupt occurs, the ISR simply sets the signal flag and terminates. This schedules the servicing thread which in turn runs the necessary code to manage the peripheral. Once finished, the thread will again hit the osSignalwait() call forcing it to block and wait for the next peripheral interrupt.

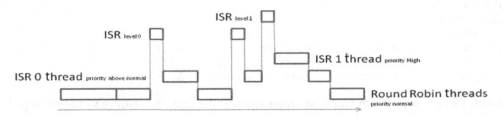

Figure 10.3
Within the RTOS, interrupt code is run as threads. The interrupt handlers signal the tasks when an interrupt occurs. The task priority level defines which task gets scheduled by the kernel.

A typical interrupt thread will have the following structure:

```
void servicing_Thread3 (void)
{
while(1)
{
osSignalWait ( isrSignal,waitForever);      // Wait for the ISR to trigger an event
        // Handle the interrupt
}        // loop round and go back sleep
}
```

The actual interrupt handler will contain a minimal amount of code.

```
void IRQ_Handler (void)
{
  osSignalSet (tsk3,isrSignal);      // Signal Thread 3 with an event
}
```

Exercise 10.1 RTOS Interrupt Exercise Handling

CMSIS-RTOS does not introduce any latency in serving interrupts generated by user peripherals. However, operation of the RTOS may be disturbed if you lock out the SysTick interrupt for a long period of time. This exercise demonstrates a technique of signaling, a thread from an interrupt and servicing the peripheral interrupt with a thread rather than a standard ISR.

Open the Pack Installer.

Select the Boards::Designers Guide Tutorial.

Select the Example tab and Copy "EX 10.1 RTOS Interrupt Handling."

In the main function, we initialize the ADC and create an ADC thread which has a higher priority than all the other threads.

```
osThreadDef(adc_Thread, osPriorityAboveNormal, 1, 0);
int main (void) {
LED_Init ();
init_ADC ();
T_led_ID1 =  osThreadCreate(osThread(led_Thread1), NULL);
T_led_ID2 =  osThreadCreate(osThread(led_Thread2), NULL);
T_adc_ID =   osThreadCreate(osThread(adc_Thread), NULL);
```

However, there is a problem when we enter main(): the code is running in unprivileged mode, so we cannot access the NVIC registers without causing a fault exception. There are several ways round this; the simplest is to give the threads privileged access by changing the setting in the RTX_Conf_CM.c (Fig. 10.4).

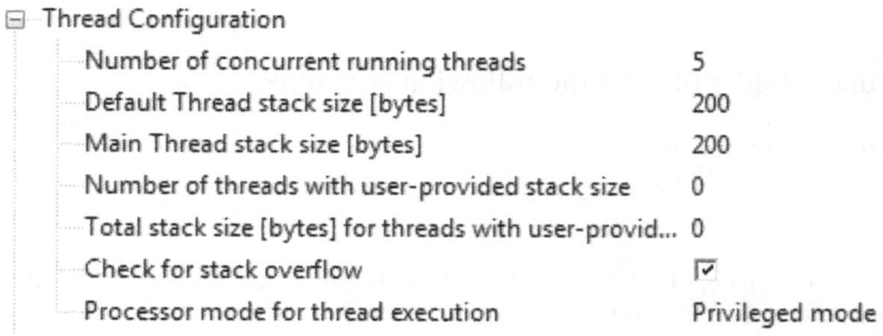

Thread Configuration	
Number of concurrent running threads	5
Default Thread stack size [bytes]	200
Main Thread stack size [bytes]	200
Number of threads with user-provided stack size	0
Total stack size [bytes] for threads with user-provid...	0
Check for stack overflow	☑
Processor mode for thread execution	Privileged mode

Figure 10.4
Configure the RTOS threads to have privileged access to the Cortex-M processor registers.

Here, we have switched the "Processor Mode for Thread Execution" to privileged which gives the threads full access to the Cortex-M processor. As we have added a thread, we also need to increase the number of concurrent running threads.

Build the code and start the debugger.

Set breakpoints in led_Thread2, ADC_Thread, and ADC1_2_IRQHandler (Fig. 10.5).

```
    57          osDelay(500);
 ⊙  58          ADC1->CR2       |=   (1UL << 22);
    59          LED_Off(2);
```

Figure 10.5
Breakpoint on "start of ADC conversion".

And in adc_Thread() (Fig. 10.6)

```
    35   osSignalWait   ( 0x01,osWaitForever);
 ⊙  36   GPIOB->ODR = ADC1->DR;
```

Figure 10.6
Breakpoint in the ADC Thread.

And in ADC1_2_IRQ Handler (Fig. 10.7)

```
    28 ⊟void ADC1_2_IRQHandler (void){
 ⊙  29 │ osSignalSet ( T_adc_ID,0x01);
```

Figure 10.7
Breakpoint in the ADC interrupt.

Run the code.

You will hit the first breakpoint which starts the ADC conversion, then run the code again and you will enter the ADC interrupt handler. The handler sets the adc_thread signal and quits. Setting the signal will cause the ADC thread to preempt any other running task, run the ADC service code, and then block waiting for the next signal.

User Supervisor Functions

In the last example, we were able to configure threads and the interrupt structure without any additional consideration because the RTOS was configured to execute thread code as privileged code. In some projects, it may be necessary to configure the RTOS so that threads are running in unprivileged mode. This means that thread code will no longer be able to write to the NVIC since the RTOS is running when we enter main(). We are stuck in Thread mode and are no longer be able to enable any interrupt source. In order for the thread code to access the NVIC, we need to be able to change our execution mode to run in Handler mode. In Handler mode we will have full privileged access to all the Cortex-M processor registers. As we saw in Chapter 5 "Advanced Architectural Features" this can be done by executing a SVC instruction.

Exercise 10.2 RTOS and User SVC Exceptions

In this exercise, we will look at using the system call exception to enter privileged mode to run "system level" code.

Open the Pack Installer.

Select the Boards::Designers Guide Tutorial.

Select the Example tab and copy "Ex 10.2 RTOS USER SVC" to your tutorial directory.

Examine the RTX_Conf_CM.c file (Fig. 10.8).

Thread Configuration	
Number of concurrent running threads	6
Default Thread stack size [bytes]	200
Main Thread stack size [bytes]	200
Number of threads with user-provided stack size	0
Total stack size [bytes] for threads with user-provided stack size	0
Check for stack overflow	☑
Processor mode for thread execution	Unprivileged mode

Figure 10.8
Thread execution is set to unprivileged mode.

In the Thread Configuration section, the operating mode for thread execution is set to "Unprivileged mode."

In the project, we have added a new file called SVC_Table.s (Fig. 10.9). This file is available as a "User Code Template" (CMSIS-RTOS User SVC) from the "Add New Item" dialog:

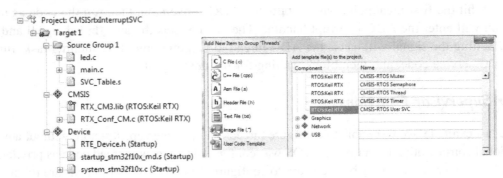

Figure 10.9
Add the user SVC template file.

Open the SVC_Tables.c file.

This is the lookup table for the SVC interrupts.

```
; Import user SVC functions here.
        IMPORT __SVC_1
        EXPORT SVC_Table
 SVC_Table
; Insert user SVC functions here. SVC 0 used by RTX Kernel.
        DCD    __SVC_1           ; user SVC function
```

In this file, we need to add the import name and table entry for each __SVC function we are going to use. **Add the __SVC_1 dentitions as shown above.**

Convert the init_ADC() function to a service call exception as shown below.

```
void __svc(1) init_ADC (void);
void __SVC_1 (void){
```

Build the project and start the debugger.

Step the code (F11) to the call to the init_ADC function and examine the operating mode in the Register window.

Here we are in Thread mode, unprivileged, and using the process stack (Fig. 10.10).

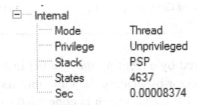

Figure 10.10
The processor is in unprivileged state using the process stack.

Now step into the function (F11) and step through the assembler until you reach the init_ADC() function.

Now, we are running in Handler mode with privileged access and are using the main stack pointer (Fig. 10.11).

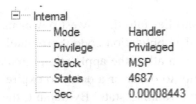

Figure 10.11
The processor has switched to privileged state using the main stack.

This allows us to set up the ADC and also access the NVIC.

Power Management

One of the advantages of developing with an RTOS is that we can consider each of the application threads to be running in parallel. This allows us to develop independent threads of code which each can perform a dedicated task and communicate together to create the desired application. This level of abstraction provides lots of benefits as projects get more complex. However, adding power management code to this multithreaded environment can seem daunting at first because have to ensure that each thread is ready to enter a low power state.

A typical Cortex-M-based microcontroller will have several low-power modes. While these will vary between different silicon manufacturers, Table 10.1 summarizes the types of low-power mode you are likely to encounter.

Table 10.1: Typical microcontroller low-power modes

Low-Power State	Power State	Wake-Up Source
CPU sleep	CPU in low-power state	Peripheral interrupt, SysTick timer
CPU and peripheral	CPU and peripherals in low-power state	Reset, wake-up pin or external interrupt pin
Deep sleep	Full power down of CPU, peripherals, SRAM, and Flash	Reset or wake-up pin

The low-power states are entered by executing the __wfi() instruction. By default, this will only affect the Cortex-M processor by placing the CPU into its sleep mode. The Cortex-M processor also has a SLEEPDEEP signal which is connected to the microcontroller power management hardware. If this is enabled in the Cortex-M "Processor System Control" register, then the SLEEPDEEP signal will be asserted when the __wfi() instruction is executed. To enter a given low-power state, the application code must configure the microcontroller power management registers to select the desired low-power state. Generally, the deeper the low-power mode the more restrictive the exit conditions.

Power Management First Steps

We need to place the microcontroller into the lowest power mode possible without disturbing the performance of the application code. Fortunately, the first step in power management is very simple. When all of the application threads are in a blocked state, for example, waiting for an interrupt to occur or a delay to expire the RTOS scheduler will place the os_Idle_demon() into a running state. By default, the os_Idle_demon() simply contains an empty for(;;) loop which is executed until an application thread becomes ready to run. In other words, we simply sit in the os_Idle_demon() burning energy.

```
void os_idle_demon (void) {
  for (;;) {
   /* HERE: include optional user code to be executed when no thread runs.*/
  }}
```

So, when we are in the os_idle_demon() there is nothing to do, we can simply enter the CPU sleep mode which will shut off the clock to the Cortex-M processor but leave the rest of the microcontroller active. Typically, the microcontroller low-power operation will default to this mode, so all we need to do is execute the __wfi() instruction when the code is in the os_idle_demon().

```
void os_idle_demon (void) {
   for (;;) {
__wfi();
   }}
```

Now as soon as the Cortex-M processor is not executing any meaningful code, the CPU clock will be halted until a peripheral generates an interrupt or the next SysTick interrupt occurs to run the RTOS scheduler.

This one line of code will stop the CPU wasting energy without affecting performance of the application code and depending on your application can have a significant impact on overall power consumption. This is an effective form of run-time power management.

Making use of the more advanced power modes does require a bit more thought. When we enter a deeper low-power mode, the wake-up methods are more restricted. This means that we have to be sure that the application code is ready to enter the low-power state. Consider the code shown below:

```
Application_thread(void)
{
while(1)
{
os_message wait();
function_1();
osDelay(10);
function_2();
}}
```

The thread starts by entering a blocked state waiting for a message to arrive. When the message data arrives, the thread will wake up and call function_1() and then it will again enter a blocked state when the osDelay() function is called. Once the delay has expired, the thread will resume and call function_2() before once again entering a blocked state until the next message arrives. This thread has two points at which it will enter a blocked state and if all the other threads are also blocked will cause the os_idle_demon to run. For the RTOS scheduler, there is no real difference between the two OS calls but for power management there is an important difference. When the thread is blocked by the os_message_wait(); call it is truly idle and cannot proceed until another part of the system gives it some data to process. When the thread is blocked by the osDelay() function for power management purposes, it is still awake in the sense that it will resume processing without input from any other part of the system. So, we need to construct our application threads so that they wait on an "event" triggered by some other part of the system, perform their function, and then wait for the next event to occur. This means that each thread now has a key blocking point (Fig. 10.12).

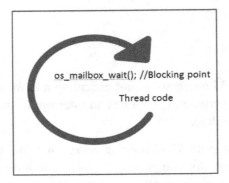

Figure 10.12
In each thread, you must identify a blocking point. At the blocking point, a thread has finished its current task and is waiting for the next event to resume processing.

If we construct all our system threads like this, we can identify when all the threads are waiting on their key blocking points. Then the application is truly idle and we can place it into a deeper low-power state.

Power Management Strategy

To build our power management strategy, we first need to declare a global variable which will hold two flags for each thread running on the system (Fig. 10.13).

uint32_t powerFlags

Task pending flags	Task running flags

Figure 10.13
Power management flags are held in a global variable and monitored in the idle thread.

Each thread has one "thread running" bit and one "thread pending" bit. Next, we need to write some simple functions that will control these bits. When the thread wakes up to start execution of its code, we need to set the thread-running bit. As multiple threads will be accessing the powerFlags variable, it should be protected by a mutex.

```
void taskIsRunning(uint16_t task) {
    osMutexWait(PowerMutex,osWaitForever);
    powertFlags |= task;
    osMutexRelease(PowerMutex);
}
```

When it has completed its task and is about to go back into the waiting state, we need to clear the thread running bit.

```
void taskIsSuspended(uint16_t task) {
    osMutexWait(PowerMutex,osWaitForever);
    powerFlags &= ~task;
    osMutexRelease(PowerMutex);
}
```

So now the application thread looks as follows:

```
#define APPLICATION_THREAD 1
Application_thread(void)
{
while(1)
{
    taskIsSuspended(APPLICATION_THREAD)
    osMessageGet(mailBox,osWaitForever);
    taskIsRunning(APPLICATION_THREAD);
    function_1();
    osDelay(10);
    function_2();
}}
```

When another part of the system is about to trigger the thread, in this case by sending a message we can set the threads pending bit. This ensures that we avoid any problems with the scheduler.

```
void taskIsPending(uint16_t task) {
    osMutexWait(PowerMutex,osWaitForever);
    powerFlags |= (task<<NUM_THREADS);   //Set Task Pending bit
    osMutexRelease(PowerMutex);
}
```

Now sending a message to our application thread looks like this:

```
taskIsPending(APPLICATION_THREAD);
osMessagePut(mailBox,appData,osWaitForever);
```

We must modify the taskIsRunning routine to clear this bit when execution of the application thread resumes.

```
void taskIsRunning(uint16_t task) {
    osMutexWait(PowerMutex,osWaitForever);
    powertFlags |= task;
    powerFlags &= ~ (task<<NUM_THREADS);
    osMutexRelease(PowerMutex);
}
```

When the application enters the os_Idle _demon () thread, we can test the powerFlags. If any of the task bits (active or pending) are set, then the system is not truly idle. If all the power flags are zero, then the system is truly idle and we can safely enter low-power mode. The following code can be placed in the os_idle_demon():

```
if (powerFlags == 0) {
    configureAndEnterDeepPowerDown();   //All tasks are idle so enter deep power down mode
    }else {
    configureAndEnterSleepMode();        //A Task is active but we can enter sleep mode
                                         for run time power saving

    }
```

So now the os_idle_Demon() will use the available power modes to minimize the run-time energy usage and also detect when the microcontroller can be placed into a deep low-power mode.

Watchdog Management

We can also use the power management flags to solve another problem. In this case, we want to enable a hardware watchdog on the microcontroller. It would be possible to refill the watchdog timer inside each thread. However, as we are continually switching between threads it is likely that the watchdog counter would be refilled by several different threads and we would be unable to detect if a given thread had failed. To make better use of the watchdog, we need to add a separate system thread that is used to monitor the thread pending bits. The system monitor runs periodically as a user timer. Each time it runs, it will check the thread pending bits. If a pending bit is set then a matching counter variable is incremented. If a threshold is exceeded then the system is in error, the watchdog will not be refilled and a hardware reset will be forced.

```
void taskMonitor (void){
uint8_t stalledTaskError = 0;
  for(loop = PEND_START_BIT,index = 0;loop<PEND_END_BIT;loop<<=1,index++)  {
    if(powerFlags &loop)){
      watchdogCounters[index]++;
      if(WatchdogCounters[index]>MAX_TASK_STALL_COUNT){
        taskIsStalled = TRUE;
```

```
}}}
if(stalledTaskError == 0){
patWatchdog();
}}
```

To complete the task monitor code, we need to reset the thread watchdog counter each time the application thread runs. This can be done in by adding another line to the taskIsRunning function.

```
void taskIsRunning(uint16_t task) {
  osMutexWait(PowerMutex,osWaitForever);
  powertFlags |= task;
  powerFlags &= ~ (task<<NUM_THREADS);
  taskNumber = 32- __clz((uint32_t)taskQuery);   //intrinsic instruction clz = count
leading zeros
  taskWatchdogCounters[taskNumber-1] = 0;
  osMutexRelease(PowerMutex);
}
```

Here, we are using the count leading zeros intrinsic to convert from a bit position to an integer value. This is very efficient and takes the minimum number of CPU cycles.

Integrating ISRs

In the case of an interrupt that will wake the processor and signal a thread to resume processing, we can use the power flags as a means of providing a timeout that allows the microcontroller to go back to sleep. For example, if the microcontroller will be in a low-power mode waiting for data to be sent to a UART. Then some random noise may wake up the processor which will then wait for a full message packet to be sent. In this situation, we can introduce a power flag for the UART ISR which is set in the ISR and cleared only when a full message has been received. In the watchdog routine, we can increment a counter if the threshold value is reached then instead of resetting the microcontroller we simply need to clear the ISR running flag to allow the processor to fall back to sleep. While the ISR power functions are essentially performing the same task as the thread power functions, we need to provide dedicated functions which do not access the mutex as the mutex cannot be acquired by an ISR routine of the same or lower priority.

```
void IsrTaskIsPending(uint16_t task) {
  powerFlags |= (task<<8);
}
void IsrIsInactive(uint16_t task) {
  powerFlags &= ~ (task);
  resetCurrentTaskWatchdogCounter(task);
}
```

Then in the watchdog task, we can test the ISR flag and increment the UART counter if it is set.

```
if( taskManagementFlags & WIRED_COMMS )
{
taskWatchdogCounters[WIRED_COMMS_COUNTER]++;
}
```

If we exceed the threshold level, we can use the mutex protected taskIsSuspended() routine to clear the ISR flag and also reset the watchdog counter.

```
if(taskWatchdogCounters[UART_ISR]>UART_TIMEOUT)
{
  taskIsSuspended(UART_ISR);
  taskWatchdogCounters[UART_ISR] = 0;
}
```

Now, if all the other threads are in a blocked state all the power management flags will be clear and the microcontroller will enter its low-power state.

Exercise 10.3 Power and Watchdog Management

We can see these examples in practice by running a simple program in the MDK-ARM simulator. This is a simple RTX program that uses several tasks to flash a group of LEDs in sequence. The start of each sequence is triggered by a hardware timer interrupt. In this example, we will start with the working program and introduce the power management and watchdog code.

Open the Pack Installer.

Select the Boards::Designers Guide Tutorial.

Select the Example tab and Copy "EX 10.3 RTOS Power Management."

Start the simulator by selecting debug\start::stop debug session.

Open the peripherals\General Purpose IO\GPOIOB debug window.

Start the code running with debug\run.

After an initial few seconds, you should see the upper byte of GPIO B pins being toggled by the application tasks.

Open the View\Analysis window\Performance analyzer.

Here, we can see that in this simple program most of the run time is being spent in the RTX_Idle_demon() (Fig. 10.14).

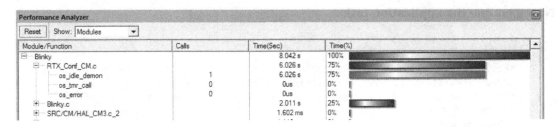

Figure 10.14
The performance analyzer show that this simple program is spending most of its time in the idle thread wasting energy.

Exit the debugger by selecting debug\start stop.

Open the RTX_Conf_CM.c file and locate the os_Idle_demon on line 141.

Uncomment the __wfi(); call and rebuild the code with project\build.

Restart the simulator with debug\start stoop and start the code running.

If we now reexamine the project in the performance analyzer, we can see a dramatic reduction in the number of cycles consumed by the os_Idle task (Fig. 10.15).

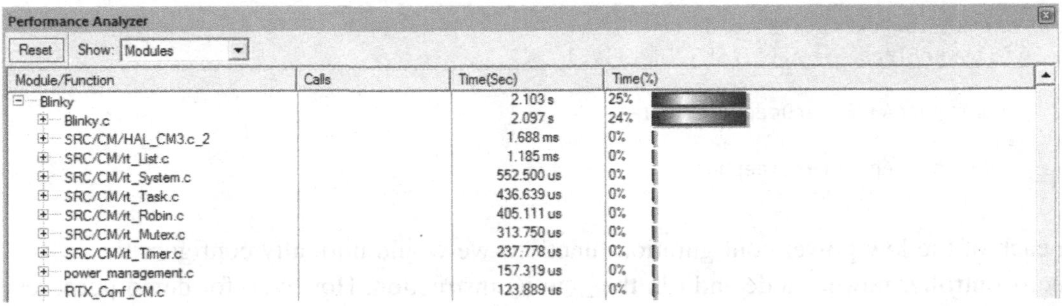

Figure 10.15
Now the processor enters sleep in the idle thread which minimizes runtime energy use.

The __wfi() instruction is placing the processor into sleep mode which is saving power in the Cortex-M processor. However, the peripherals of the microcontroller are still consuming energy. To take advantage of the deeper low-power modes provided by a chip manufacturer, we need to detect when the application is idle and then place the device into a deeper low-power mode until it is ready to restart processing.

Now examine the application tasks.

```
__task void phaseA (void) {
  for (;;) {
  taskIsSuspended(p_phaseA);              //Task has finished prepare for sleep
    osSignalWait(0x0010,osWaitForever);   //Task blocking point
  taskIsRunning(p_phaseA);                //Task has woken up set running bit
    LED_On (LED_A);
    signal_func ();
  taskIsPending(p_phaseB);                //set the pending bit on Task B
  osSignalSet(0x0001,t_phaseB);
  osDelay(5);
  LED_Off(LED_A);
 }}
```

In each thread, we identify the key blocking point and surround it with the taskIsSuspended() and taskIsRunning() functions. If the task communicates with another thread, we must first call the taskIsPending() function.

Now go back to the os_Idle_Task() and comment out the __wfi() instruction and uncomment the enterLowPowerState() function.

The enterLowPowerState() function checks the power flags and decides which power mode to enter.

```
void enterLowPowerState(void) {
  if (powerFlags == ALL_TASKS_INACTIVE)
  {
    configureAndEnterDeepSleepMode();
  }else{
    configureAndEnterSleepMode();
 }}
```

In each of the low-power configuration functions we would normally configure the microcontroller power mode and run the __wfi() instruction. However, for demo purposes we will sit in the function and simply toggle a debug variable.

```
void configureAndEnterDeepSleepMode(void)
{
  uint32_t delay;
//  SCB->SCR |= (1<<SCB_SCR_SLEEPDEEP_Pos);   //enable the deep sleep mode
//  __wfi();                                  //Place the micro in deep sleep
  for(delay = 0;delay<0x100;delay++);         //Simple delay
  debugDeepSleep ^= 0x01;                     //Toggle the debug variable
}
```

Rebuild the code and restart the debugger.

Open the Logic analyzer window with view\analysis window\Logic analyzer.

Start the code running.

After the code has been running for a few seconds press the Min/Max Auto button and press the zoom out button until you get a view similar to the Fig. 10.16.

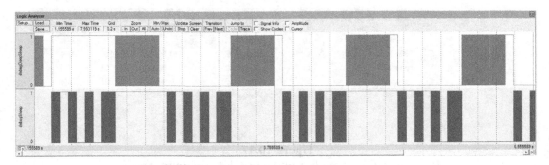

Figure 10.16
The logic analyzer can be used to show entry into sleep modes.

The upper trace is showing the deep-sleep debug variable. The solid blocks indicate where we are in the low-power mode. The lower trace is showing the sleep debug variable again the solid blocks are showing where we are in low-power mode. Here, we can see how the code is saving energy in both its standby mode and also during the run-time execution of the application code.

In blinky.c uncomment the os_tsk_create (Monitor, 10); call on line 171.

This will enable the watchdog monitor task which will ensure each of the application tasks is running correctly.

To create an error, uncomment the code in blinky.c 131 to 136 this will let the application run a few times and then delete task A.

Build the code and start the simulator.

Open watchdog.c and add the taskWatchdogCounters array to the watch window.

Add a breakpoint on the taskIsStalled = TRUE; line 86.

Open the debug\os support\RTX Task and system window.

Start the code running.

After the code has run, the LED flashing sequence five times will delete task phase_A() from the system (Fig. 10.17).

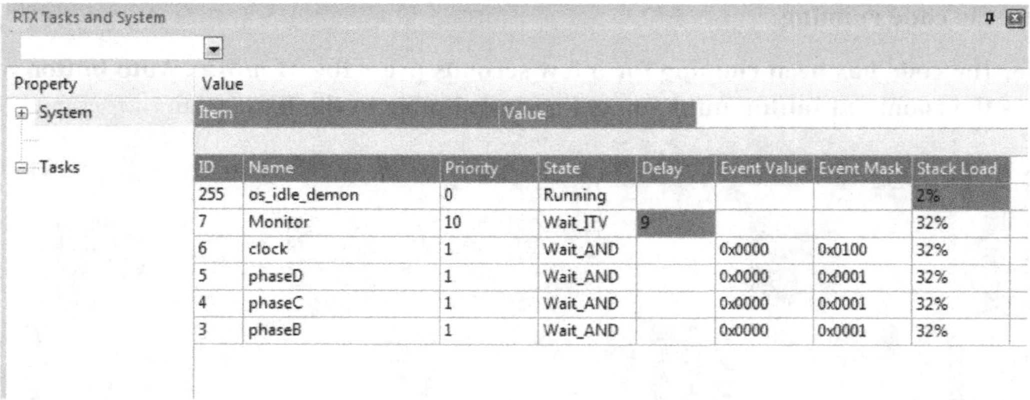

Figure 10.17
The application threads and the watchdog monitor thread.

Now the task monitor will detect that the task A pending bit is set and increment the taskWatchdogCounters associated with task phaseA.

Observe the taskWatchdogCounters in the watch window (Fig. 10.18).

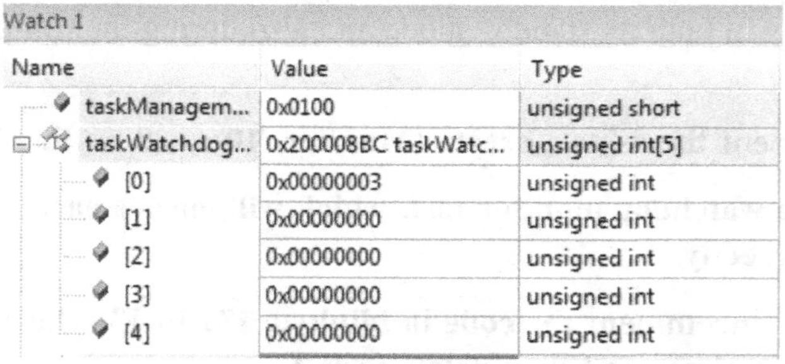

Figure 10.18
The watchdog counters for each thread.

Counter 0 will start to increment. When it hits the MAX_TASK_STALL_COUNT (five), we will hit the breakpoint.

Start the code running again, the taskIsStalled variable will be set. This will stop the monitor task from patting the watchdog and the microcontroller will be reset causing the application to restart.

This approach to power and watchdog management in and RTOS should be considered when designing the structure of your low power application. It can then be adapted to meet your specific needs. All of the code is contained in two files making it easy to reuse across multiple projects.

Startup Barrier

If you are starting work with a new microcontroller or about to start adding low-power code, it is good practice to add a "startup barrier" to your code. In the early stages of development when you are configuring the system peripherals, it is possible to accidently put the microcontroller into a disturbed state where it becomes difficult or impossible to connect via the CoreSight debug. Also, it is possible to add code which places the chip into a power down state which again makes it impossible to connect the debugger. If this happens, you can no longer erase the FLASH and program the memory. If you only have a limited supply of development boards, this can be an embarrassing problem! A "startup barrier" is simply a small function that will only allow the code to continue if a specific pattern is in a specific RAM memory location.

```
volatile uint32_t debuggerAttached __attribute__((at(0x2000FFFC)));
void startupBarrier (void)
{
  if(debuggerAttached != 0x55555555){
    __BKPT(0);
  }
}
```

This function is called on the very first line of system_init() before any code that may upset the Cortex-M processor has been called. If we build a program with such a "startup barrier" and download it into the FLASH memory, it will start execution and then become trapped in the barrier function. This will guarantee that we can erase and reprogram the Flash memory. To get past the "startup barrier," we can add a script file to the debugger which programs the barrier pattern into the RAM (Fig. 10.19).

```
_WWORD (0x2000FFFC, 0x55555555);
```

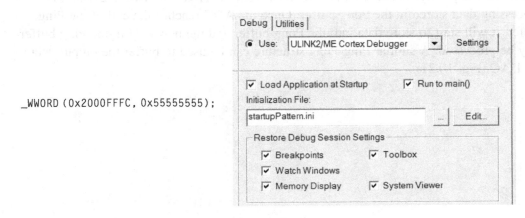

Figure 10.19
Add the startup barrier script as the debugger initialization file.

Now, when we start the debugger, the barrier pattern will be present in the RAM memory location and the processor will begin execution of our application code. If the code does something to latch up the processor, we can repower the board and this time it will stop at the barrier allowing the CoreSight debugger to get control of the Cortex-M processor so we can erase the flash memory, modify the source code, and try again.

Designing for Real Time

So far we have looked at the DSP features of the Cortex-M4 and the supporting DSP library. In the next section, we will look at developing a real-time DSP program that can also support non real-time event driven code like a user interface or communication stack without disturbing the performance of the DSP algorithm.

Buffering Techniques—The Double or Circular Buffer

When developing your first real-time DSP application, the first decision you have to make is how to buffer the incoming blocks of data from the ADC and the outgoing blocks of data to the DAC. A typical solution is to use a form of double buffer as shown in Fig. 10.20.

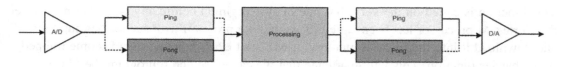

Figure 10.20
A simple DSP system can use a double buffer to capture the stream of data.

While the Ping buffer is being filled with data from the ADC, the DSP algorithm is processing data stored in the Pong buffer. Once the ADC reaches the end of the Ping buffer, it will start to store data into the Pong buffer and the new data in the Ping buffer may be processed. A similar Ping-Pong structure can be used to buffer the output data to the DAC (Fig. 10.21).

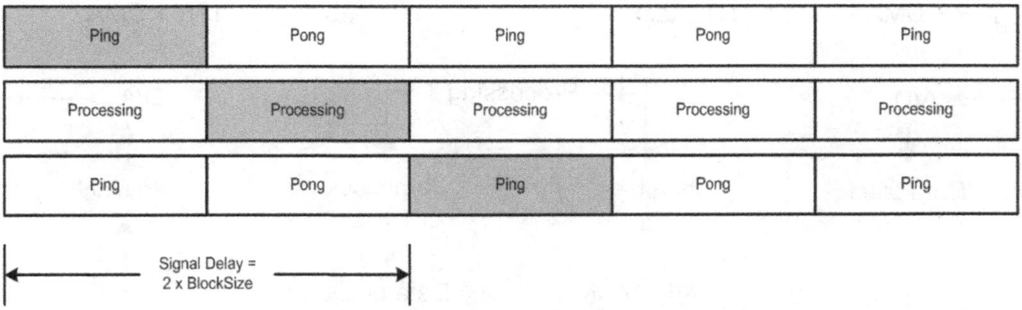

| Ping | Pong | Ping | Pong | Ping |

| Processing | Processing | Processing | Processing | Processing |

| Ping | Pong | Ping | Pong | Ping |

Signal Delay =
2 x BlockSize

Figure 10.21
The double buffer causes minimum signal latency but requires the DSP algorithm to run frequently. This makes it hard to maintain real-time performance when other software routines need to be executed.

This method works well for simple applications; it requires the minimum amount of memory and introduces the minimal latency. The application code must do all of the necessary buffer management. However, as the complexity of the application rises, this approach begins to run into problems. The main issue is that you have a critical deadline to meet, in that the DSP algorithm must have finished processing the Pong buffer before the ADC fills the Ping buffer. If, for example, the CPU has to run some other critical code such as user interface or communications stack then there may not be enough processing power available to run all the necessary code and the DSP algorithm in time to meet the end of buffer deadline. If the DSP algorithm misses its deadline, then the ADC will start to overwrite the active buffer. You then have to write more sophisticated buffer management code to "catch up" but really you are onto a losing game and data will at some point be lost.

Buffering Techniques FIFO Message Queue

An alternative to the basic double buffer approach is to use an RTOS to provide more sophisticated buffering structure using the mailbox system to create a FIFO queue (Fig. 10.22).

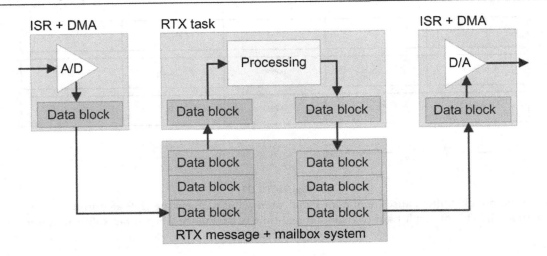

Figure 10.22
Block processing using RTOS mail queues increases the signal latency but provides reliable integration of a DSP thread into a complex application.

Using the RTX RTOS, we can create a task which is used to run the DSP algorithm. The inputs and outputs of this task are a pair of mailboxes. The ADC interrupt is used to fill a mailbox message with a block of ADC data and then post that message to the DSP task. These message blocks are then placed in a FIFO queue ready to be processed. Now if other critical code needs to run, the DSP task will be preempted and rather than risk losing data as in the double buffer case the blocks of data will be sitting in the message queue. Once the critical code has run, the DSP task will resume running and can use all the CPU processing power to "catch up" with the backlog of data. A similar message queue is used to send blocks of data to the DAC. The use of the RTOS message queues provides a simple and elegant buffering system without the need to write any low level code.

In the first example program, the ADC is sampling at 40 kHz and the sampled data is placed in a 32 byte message buffer.

The message queue is defined by allocating a block of memory as the message space. This region is formatted into a group of mailslots. Each mailslot is a structure in the shape of the data you want to send.

The shape of the mailslot can be any collection of data declared as a structure. In our case, we want an array of floating point values to be our block of data.

```
typedef struct {
float32_t dataBuffer[BLOCK_SIZE];   //declare structure for message queue mailslot
} ADC_RESULT;
```

Next, a block of memory is reserved for the mail queue. Here memory is reserved for a queue of 16 mailslots. Data can be written into each mailslot, the mailslot is then locked until the message is received and the buffer released.

```
osMailQDef(mpool_ADC, 16, ADC_RESULT);
```

In the mailbox system the data is loaded into a buffer and is accessed via a pointer. The pointer is placed in a queue which is read as a FIFO. It is important to note that the pointer is moved while the actual data stays static.

```
os_mbx_declare (MsgBox_ADC,16);   //declare message pointers
```

Then the mailbox can be instantiated by formatting the message box.

```
osMailQId MsgBox;
```

And then initializing the message pointer queue.

```
MsgBox = osMailCreate(osMailQ(MsgBox), NULL);
```

Once this is done, the mailbox is ready to send data between threads or ISR and threads. The code below shows the ADC handler logging 32 conversions into a mailslot and then posting it to the DSP routine.

```
void ADC_IRQHandler(void) {
static ADC_RESULT *mptr;                                      //declare a mailslot
                                                                  pointer

static unsigned char index = 0;
if(index == 0) {
mptr = _alloc_box (mpool_ADC);
mptr = (ADC_RESULT*)osMailAlloc(MsgBox, osWaitForever);       //alocate a new
                                                                  mailslot
}
mptr->dataBuffer[index] = (float) ((LPC_ADC->ADGDR>>4)& 0xFFF) ;   //place ADC data into
                                                                  the mailslot
index++;
if(index == BLOCK_SIZE ){
index = 0;
osMailPut(MsgBox, mptr);                                      //when mailslot is
                                                                  full send
}}
```

In the main DSP thread, the application code waits for the blocks of data to arrive in the message queue from the ADC. As they arrive each one is processed and the results placed in a similar message queue to be fed into the DAC.

```
void DSP_thread (void){
ADC_RESULT *ADCptr;                                      //declare mailslot pointers
                                                           for ADC and DAC mailboxes

ADC_RESULT *DACptr;
while(1){
evt = osMailGet(MsgBox_ADC, osWaitForever);              //wait for a message from the
                                                           ADC
if (evt.status == osEventMail)  ADCptr = (ADC_RESULT*)evt.value.p;
DACptr = (ADC_RESULT*)osMailAlloc(MsgBox, osWaitForever); //Allocate an mailslot for
                                                           the DAC message queue
DSP_Alogrithm(ADCptr->dataBuffer, DACptr->dataBuffer);   //Run the DSP algorithm
osMailPut(MsgBox,DACptr );                               //Post the results to the DAC
message queue
osMailFree(MsgBox, ADCptr);                              //Release the ADC mailslot
}}
```

In the example code, a second thread is used to read the DAC message and write the processed data to the DAC output register.

```
void DAC_task (void){
unsigned int i;
ADC_RESULT *DACrptr;
while(1){
  evt = osMailGet(MsgBox, osWaitForever);                //Wait for a mailslot of
                                                           processed data
  if (evt.status == osEventMail) DACrptr = (ADC_RESULT*)evt.value.p;
  for(i = 0;i<(BLOCK_SIZE);i++){
    osSignalWait(0x01,osWaitForever);                    //synchronise with the
                                                           ADC sample rate
    LPC_DAC->DACR = (unsigned int) DACrptr->dataBuffer[i]<<4; //write a processed data
                                                           value to the DAC register

  }
  osMailFree(MsgBox, DACrptr);                           //free the mailslot
}}
```

The osSignalWait() RTX call causes the DAC task to halt until another thread or ISR sets its event flags to the pattern 0x0001. This is a simple triggering method between thread that can be used to synchronize the output of the DAC with the sample rate of the ADC. Now, if we add the line

```
osSignalSet (tsk_DAC,0x01); //set the event flag in the DAC task.
```

to the ADC ISR, this will trigger the DAC thread each time the ADC makes a conversion. The DAC task will wake up each time its event flag is triggered and write an output value to the DAC. This simple method synchronizes the input and output streams of data.

Balancing the Load

The application has two message queues, one from the ADC ISR to the DSP thread and the other from DSP thread to the DAC thread. If we want to build a complex system say with a GUI and a TCP/IP stack, then we must accept that during critical periods the DSP thread will be preempted by other threads running on the system. Provided that the DSP thread can run before the DAC message queue runs dry, then we will not compromise the real-time data. To cope with this, we can use the message queues to provide the necessary buffering to keep the system running during these high-demand periods. All we need to do to ensure that this happens is post a number of messages to the DAC message queue before we start the DAC thread running. This will introduce an additional latency but there will always be data available for the DAC. It is also possible to use the timeout value in the osMessageGet() function. Here, we can set a timeout value to act as a watchdog.

```
evt = osMailGet(MsgBox, WatchdogTimeout);
if(evt.status == osEventTimeout)
  osThreadSetPriority(t_DSP,osPriorityHigh);
```

Now, we will know the maximum period at which the DAC thread must read a data from the message queue. We can set the timeout period to be three quarters of this period. If the message has not arrived then we can boost the priority of the DSP thread. This will cause the DSP thread to preempt the running thread and process and delayed packets of data. These will then be posted on to the DAC thread. Once the DAC has data to process, it can lower the priority of the DSP thread to resume processing as normal. This way we can guarantee the stream of real-time data in a system with bursts of activity in other threads.

The obvious downside of having a bigger depth of message buffering is the increased signal latency. However, for most applications this should not be a problem. Particularly when you consider the mailbox ease of use combined with sophisticated memory management of the RTX RTOS.

Exercise 10.4 RTX Real Time

This exercise implements an Infinite Impulse Response (IIR) filter as an RTX task. A periodic timer interrupt takes data from a 12-bit ADC and builds a block of data which is then posted to the DSP task to be processed. The resulting data is posted back to the timer ISR. As the timer ISR receives processed data packets, it writes an output value per interrupt to the 10-bit DAC. The sampling frequency is 32 kHz and the sample block size is 256 samples (Fig. 10.23).

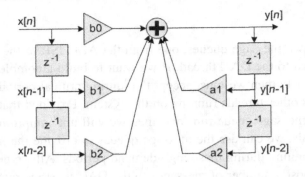

Figure 10.23
An IIR filter is a feedback filter which is much faster than an Finite Impulse response (FIR) filter but can be unstable if incorrectly designed.

Open the Pack Installer.

Select the Boards::Designers Guide Tutorial.

Select the Example tab and Copy "EX 10.4 RTOS Real Time" (Fig. 10.24).

Figure 10.24
Real-time DSP project.

First examine the code, particularly the DSP_App.c module.

The project is actually designed for a Cortex-M3 based microcontroller which as a simulation model that includes an ADC and DAC. The modules ADC.C, Timer.c, and

DAC.c contain functions to initialize the peripherals. The module DSP_APP.c contains the initializing task which sets up the mail queues and creates the DSP thread called "sigmod ()." An additional "clock()" thread is also created; this thread periodically flashes GPIO pins to simulate other activity on the system. DSP_App.c also contains the timer ISR. The timer ISR can be split into two halves, the first section reads the ADC and posts sample data into the DSP thread inbound mail queue. The second half receives messages from the outbound DSP thread mail queue and writes the sampled data to the DAC. The CMSIS filter functions are in DSP_IIR.c.

Build the project and start the simulator.

When the project starts, it also loads a script file that creates a set of simulation functions. During the simulation these functions can be accessed via buttons created in the toolbox dialog (Fig. 10.25).

Select view\toolbox

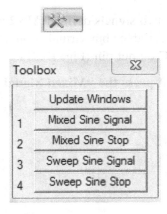

Figure 10.25
The toolbox buttons are created by the script file.

The simulation script generates simulated analog signals linked to the simulator time base and applies them the simulated ADC input pin (Fig. 10.26).

Open the Logic Analyzer window.

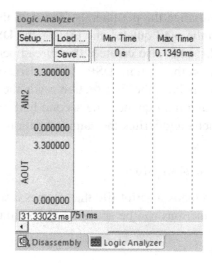

Figure 10.26
Logic analyzer window.

The Logic Analyzer should have two signals defined AIN2 and AOUT each with a range of 0−3.3V. These are not program variables but virtual simulation registers which represent the analog input pin and the DAC output pin (Fig. 10.27).

Start the simulator running and press the "Mixed Signal Sine" button in the toolbox dialog.

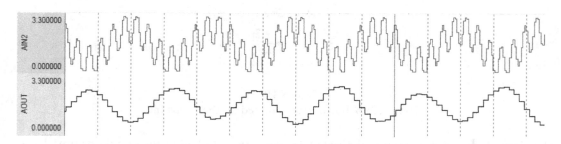

Figure 10.27
The input mixed sine wave and the output filtered waveform.

This will generate an input signal which consists of a mixed high-frequency and low-frequency sine wave. The filter removes the high-frequency component and outputs the processed wave to the DAC.

Now open the debug\os support\event viewer window.

In the event viewer window, we can see the activity of each thread in the system. Tick the cursor box on the event viewer toolbar. Now, we can use the red and blue cursors to make thread timing measurements. This allows us to see how the system is working at a thread level and how much processing time is free (ie, time spent in the idle thread) (Fig. 10.28).

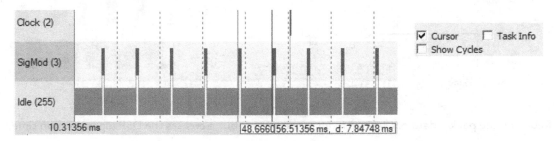

Figure 10.28
The SigMod function contains the DSP processing function.

Stop the simulation and wind back the Logic Analyzer window to the start of the mixed sine wave (Fig. 10.29).

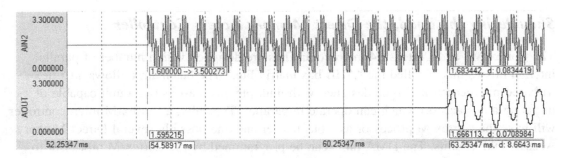

Figure 10.29
Processing of the first packet of samples introduces a latency between input and output waveform.

This shows a latency of around 8.6 ms between signal data being sampled and processed data being output from the DAC. While part of this latency is due to the DSP processing time, most of the delay is due to the sample block size used for the DSP algorithm.

Exit the simulator and change the DSP block size from 1 to 256.

The DSP blocksize is defined in IIR.h.

Rebuild and rerun the simulation (Fig. 10.30).

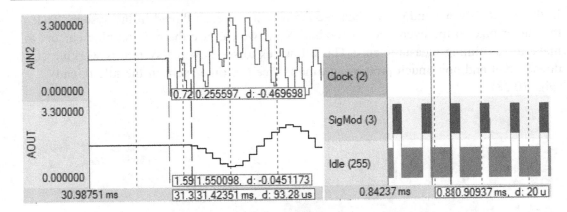

Figure 10.30
Reducing the packet data packet size reduces the latency but increases the DSP algorithm run time.

Now, the latency between the input and output signal is reduced to around 90 µs. However, if we look at the event viewer, we can see that the SigMod thread is running for every sample and is consuming most of the processor resources and is in fact blocking the clock task.

Shouldering the Load, the Direct Memory Access Controller

The Cortex-M3 and Cortex-M4 microcontrollers are designed with a number of parallel internal busses this is called the "AHB bus matrix lite." The bus matrix allows a Cortex-M-based microcontroller to be designed with multiple bus masters (ie, a unit capable of initiating a bus access) which can operate in parallel. Typically, a Cortex-M microcontroller will have the Cortex-M processor as a bus master and one possibly several Direct Memory Access (DMA) units. The DMA units can be programmed by the Cortex-M processor to transfer data from one location in memory to another while the Cortex-M processor performs another task. Provided that the DMA unit and the Cortex-M processor are not accessing resources on the same arm of the bus matrix then they can run in parallel with no need for bus arbitration. In the case of peripherals, a typical DMA unit can, for example, transfer data from the ADC results register to incremental locations in memory. When working with a peripheral, the DMA unit can be slaved to the peripheral so that the peripheral becomes the "flow controller" for the DMA transfers. So each time there is an ADC result, the DMA unit transfers the new data to the next location in memory. Going back to our original example by using a DMA unit, we can replace the ADC ISR with a DMA transfer that places the ADC results directly into the current mailslot. When the DMA unit has completed its 32 transfers it will raise an interrupt. This will allow the application code to post the message and configure the DMA unit for its next transfer.

The DMA can be used to manage the flows of raw data around the application leaving the Cortex-M processor free to run the DSP processing code.

Designing for Debug

When we use an RTOS in a project, the RTOS API functions start to become the scaffold on which the application rests. Many of these API calls have a return code that will tell us if the function completed successfully or will indicate the reason for its failure. As the code executes, we can report any errors and also provide informational messages to aid the debug process. A simple way to do this is to add printf() statements that report failings within the project.

```
status = osMessagePut(Q_LED,ledData,0);
if(status != osOK)
{
  printf("error Q_LED %d\n", status);
}
```

Now as the code runs any problems with posting messages into the Q_LED, message queue will be displayed on the STDIO channel (Fig. 10.31).

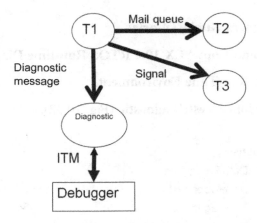

Figure 10.31
Creating a low-priority diagnostic thread allows application threads to send error and informational messages. The diagnostic thread can write these to the debugger via the ITM and/ or log them to nonvolatile memory.

As we saw in Chapter 7 "Debugging with CoreSight," the Instrumentation trace (ITM) provides a serial communications channel between the Cortex-M processor and the debugger. This allow us to take diagnostic reporting a step further by designing a dedicated diagnostic thread which sends messages to the debugger using the ITM. If we run this thread at a low priority, messages can be sent to the CPU using run-time not required by the rest of the system. Any of the application threads can post a message to this thread which in turn reports them to the debugger. The diagnostic thread can also be developed to log messages into the microcontroller FLASH memory or a file system if one is available. When you start a project, you should add some form of diagnostic messaging from the outset. You can add a range of error and informational messages that will give you a run-time application level trace. This helps build confidence in your code and allows you to spot run-time errors very quickly. Once you have developed a suitable system it can then be reused in future projects.

Exercise 10.5 Run-Time Diagnostics

In this example, we will look at creating a diagnostic thread and will structure a message queue so that we can send error and informational messages from the application threads to the debugger.

For this exercise you will need to have the Utility pack installed. See Chapter 2 "Developing Software for the Cortex-M Family" for installation instructions.

Open the Pack Installer.

Select the Boards::Designers Guide Tutorial.

Select the Example tab and Copy "EX 10.4 RTOS Run-time Diagnostics."

In the project, select the Run-Time Environment.

Select the Utilities::Developer Test:Diagnostic (Fig. 10.32).

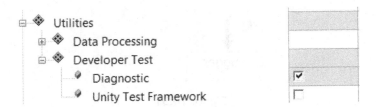

Figure 10.32
Add the diagnostic framework.

This adds a diagnostic thread to the project. This thread implements a mail queue which has the mailslots defined as shown below:

```
typedef struct {
  char message[30];
  uint8_t level;
  uint32_t line;
  char file[20];
  uint32_t parameter;
} mail_format;
```

Here, we can send a text message together with a diagnostic level (1 = Error, 2 = Informational). We can also pass in the "C" source line, file, and a user defined parameter.

The diagnostic thread receives these messages and can print them to the ITM and log them to a file.

Open the diagnostic.h file (Fig. 10.33).

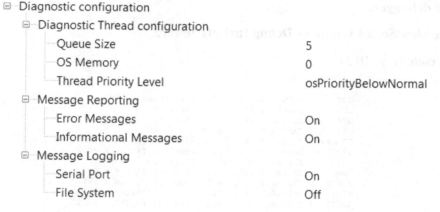

Figure 10.33
Diagnostic framework configuration options.

This is a template file that allows us to configure the diagnostic thread parameters, diagnostic message queue size, thread memory allocation, and the thread priority level. We can also decide what level of diagnostic message to process and finally how to display it. Also, in this file are two macros that define ERROR and INFO macros.

```
#define INFO(message,informational) diagnosticMessage(message,informational,
INFO_MSSG,__LINE__,__FILE__)
#define ERROR( message,informational ) diagnosticMessage(message,informational,
ERROR_MSSG,__LINE__,__FILE__)
```

In main_app.c, we can start the diagnostic thread.

```
#include "diagnostic.h"
int main (void)
{
  tid_mainThread = osThreadGetId ();
  LED_Initialize ();
  Init_diagnosticThread ();
```

Now in our code, we can add diagnostic messages at any key points.

```
txResult = osMessagePut(Q_LED,ledData,0);
if(txResult ! = 0)
{
  ERROR("Mail Queue Put failed",txResult);
}
```

Here we are sending an error message if posting data to a mail queue fails. The error function sends a text message and passes the txResult return code as a user defined parameter.

Build the code.

Start the debugger.

Open the view\Serial Windows\Debug (printf)viewer.

Run the code (Fig. 10.34).

```
INFO    : main_app.c 29 : Threads Started : 0x0
INFO    : Thread.c 37 : LED's Off : 0x0
INFO    : lookup.c 21 : Extended Delay : 0xb
INFO    : Thread.c 37 : LED's Off : 0x0
ERROR--> Thread.c 76 : Mail Queue Put failed : 0x81
ERROR--> Thread.c 76 : Mail Queue Put failed : 0x81
INFO    : Thread.c 37 : LED's Off : 0x0
ERROR--> Thread.c 76 : Mail Queue Put failed : 0x81
ERROR--> Thread.c 76 : Mail Queue Put failed : 0x81
ERROR--> Thread.c 76 : Mail Queue Put failed : 0x81
ERROR--> Thread.c 76 : Mail Queue Put failed : 0x81
```

Figure 10.34
The diagnostic messages reveal a mail queue overflow.

After the code has been running for about 40ms, the application mail queue will fail and the error message will be displayed.

Exit the debugger.

Uncomment line 27 in main_app.c.

Rebuild the code.

Run the code and observe the output in the Debug Viewer (Fig. 10.35).

```
INFO    : main_app.c 29 : Threads Started : 0x0
INFO    : Thread.c 114 : LED On : 0x0
INFO    : Thread.c 114 : LED On : 0x1
INFO    : Thread.c 114 : LED On : 0x2
INFO    : Thread.c 114 : LED On : 0x3
INFO    : Thread.c 114 : LED On : 0x4
INFO    : Thread.c 114 : LED On : 0x5
INFO    : Thread.c 114 : LED On : 0x6
INFO    : Thread.c 114 : LED On : 0x7
INFO    : Thread.c 37 : LED's Off : 0x0
INFO    : Thread.c 114 : LED On : 0x0
INFO    : Thread.c 114 : LED On : 0x1
INFO    : lookup.c 21 : Extended Delay : 0xb
INFO    : Thread.c 114 : LED On : 0x2
```

Figure 10.35
Informational messages from the corrected code.

Conclusion

One of the chief reasons made by developers for not adopting the use of an RTOS is that it represents too much overhead in terms of processing time for a small microcontroller. This no longer stacks up, just about every Cortex-M-based device is capable of supporting an RTOS and should be used for all but the simplest projects. Used correctly the benefits of using an RTOS far outweigh the negatives. Once you have invested the time in learning how to use an RTOS, you will not want to go back to bare metal projects.

CHAPTER 11

Test Driven Development

Introduction

In this section we are going to look at a new technique called "Test Driven Development" (TDD) for developing code. When I say new I really mean new to embedded system developers. TDD has been established in other areas of computing since around 2003. Traditionally most embedded systems developers will tend to write a relatively small amount of code, compile it, and then test it on the hardware using a debugger. Any formal testing is then done by a separate team once all of the codebase has been written. Unit testing would then be a long phase of testing, debug, and fix until the final production code is deemed fit to be released. TDD is not a replacement for the formal unit testing phase but introduces testing as a tool which is part of the code development process. Put simply, TDD requires you to add tests for every function that you write but crucially you write the tests first then write the application code to pass the tests. At first this seems unnecessarily time consuming and complicated, however, TDD does have a number of very positive benefits.

As we develop our application code tests are written for all the major functions. As the project grows this provides us with a fully recursive test suite. Now as new code is added or existing code is modified we can run all the tests and be confident that the new code works and does not have any unexpected side effects.

By writing tests for each incremental addition to our production code we will catch bugs early. This is the holy grail of project development. Trapping and removing errors as early as possible saves a huge amount of debug time. There is nothing worse than trying to catch an elusive bug buried deep in the final production code with a deadline looming (well possibly there are worse things).

If we develop the discipline of testing code as we go it increases our overall confidence that the code works. Plus if you are working within a team of developers a full range of software tests will help speed up any software integration phase.

In order to be able to successfully write unit tests we need to write the production code so that it is testable. This means that we must be able to decouple our production code functions from the main application. This encourages the writing of well-structured production code.

As our application codebase grows a traditional compile and debug cycle will begin to take longer and longer to complete. Also the system complexity grows and it can be harder to debug a portion of the system because it is dependent on another part of the system being in the correct state. The TDD approach only requires us to compile our test code and the module under test. This leads to a very fast compile and test cycle, typically this will be under 30 seconds. Over the design cycle of the project this can really mount up. This also means that in order to apply TDD effectively we must have an automated compile and test process that can be launched by a single action in the IDE.

Software has two aspects, the first and most obvious it must fulfill the required functionality. The second is more subtle, software must also be well crafted and expressive so that it may be maintained over the lifetime of the project and ideally reused in future projects. Once we have written the code to be functionally correct it can then be refactored to become releasable production code. With TDD once the code is functionally correct we will have a set of passing tests. We can now refractor the code and go through a very fast compile and test cycle to ensure that we have not introduced any bugs. TDD can save a huge amount of time during this stage of code development.

The written tests are also a very good form of living documentation. You may well write a formal document describing how the code is structured and how it works but most programmers hate to read it, instead we all prefer to look at the source code. In this context the test cases describe what each function does and how to use it. Examining the test cases alongside source code is an express route to understanding how the application works.

Many microcontroller-based embedded systems are developed by small teams or individual developers. In such an environment formal testing can really be a luxury and often simply does not happen. In such an environment TDD is a vast improvement to the development process and also has the benefit of being free to adopt.

Finally, any kind of software development is a marathon rather than sprint. As you get deeper into the project it can become hard to see the wood for the trees. If you are a sole developer you will be holding most of the project details in your head day after day. Using TDD does breed a confidence that your codebase is working and gives a feeling of progress and accomplishment through this daily grind.

The TDD Development Cycle

The philosophy of TDD may be summed up in three rules formulated by Robert C. Martin (no relation):

1. You are not allowed to write any production code unless it is to make a failing unit test pass.
2. You are not allowed to write any more of a unit test than is sufficient to fail; and compilation failures are failures.
3. You are not allowed to write any more production code than is sufficient to pass the one failing unit test.

While this feels very awkward at first with a little practice it becomes second nature. Once you have invested the time and discipline to work with TDD in a project you will start to reap the benefits.

To use TDD in a real project we can go through the following cycle:

1. Write the code to add a test.
2. Build the code and run the test to see it fail.
3. Add the minimal amount of application code to pass the test.
4. Run all the test code and check all the tests are passed.
5. Refractor the code to improve its expressiveness.
6. Repeat.

Test Framework

In order to make the TDD approach work in practice we need to be able to create tests and run them as quickly as possible. Fortunately, there are a number of test frameworks available that can be added to our project to create a suitable test environment. Two of the most popular frameworks are Unity and CppUtest:

> Unity   unity.sourceforge.net
> CppUtest   www.cpputest.org

In the example below we will use the Unity framework. Unity is designed to test "C"-based applications. It has a small footprint and is also available as a software pack so is very easy to add to a project. CppUtest is a similar framework designed for both the "C" and "C++" languages.

Test Automation

As well as being able to quickly add tests to our framework we need to be able to automate the build and test cycle. So ideally within the IDE we will be able to build the test project, download to the target, run the tests, and see the results with one button press. This should then give us a development cycle of under 30 seconds. In the next two examples we will look at setting up and automating a TDD framework with a simple project.

This project is configured to use the simulator so we can experiment with the TDD approach without the need for specific hardware.

Installing the Unity Framework

For this example you will need to have the Utility pack installed. See Chapter 2 "Developing Software for the Cortex-M Family" for installation instructions.

You will also need to install the following packs using the pack installer (Table 11.1).

Table 11.1: Required software packs

Pack	Minimum Version Number
Keil::LPC1700_DFP	2.1.0
Keil::ARM_Compiler	1.0.0
ARM::CMSIS	4.5.0

As tests run the results are sent to the STDIO channel. The Unity framework component requires the Compiler IO options to be configured. In most projects the test results are best sent via the ITM to the console window in the debugger. A Cortex-M0 does not have the ITM trace so you would need to use a different IO channel to display the results, typically this will be a USART.

Exercise 11.1 Test Driven Development

Open the μVision pack installer (Fig. 11.1).

Figure 11.1
Pack installer Boards and examples tabs.

Select the Boards tab and the Examples tabs (Fig. 11.2).

Figure 11.2
Select the MCB1700 and then the Unity test framework.

Select the MCB1700.

Select the example tab and Copy "Unity Developer Test Framework".

This will copy the project to your selected development directory.

The project contains a subdirectory called "TDD Setup PreBuilt" which has a fully configured version of the project for reference.

The application code is an RTOS-based project that is used to switch on a bank of LEDs one at a time. Once all of the LEDs have been switched on they will be simultaneously switched off and the cycle will repeat. We are going to develop it further by adding a function that will introduce different period delays at selected points in the LED cycle. While this is easy enough to do we will use this example to see how to add a TDD approach to our project.

In our project we have our new code will be placed in a file called lookup.c (Fig. 11.3).

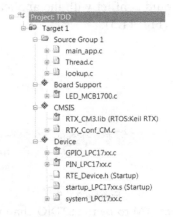

Figure 11.3
Initial project with code under test in lookup.c.

Adding the Unity Test Framework

Lookup.c contains the skeleton of a function called checkTick(). This function is passed an integer value, it must then scan an array to see if this value is in the array. If there is a matching value it returns the location of that value in the array. If there is no matching value in the array zero is returned.

In the TDD project open the Run-Time Environment (RTE). ❖ (Fig. 11.4)

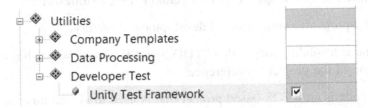

Figure 11.4
The sel column will be colored orange (light gray in print version) if a subcomponent is required.

Select the Utilities::Developer Test:Unity Test Framework.

The test framework used STDIO as an output for its results so we must enable this and select the low level output channel. Using the ITM will allow us to run on any Cortex-M core which has the ITM fitted. The ITM will give us an IO channel that does not use any microcontroller resources that would conflict with the application code and does not need any external hardware resources (Fig. 11.5).

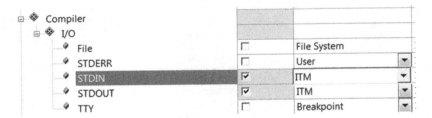

Figure 11.5
Select ITM to be the STDIO channel.

Finally we just need to make sure that the ITM support in the debugger is correctly configured. This example is using the simulator and does not need any further configuration but for a project using a hardware debugger the ITM is configured as follows.

Only change the following debugger settings if you are using a hardware debugger.

Open the options for target\debug menu and press the settings button (Fig. 11.6).

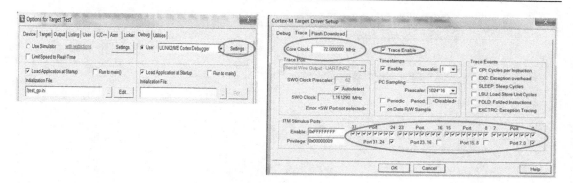

Figure 11.6
For hardware debug you must enable the ITM Trace Unit to get the test results.

Select the trace menu.

Set the core clock speed to match the processor CPU frequency.

Enable the trace and enable ITM channel 31.

It is also advisable to disable any other trace features in order to minimize the amount of data being sent through the Serial Wire trace channel.

The ITM data is sent to the debugger through the SWO debug pin. On some microcontrollers this is multiplexed with a GPIO function so it may be necessary to use a debugger script file to enable this pin as the debug SWO pin.

Configuring the Project Build Targets

Our project should now have the Unity component added alongside the application code. Now we need to configure the project so we can easily switch between building the full application code and building a test version.

Select project\manage\project items (Fig. 11.7).

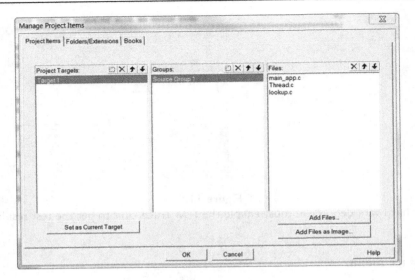

Figure 11.7
Manage projects allows you to define project folders and build versions.

Currently we have a single build version of the project which builds the application code. Using the project window we can define different build versions and additional project folders (Fig. 11.8).

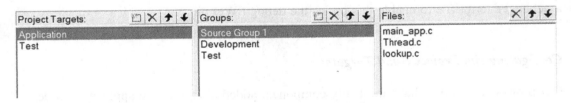

Figure 11.8
We can define a test target and additional project folders.

In the "Project Targets" window press the insert button and add a Test build.

In the "Groups" section add two new groups "Development" and "Test" then press ok to return to the IDE (Fig. 11.9).

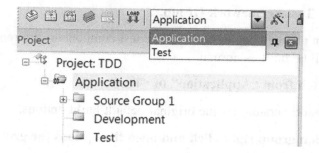

Figure 11.9
Switch to the application build using the tool bar drop down.

Now we can switch between build versions using the drop down test box on the toolbar. We also have additional project folders to layout the project.

With the project build set to Application select the Test folder.

Right click and select options for Test folder.

In the properties menu uncheck the "Include in Target Build" and "Always Build" options (Figs. 11.10 and 11.11).

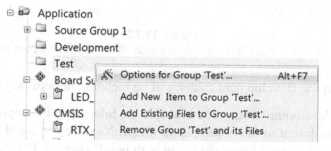

Figure 11.10
Open the local options for the test group folder.

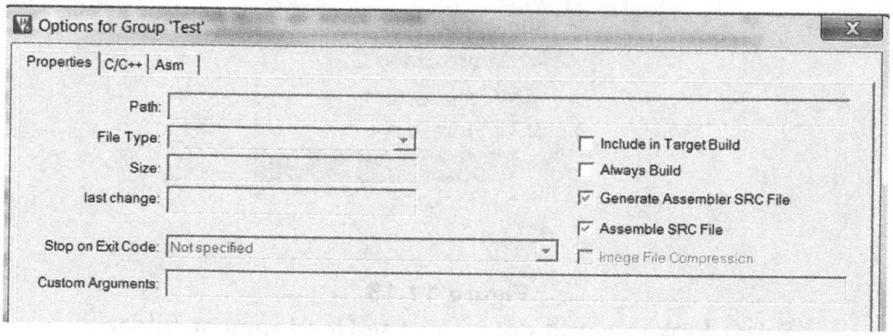

Figure 11.11
Uncheck the build options to remove this folder from the build.

Repeat this for the Test Framework Group.

Now when we are in the application build all the test components are removed and the application code will be built as normal.

On the toolbar switch from "Application" to "Test".

Currently the Test build version has the original default build options.

Select the application group right click and open the options for group "Application" build.

Uncheck the "Always Build" and "Include in Target Build" options and click ok.

Now when we select the "Test Target" the application code will be removed and the test components will be active in the project (Fig. 11.12).

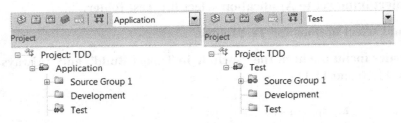

Figure 11.12
The "stop signs" on the project folders indicate that the folder is not part of the Target build.

Any code that is in the Development folder will be included in both projects.

We are currently developing code in the lookup.c module. Open the application folder, select lookup.c and drag it to the development folder. Now we can build the project in either Target and lookup.c will be included in both builds (Fig. 11.13).

Figure 11.13
Select lookup.c and drag it to the Development folder.

Adding the Test Cases

Select the Test folder, right click and select "Add New Item to Group Test"
(Fig. 11.14).

Component	Name
⊞ ❖ CMSIS	
⊟ ❖ Utilities	
Developer Test:Unity Test Framework	Test Group
Developer Test:Unity Test Framework	Test Main

Figure 11.14
User code templates for the Unity Test harness.

Select "User Code Templates" and in Utilities add Test Group and Test Main
(Fig. 11.15).

```
⊟ ♘ Project: TDD
  ⊟ 🔊 Test
    ⊟ 📚 Source Group 1
      ⊞ 📄 main_app.c
      ⊞ 📄 Thread.c
    ⊟ 📂 Development
      ⊞ 📄 lookup.c
    ⊟ 📂 Test
      ⊞ 📄 TestGroup.c
      ⊞ 📄 TestMain.c
```

Figure 11.15
The project with the Test harness added.

Open Test Main.c.

This file provides the main() function for the Test Target and calls the Unity test framework. When the tests are finished execution will reach the stop() function.

Once Unity starts to run it will start to execute any tests that have been defined. These are declared as test cases which are then collected into test groups. We can create this hierarchy of tests by modifying the runAllTests() and TEST_GROUP_RUNNER() functions.

In the runAlltest() function rename the test group to lookup.

In the TEST_GROUP_RUNNER() function rename the "testgroup" to "lookup" and add the test case "correctTickValues" and "failingTickValues".

```
TEST_GROUP_RUNNER(lookup)
{
  RUN_TEST_CASE (lookup, correctTickValues);
  RUN_TEST_CASE (lookup, failingTickValues);
}
static void RunAllTests(void)
{
  RUN_TEST_GROUP(lookup);
}
```

Open TestGroup.c.

In this file we can define a group of tests for our lookup.c function. This file defines the test group name and provides setup and tear down functions which run before and after the tests. The TEST_SETUP() function runs once before each test in the test group and is used to prepare the application hardware for the test cases. The tear down function is used to return the project to a default state so that any other test groups can run. The final function is our test case which allows us to create an array of tests for our new function.

```
TEST_GROUP(lookup);
TEST_SETUP(lookup)
{
}
TEST_TEAR_DOWN(lookup)
{
}
```

For this simple example we do not need to provide any setup or teardown code.

Copy and paste the TEST function so we have two test functions.

Rename the first one as follows   **TEST(lookup, correctTickValues).**

And the second as shown below   **TEST (lookup,failingTickValues).**

Now we can create test cases in both of these functions as shown below.

```
TEST(lookup, correctTickValues)
{
TEST_ASSERT_EQUAL(1, checkTick(34));
TEST_ASSERT_EQUAL(2, checkTick(55));
TEST_ASSERT_EQUAL(3, checkTick(66));
TEST_ASSERT_EQUAL(4, checkTick(32));
TEST_ASSERT_EQUAL(5, checkTick(11));
TEST_ASSERT_EQUAL(6, checkTick(44));
TEST_ASSERT_EQUAL(7, checkTick(77));
TEST_ASSERT_EQUAL(8, checkTick(123));
}
```

The first set of test cases will call the function with valid input values and check that the correct value is returned.

```
TEST (lookup,failingTickValues)
{
TEST_ASSERT_EQUAL(0, checkTick(22));
TEST_ASSERT_EQUAL(0, checkTick(93));
}
```

The second set of test cases will call the function with invalid values and check that the correct error value is returned.

So far we have added the test framework and defined some test cases. We can now build the code and execute the code and the test cases should fail as there is no actual application code.

Build the code.

Start the debugger.

Open the view\serial windows\printf window.

Run the code.

A test fail will be reported in the serial window (Fig. 11.16).

```
Unity test run 1 of 1
.TestGroup.c:18:TEST(lookup, correctTickValues):FAIL: Expected 1 Was -1

.TestGroup.c:30:TEST(lookup, failingTickValues):FAIL: Expected 0 Was -1


-----------------------
2 Tests 2 Failures 0 Ignored
FAIL
```

Figure 11.16
A failing test is reported in the ITM console window.

Automating the TDD Cycle

Before we go and start writing the application code we can further automate this testing procedure so we can go around the build and test cycle with one click in the IDE.

Set the project target to be the "Test" build.

Open the options for target menu and select the user menu and enable the "Start Debugging" option (Fig. 11.17).

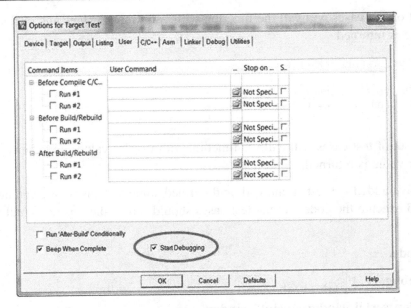

Figure 11.17
The User tab allows you to start the debugger as soon as the build is finished.

Now after a build the code will be downloaded into the target and the debugger will be started.

Next select the options for target\debug menu and add the test_go.ini script file which is located in the <project>\RTE\utilities directory as the simulator initialization file. (Fig. 11.18).

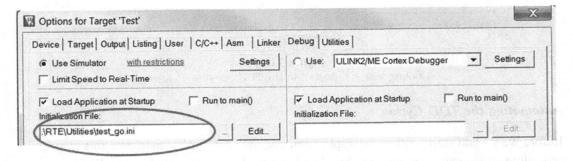

Figure 11.18
Add the automation script as the simulator initialization file.

Next uncheck the "Run to main()" box. Or this will stop the script running correctly.

Now this script will be run as soon as the debugger is started.

```
coverage clear
SLOG≫testResults.txt
g,stop
SLOG OFF
LOG≫testResults.txt
coverage \testgroup DETAILS
LOG OFF
```

The script file starts the code running as soon as the debugger is started. The code will run until it has reached the stop function. All of the test results are printed to the ITM console and also logged to a file on the PC hard disk. At the end of the tests the code coverage information for the testgroup.c module is also saved to the log file. The code coverage information proves that all the tests have been executed. You could also add the coverage information for the modules under test.

Now if you rebuild the code it will compile link, start the debugger, and execute the tests.

Since we are only building and downloading the test framework and our new code is a very fast development cycle. In a large project this can be a big time saver.

Now open the lookup.c file. Read the function description and write the code required to make the function work.

Build the code and check all the tests pass. If there is a failure correct the error and repeat until the tests pass (Fig. 11.19).

```
Unity test run 1 of 1
..
-----------------------
2 Tests 0 Failures 0 Ignored
OK
```

Figure 11.19
Once the correct code is added we have passing tests.

Once we have reached this happy state we can switch to the application build and add the new function to the code in thread.c.

```
while(1)
{
    ledData = 0;
    for(i = 0;i<8;i++)
    {
        extendedDelay = checkTick(tick);      //call checkTick to calculate an extended delay
        totalDelay = 100 + (100*extendedDelay);
        osDelay(totalDelay);
        osMessagePut(Q_LED,ledData,0);
        tick = (tick + 1) & 0xDFF;
        ledData = ledData + 1;
    }
}
```

Switch to the Application Target and build the code.

Start the debugger and run the new version of the blinky program.

Testing RTOS Threads

This form of testing also works very well with RTOS threads. In our example we have four threads, main plus the three application threads. Each of the threads is a self-contained "object" with very well-defined inputs and outputs. This allows us to test each of the threads in isolation since we can start each thread individually and build a test harness that "mocks" the rest of the system (Fig. 11.20).

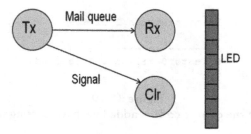

Figure 11.20
An RTOS-based project is composed of thread objects with well-defined inputs and outputs. A thread is an excellent target for testing.

Our application threads consist of a Tx thread that writes values into a message queue, an Rx thread that reads values from the message queue and writes these to the bank of LEDs. When all the LEDs have been switched on, the Tx thread signals the Clr thread which switches off all the LEDs (Fig. 11.21).

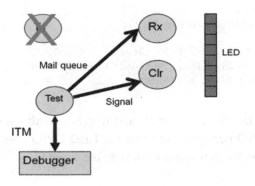

Figure 11.21
A test thread can use the RTOS API to control the application threads. Here we can terminate a thread and apply test cases to the remaining thread using the mail queue and signal flags.

We can test the application at the thread level by using the same testing framework. The test thread can start each thread individually and then we can apply test data using the RTOS calls (message queue and signals). In our test project main() is running as an RTOS thread, so in our test cases we can also access all of the RTOS API calls to create thread level test cases.

```
TEST(thread, rxThread)
{
        LED_Initialize ();
        Init_queue();
        Init_rxThread();
        ..............................
```

Here we are creating a test group that initializes the LEDs and the application message queue. We can then start the rxThread() running. This thread is now waiting for messages to be posted into the mail queue. This allows us to create test cases that post messages into the mail queue in place of the txThread(). We can then check the behavior of the rxThread().

```
osMessagePut(Q_LED,(0x00),0);
osDelay(10);
TEST_ASSERT_EQUAL((1≪28),(GPIO_PortRead (1) & 0xB0000000));
```

In our test case we can post some data into the rxThread() message queue, and then read the LED Port settings to check that the correct LEDs are switched on. The osDelay() call ensures the RTOS scheduler will task switch and run the rxThread() code. At the end of the test cases we can kill the rxThread() and continue on to create and test another of the application threads.

```
TEST(thread, clrThread)
{
    Init_clrThread ();
    osSignalSet  (tid_clrThread,0x01);
    osDelay(10);
    TEST_ASSERT_EQUAL(0,(GPIO_PortRead (2) & 0x0000007C));
    TEST_ASSERT_EQUAL(0,(GPIO_PortRead (1) & 0xB0000000));
    osThreadTerminate(tid_clrThread);
}
```

In the next test case we start the clrThread() and trigger it with the osSignalSet() function. We can then read the GPIO port to ensure that the LED GPIO pins have been written to zero; at the end of the test we can again terminate the thread.

Exercise 11.2 Testing RTOS Threads

Return to the 11.1 TDD project used in the last example.

Highlight the Test folder and select "Add existing item to project".

Add the file testGroupRTX.c.

Open TestMain.c.

Add the thread test group and call it from RunAllTests() as shown below.

```
TEST_GROUP_RUNNER(thread)
{
RUN_TEST_CASE (thread, rxThread);
RUN_TEST_CASE (thread, clrThread);
RUN_TEST_CASE (thread, txThread);
}
static void RunAllTests(void)
{
printf("\nRunning lookup test case \n");
RUN_TEST_GROUP(lookup);
printf("\nRunning Thread test case \n");
RUN_TEST_GROUP(thread);
}
```

Build the project and let it run through the compile link and test loop (Fig. 11.22).

```
-------------------------------Build Time and Date 15:31:39, Nov 19 2015--

Unity test run 1 of 1

Running lookup test case
..
Running Thread test case
...
-----------------------
5 Tests 0 Failures 0 Ignored
OK
```

Figure 11.22
Now the original and new RTOS tests run and pass.

This will execute the original set of test cases on the code in lookup.c. Provided these pass ok, the new RTOS test group will be executed to test each of the three RTOS threads.

Decoupling Low Level Functions

Ideally we do not want to modify the code under test. However, at some point we will encounter the problem that the function under test calls a routine that requires a result from a peripheral.

```
int detectPacketHeader (void)
{
while ( !packetHeader)
{
cmd = receiveCharacter(USARTdrv);
..........
```

In this case the receiveCharacter() routine is waiting to receive a character from the microcontroller USART. This makes automated testing the detectPacketHeader() function difficult to test because it is reliant on the receiveCharacter() routine. To get round this we need to make a "mock" version of receiveCharacter() that can be called during testing in place of the real function. We could comment out the real receiveCharacter() function but this means changing the application code or adding conditional compilation using #ifdef pre-processor commands. Both of these approaches are undesirable as they can lead to mistakes when building the full application. One way round this problem is to use the __weak directive provided by the compiler.

```
__weak char receiveCharacter (ARM_DRIVER_USART *USARTdrv)
```

When the application code is built this function will be compiled as normal. However, during testing we can declare a "mock" function with the same function prototype minus the __weak pragma. The linker will then "overload" the original function with the weak declaration and use the "mock" version in its place.

```
int8_t mockSerialData[] = "Hello World";
char receiveCharacter (ARM_DRIVER_USART *USARTdrv)
{
int8_t val,count = 0;
          val = mockSerialData[count++];
          if(count > sizeof(mockSerialdata)) count = 0;
return val;
}
```

Now we can build a test for the detectPacketHeader() function and use the "mock" version of receiveCharacter() to provide appropriate data during the test case.

Testing Interrupts

An embedded microcontroller project will tend to have code that is intimately interconnected with the microcontroller hardware. Some functions within the application code will be reliant on a microcontroller peripheral. In most projects we may also have a number of peripheral interrupts active. We can use the same method of overloading functions to test the application functions that are dependent on hardware interrupts. If we have an ISR function we can declare it as __weak and then overload it with a "mock" function that is not hardware dependent.

```
__weak void ADC_IRQHandler(void) {
  volatile uint32_t adstat;
  adstat = LPC_ADC->ADSTAT;            /* Read ADC clears interrupt   */
  AD_last = (LPC_ADC->ADGDR >> 4) & 0xFFF;   /* Store converted value  */
  AD_done = 1;
}
```

So the above ADC interrupt handler which is dependent on the microcontroller ADC peripheral can be replaced with a simplified "mock" function.

```
void ADC_IRQHandler(void)
{
        AD_last = testValue;
        AD_done = 1;
}
```

The "mock" function mimics the functionality of the original ISR but is in no way dependent on the actual ADC hardware. As we are naming the function using the prototype reserved for the CMSIS-ADC interrupt handler the function will be triggered by an interrupt on the ADC NVIC channel. To use this function during testing we can also overload the original ADC_StartConversion() function.

```
__weak int32_t ADC_StartConversion (void) {
    LPC_ADC->ADCR &= ~( 7 << 24 );        /* stop conversion      */
    LPC_ADC->ADCR |= ( 1 << 24 );         /* start conversion     */
    return 0;
}
```

Again the overload "mock" function does not address the hardware ADC peripheral registers but uses the CMSIS-NVIC_setPending() function to cause an interrupt on the NVIC ADC channel.

```
int32_t ADC_StartConversion (void)
{
NVIC_SetPendingIRQ(ADC_IRQn);        //trigger the ADC interrupt channel
}
```

Now we can build test cases around functions that use results from the ADC while using the overload functions to mock the activity of the hardware ADC. It is also important to note that we have made minimal changes (two __weak directives) to the application source code.

Exercise 11.3 Testing with Interrupts

In this example we will enable the 12-bit ADC and its interrupt. Then in a background function we will start an ADC conversion. Once the ADC conversion has finished we will read the result, shift the result 4 bits to the right and then copy the result to a bank of 8 LEDs.

Open the Pack Installer.

Select the Boards:: MCB1700.

Select the example tab and Copy "Unity Framework Interrupt Testing".

The project is constructed with the application and test builds we setup in the first example in this chapter (Fig. 11.23).

Figure 11.23
The project is setup to work with the test framework.

Open testGroup.c.

In this file we have the overload versions of the ADC_IRQHandler() and the
ADC_StartConversion() functions.

The code under test is the function displayADC() in DisplayADC.c

```
void displayADC (void)
{
uint32_t result;
ADC_StartConversion();
while(ADC_ConversionDone() != 0);
result = ADC_GetValue();
result = result>>4;
LED_SetOut(result);
}
```

In the test case we can load an ADC result value and then call the displayADC() function.

```
testValue = 0x01<<4;
displayADC();
TEST_ASSERT_EQUAL(1<<28,(GPIO_PortRead (1) & 1<<28));
```

displayADC() will use the value from the ADC and write this value to the LED port
pins. We can then check that the correct bits have been set by reading the state of the
appropriate port.

Examine the code in the displayADC.c and ADC_IRQ.c.

Build the code and execute the test cases (Fig. 11.24).

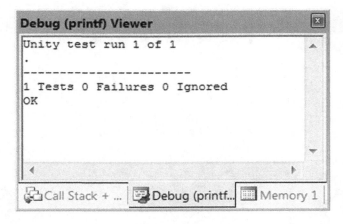

Figure 11.24
The test cases run with the mock overloaded functions.

Conclusion

Hopefully this has been a useful chapter. If you are new to TDD it should give you some food for thought. The only real way to get a real sense of how to use TDD in a real design is to actually do it. Initially you will go slower as you have to build experience in how to construct suitable tests. Once you have built some useful testing "patterns" you will start to make more rapid progress. With practice and a degree of discipline you will start to see some real benefits by adopting a TDD approach to code development.

CHAPTER 12

Software Components

Introduction

Over the last 25 years, the focus of embedded systems development tools has seen the move from application firmware written entirely in assembler, to the introduction of compilers for the "C" and "C++" languages. This was followed by increasingly sophisticated development environments and debuggers. Today with the advent of ever-more complex microcontrollers, a project may require several software "stacks" to drive peripherals such as USB Device and Host, Ethernet, Multimedia card, and LCD. Writing the necessary code for any one of these peripherals is a project in its own right. Consequently, it is becoming a fact of life that you will have to use third party code in your project, be it open source, Silicon Vendor library or a commercial library. Furthermore, you will need to be able to integrate several of these libraries, possibly from different sources, into a single project and make them all work happily together (Fig. 12.1).

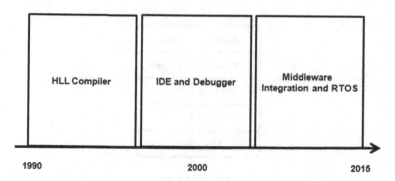

Figure 12.1

Over the last quarter century, microcontroller-based systems have moved from the introduction of high-level languages to the introduction of high-level software components.

As we saw in Chapter 4 "Cortex Microcontroller Software Interface Standard," CMSIS-Core standardizes how "C" is used on a Cortex-M processor. In this chapter, we are going to look at the next two Cortex Microcontroller Software Interface Standard (CMSIS) specifications, CMSIS Driver and CMSIS Pack. CMSIS Driver defines a standard application programming interface (API) to access common microcontroller peripherals.

The Designer's Guide to the Cortex-M Processor Family.
DOI: http://dx.doi.org/10.1016/B978-0-08-100629-0.00022-0

The intention of a CMSIS Driver is to provide a standardized low-level interface for middleware libraries. So, a company that produces a middleware stack such as a USB or TCP \IP library can use the CMSIS Driver API, this then enables their software to run on any Cortex microcontroller which supports a matching CMSIS Driver. For an application developer, CMSIS Drivers can be used to design generic software components which can be reused across future projects even if they are using a different microcontroller family. The CMSIS Pack specification allows us to take any code that we want to reuse and bundle it into a software pack so that it can be easily distributed to other users. Once installed, the CMSIS Pack component is available to add to your project through the Run-Time Environment (RTE). To see how this works in practice, we will go through the design of a simple software component that receives serial data from a GPS receiver via a Universal Synchronous Asynchronous Receiver Transmitter (USART). The component will verify and parse the data and make it available to other parts of our application. We will use the CMSIS USART driver to make the GPS code device independent. Once we have some functioning code, we can make it into a reusable software component using the CMSIS Pack specification.

CMSIS Driver

The CMSIS Driver specification defines a generic peripheral driver interface. This creates an abstraction from the microcontroller hardware. Any code that is written using CMSIS Driver can be reused on any Cortex-M microcontroller that has a matching CMSIS Driver (Fig. 12.2).

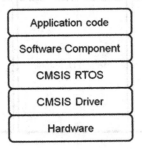

Figure 12.2
CMSIS Driver and CMSIS RTOS allow you to create a software component that can be reused across different hardware platforms.

This has lots of advantages in that once you are familiar with the CMSIS Driver API, you can reuse that knowledge across many microcontrollers and projects. You can also move your code between different microcontrollers and even different toolchains. The downside is that as you are using a generic interface, that has to be able to work on any microcontroller. This means that the features offered by the CMSIS Driver API are limited. As the CMSIS Driver may be implemented as a wrapper over a Silicon Vendor peripheral library, the code size and performance may not be optimal. Lastly, if a Silicon Vendor has added a unique hardware feature to one of their peripherals, then it may not be used by the CMISIS Driver code (Fig. 12.3).

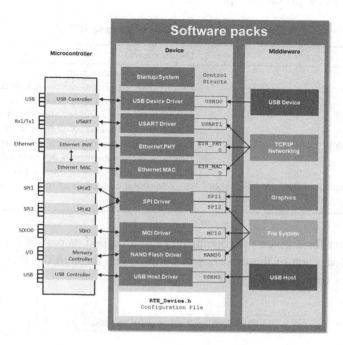

Figure 12.3
CMSIS Driver provide a standard API for a range of peripherals common to many microcontrollers.

CMSIS Driver currently supports communication peripherals such as Inter Intergrated communications (I2C), Serial Peripheral interface (SPI), universal synchronous asynchronous receiver transmitter (USART), controller area network (CAN), and Serial Audio interface (SAI) as well as more complex peripherals such as Ethernet, USB Device, and USB Host. CMSIS Driver also has specifications for Multi Media Card interface and specifies drivers for NAND and NOR Flash memory to support embedded file systems.

CMSIS Driver API

Each CMSIS Driver contains a set of API functions that are used to configure each driver. The CMSIS Driver functions have a common structure across each of the supported peripherals, once you are familiar with how to use one type of driver the same logic applies to all the others. Each CMSIS Driver is capable of supporting multiple instances, so within a project it is possible to instantiate several drivers to support multiple peripherals of a given type, for example, three SPI peripherals. The common CMSIS Driver functions are shown below in a generic form. These functions vary a little between different driver types but once you are familiar with how the CMSIS Driver API works moving between different peripherals presents few problems (Table 12.1).

Table 12.1: CMSIS Driver generic API

Function	Description
get Version()	Returns the driver version
get Capabilities()	Returns the supported driver capabilities
getStatus()	Returns the current peripheral status
initialize()	Initial driver setup and registers the driver callback
uninitialize()	Return the peripheral to its reset state
powerControl()	Enable/disable the peripheral power state
control()	Configures the peripherals operational parameters
signalEvent	A user define callback to handle peripheral events
Peripheral data transfer functions send()	An additional collection of functions to control
receive()	data transfer

The CMSIS Drivers for a given microcontroller are provided as part of the device family pack. Once installed, each of the drivers will be available through the RTE. A typical calling sequence is shown in Fig. 12.4.

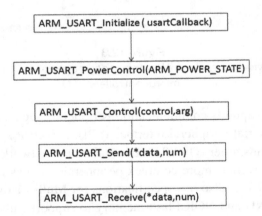

Figure 12.4

The CMSIS Driver API is (fairly) orthogonal. Once you have used the CMSIS USART driver, you will be able to use any other CMIS driver.

Exercise 12.1 CMSIS-Driver

In the following example, we are going to look at how to use the CMSIS USART driver. The principles learned with this driver can then be applied to any of the other peripheral drivers.

This example is a multiproject workspace. One project is setup for an NXP microcontroller, the second is for an ST Microelectronics microcontroller. We are going to use the same code on both microcontrollers to demonstrate the use of the CMSIS USART driver (Fig. 12.5).

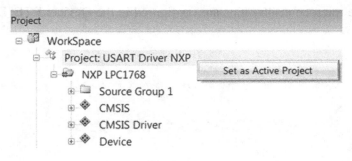

Figure 12.5
Make "USART Driver NXP" the active project.

Select the NXP project. Right click and select "Set as Active project."

First, we can add the CMSIS USART driver to our project by selecting it in the RTE then resolve any required subcomponents and add it to our project (Fig. 12.6).

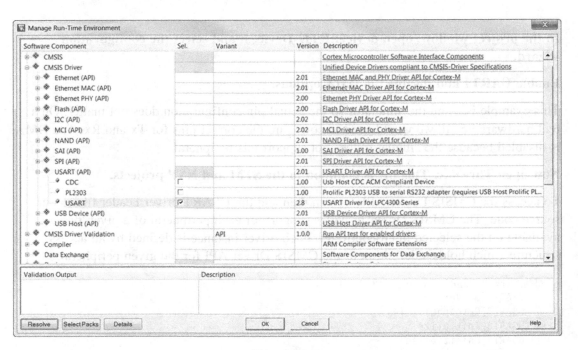

Figure 12.6
Select the CMSIS USART Driver in the Run-Time Environment Manager.

Open the RTE and select the CMSIS Driver: USART::UART.

Press the resolve button and then OK to add the UART driver to the project.

First, we need to configure the microcontroller pins to switch from GPIO to the USART Tx and Rx pins. The microcontroller pins are configured in the Device::RTE_Device.h file (Fig. 12.7).

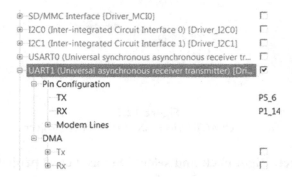

Figure 12.7
Configure the USART pins in the RTE_Device.h file.

In the project window, open device::RTE_Device.h and select the configuration wizard.

Enable UART1 and disable the DMA options.

This example is using the simulator so the actual pin configuration does not matter. On a real hardware board, we would need to configure the correct pins for Tx and Rx. The DMA is disabled because the simulator does not support this peripheral.

Now open Thread.c. This file is common to the STM and NXP projects.

To access the CMSIS Driver API, we can then add the USART driver header file to our source code. The CMSIS Driver provides support for each peripheral of a given type available on the selected microcontroller. Each driver instance is defined by an access structure which holds a definition of the CMSIS Driver API for the given peripheral.

```
ARM_DRIVER_USART Driver_USART3 = {
    USARTx_GetVersion,
    USART3_GetCapabilities,
    USART3_Initialize,
    USART3_Uninitialize,
    USART3_PowerControl,
    USART3_Send,
```

```
        USART3_Receive,
        USART3_Transfer,
        USART3_GetTxCount,
        USART3_GetRxCount,
        USART3_Control,
        USART3_GetStatus,
        USART3_SetModemControl,
        USART3_GetModemStatus
    };
```

To install the driver support, we need to add the driver include file to our source code and add an "extern" declaration for the Universal Synchronous Asynchronous Receiver Transmitter (USART) peripheral we want to access.

```
#include "Driver_USART.h"
  extern ARM_DRIVER_USART Driver_USART1;
```

In thread.c, we can make the first API call to initialize the driver.

```
Driver_USART1->Initialize(myUSART_callback);
```

The initializing call starts up the driver and registers a callback function. Once the callback function has been registered with the driver, it will be triggered by hardware events in the USART peripheral and driver interrupt service routines. Each time this function is triggered, it will pass a parameter which indicates a specific USART hardware event (Table 12.2).

Table 12.2: USART callback events

USART Hardware Events
ARM_USART_EVENT_SEND_COMPLETE
ARM_USART_EVENT_RECEIVE_COMPLETE
ARM_USART_EVENT_TRANSFER_COMPLETE
ARM_USART_EVENT_TX_COMPLETE
ARM_USART_EVENT_TX_UNDERFLOW
ARM_USART_EVENT_RX_OVERFLOW
ARM_USART_EVENT_RX_TIMEOUT
ARM_USART_EVENT_RX_BREAK
ARM_USART_EVENT_RX_FRAMING_ERROR
ARM_USART_EVENT_RX_PARITY_ERROR

In the callback function, we can now provide code for any of the USART hardware events we want to handle.

```
void myUSART_callback(uint32_t event)
{
    switch (event)
```

```
    {
        case ARM_USART_EVENT_RECEIVE_COMPLETE:
        case ARM_USART_EVENT_SEND_COMPLETE:
                osSignalSet(tid_Thread, 0x01);

        break;

    }

}
```

When the USART successfully sends or receives some data the callback will be triggered and we can send an RTOS signal to one of our application threads to indicate the event.

Once we have initialized the driver and installed the callback, we can switch on the peripheral by calling the power control function.

```
USARTdrv->PowerControl(ARM_POWER_FULL);
```

This function provides the necessary low-level code to place the USART into an operating mode. Typically, this involves configuring the system control unit of the microcontroller to enable the clock tree and release the peripheral from reset. Once the USART has been powered up, we can configure its operating parameters using the control function.

```
USARTdrv->Control(ARM_USART_MODE_ASYNCHRONOUS |

        ARM_USART_DATA_BITS_8 |

        ARM_USART_PARITY_NONE |

        ARM_USART_STOP_BITS_1 |

        ARM_USART_FLOW_CONTROL_NONE, 9600);
```

Now that the USART's operating parameters have been configured, we can use the send and receive functions to transfer data.

```
USARTdrv->Send("A message of 26 characters", 26);
USARTdrv->Receive(&cmd, 1);
```

Depending on the driver configuration, the underlying data transfer may be performed by the CPU or by an internal DMA unit provided by the microcontroller.

Examine the CMSIS Driver code in Thread.c.

Select project "Batch Build" to build both projects.

Start the debugger.

In the NXP project, open the "view\serial window\serial 2."

Open the peripherals\usart\usart1 window.

Run the code.

Check the USART configuration in the peripheral window.

The initial "Hello World" message will be displayed in the serial console window.

Exit the debugger.

Switch to the second project and repeat the above debugger session.

Due to the different naming conventions used by the silicon manufacturers the STM serial data will be output to "view\serial window\serial 1."

We are now running exactly the same USART code on two different Silicon Vendors' microcontrollers.

The CMSIS Driver specification provides us with a way to very quickly bring up key peripherals on a microcontroller without having to spend a lot of time writing the low-level configuration code. Whilst configuring a USART may seem like a trivial matter many of the higher end Cortex-M-based microcontrollers are becoming quite complex. Take a look at the LPC1768 clock tree and fractional baud rate generator if you are in any doubt.

Driver Validation

The CMSIS Driver specification is an open specification that anyone can download and use to implement their own driver. Ideally a Silicon Vendor will provide a set of CMSIS Drivers for their microcontroller as part of a Device Family Pack. Wherever the CMSIS Driver code comes from, it is still alien third party code which is being added to your project and as such may contain its own bugs. Before we trust this code and build it into our application, we need to perform some initial tests to check it works correctly. Fortunately, there is an easy way to do this in the form of a CMSIS Driver Validation Pack. As its name implies, the Validation Pack is designed to test the capabilities of each CMSIS Driver and output a report on the results.

Exercise 12.2 Driver Validation

In this exercise, we will set up the CMSIS Driver Validation Pack and create a validation project to test the capabilities of the I2C, SPI, and USART drivers.

Open the pack installer and select the ARM::CMSIS-Driver_Validation pack (Fig. 12.8).

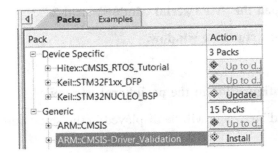

Figure 12.8
Select and install the CMSIS Driver Validation pack.

In the tutorial examples, open the project "Exercise 12.1 CMSIS-Driver."

This project has the SPI, I2C, and USART drivers selected and configured in the project (Fig. 12.9).

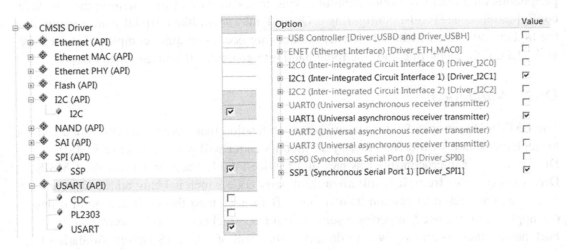

Figure 12.9
Enable the drivers you want to test in the RTE Manager and RTE_Device.h.

The validation report will be send the STDIO channel. The ARM::compiler:IO setting in the RTE is configured to redirect the STDIO output to the ITM so now all our results will appear in the view\serial windows\printf viewer (Fig. 12.10).

Figure 12.10
Setup the ITM to be the STDIO channel.

In the RTE, select the CMSIS Driver validation and enable the Framework, I2C, SPI, and USART options (Fig. 12.11).

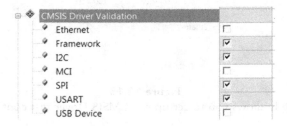

Figure 12.11
In the RTE Manager, select the test framework and the drivers to test.

Click OK to add the validation components to the project (Fig. 12.12).

Figure 12.12
The project now has the validation framework added.

Open the CMSIS Driver Validation::DV_Config.h and select the Configuration Wizard. For each of the CMSIS Drivers, select Driver instance 1 (Fig. 12.13).

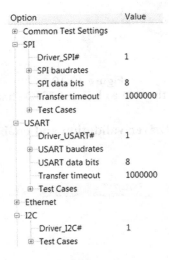

Figure 12.13

DV_Config.h allows you to set up the CMSIS Driver test configuration.

Next open the test cases section for each driver and enable some of the test cases (Fig. 12.14).

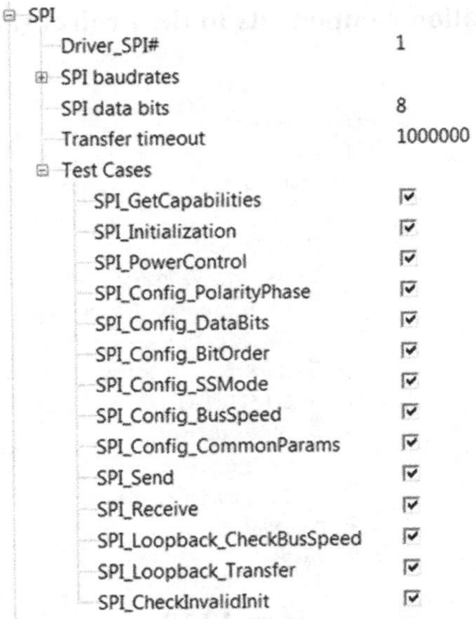

Figure 12.14

Each CMSIS Driver has a range of predefined tests.

This configures the validation tests, to run the tests we have to call the test framework.

Open main.c in the project and the cmsis_dv() function to call the validation framework.

```
int main (void) {
        osKernelInitialize();
        cmsis_dv();
        osKernelStart();
}
```

Build the project.

Start the debugger and run the code.

Each of the drivers will be exercised in turn and the results will be shown in the view\serial windows\debug(printf) window (Fig. 12.15).

```
Debug (printf) Viewer
CMSIS-Driver Test Suite    Nov 14 2015    14:17:12

TEST 01: SPI_GetCapabilities                    PASSED
TEST 02: SPI_Initialization                     PASSED
TEST 03: SPI_PowerControl
  DV_SPI.c (244) [WARNING] Low power is not supported
TEST 04: SPI_Config_PolarityPhase               PASSED
TEST 05: SPI_Config_DataBits                    PASSED
TEST 06: SPI_Config_BitOrder
  DV_SPI.c (315) [WARNING] Bit order LSB_MSB is not supported
```

Figure 12.15
The test results are output to the ITM console window. The local options allow you to save the results to a file.

In the dv_config.h configuration file, the "Common Test Settings" allow this report to be generated as plain text or XML (Fig. 12.16).

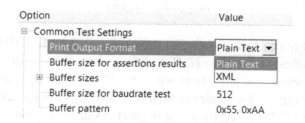

Figure 12.16
The test results can be output as plain text or XML.

Designing a Software Component

Now that we have the CMSIS USART driver configured, we can design a software component that uses the USART to receive data from an external low-cost GPS unit. We can then bundle the component into a software pack so that is can be installed into a tool chain and reused across different projects. We will be able to reuse it on any project that has an appropriate CMSIS Driver. So as a developer, we have written this code once but can now use it on many different microcontrollers.

Exercise 12.3 GPS Component

Open the Pack Installer.

Select the Boards::Designers Guide Tutorial.

Select the Example tab and Copy "EX 12.2 GPS Component."

This project uses the CMSIS USART Driver to receive data from an external GPS unit. The GPS unit used in this example is a "Ublox NEO-6M GPS" module. This is a low-cost module which is widely available (Fig. 12.17).

Figure 12.17
The UBLOX NEO-6M is a low-cost GPS module that can be connected to a microcontroller through an USART.

The GPS data is sent in the form of a serial protocol called NMEA 0183 (National Marine Equipment Association). The NMEA data is sent in the form of an ASCII serial string. Each NMEA serial sentence starts with "$GP" followed by a set of letters to identify the sentence type. This header is then followed by comma delimited data. Each sentence ends with a "*" followed by a two-digit checksum. A typical sentence is shown in Fig. 12.18.

$GPRMC,015509.00,-0.031,-0.186,0.219,19,0.000,-0.354,6.972*4D

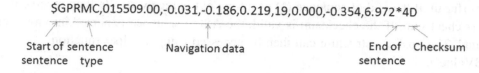

Start of   Navigation data   End of Checksum
sentence   sentence
sentence type

Figure 12.18
A GPS receiver will typically output an ASCII text sentence using the NMEA 0183 protocol.

All of the GPS code is in the module gpsThread.c. The configuration options are in gps_config.h. The user API functions are declared in gps.h (Fig. 12.19).

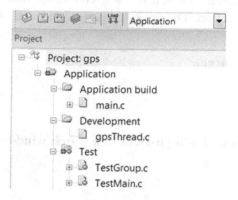

Figure 12.19
The GPS component project has application and test builds.

As discussed in Chapter 11 "Test-Driven Development," the code has been developed using the "Test Driven Development" method so we can switch from an application build to a test build to examine how the code works.

```
static void GPS_Thread (void const *argument) {
static ARM_DRIVER_USART * USARTdrv = &SELECTED_USART;
char valid;
gpsMutexCreate();
initilizeUSART(USARTdrv);
        while (1){
                detectGpsHeader(USARTdrv);
                valid = receiveNmeaData(USARTdrv);
                if(valid == 0){
                        processNmeaData();
}}}
```

The GPS component works by first initializing the CMSIS USART Driver. The component then listens for the GPS sentence header ($GP). When this is received, characters are read into a buffer until the end of sentence (*) or a time out is received. We then receive the two character checksum. If the checksum is valid, the ASCII sentence is parsed into a structure "gpsData." The gpsData structure can then be accessed using a helper function getGPSValue().

You can use the test build to understand how each of the key functions are called. The application build will also run in the simulator. A script is included to simulate GPS data arriving at the serial port.

Examine the GPS component project.

The GPS code has been written so that it is a self-contained module that is decoupled from any other parts of the application. This makes it easy to test and also easy to reuse (Fig. 12.20).

Switch to the application target.

Build the project.

Start the simulator.

Open the Symbol window and add gpsData to a watch window.

Open the View\Toolbox.

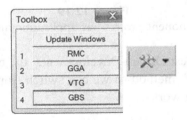

Figure 12.20
Functions in the simulation scripts can be triggered by used define buttons.

Pressing one of these buttons will send a matching NMEA sentence to the simulated microcontroller serial port.

Once the NMEA sentence has been received, the parsed data will be updated in the gpsData structure.

Now, we have a generic software module which can be used on any Cortex-M microcontroller that has a CMSIS USART Driver.

Creating a Software Pack

In the last example, we have seen how we can use a CMSIS Driver to design a software component that can be reused across multiple microcontroller families. We have also seen how software packs are used to install support for microcontrollers. Software packs can also be used to install software libraries and other software components into our toolchain (Fig. 12.21).

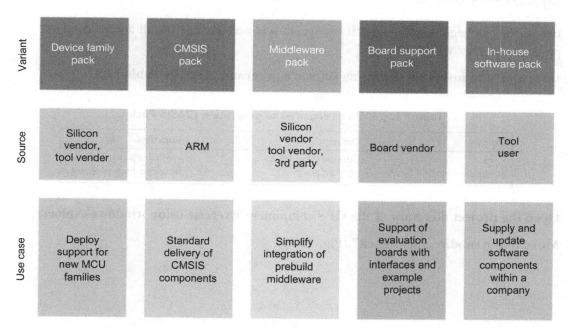

Figure 12.21

The CMSIS Pack system can be used to install software components into a toolchain. This can be support for a particular microcontroller family, other uses are to install middleware or board support libraries. In house software packs can also be created to reuse code.

In this next section, we are going to create a software pack based on the GPS component. This will allow our code to be easily distributed between developers and reused in new projects. Finally, we will have a look at how to deploy one within your organization to support a team of developers.

Software Pack Structure

A CMSIS software pack is simply a collection of software files that you want to add to the RTE Manager plus additional supporting files such as documentation, license information, examples, and templates. To make the software component, you need to generate only one additional file. This is an XML "pack description file" which describes the contents of the pack and how it is to be used. Once you have created the pack description file, a simple utility can be run to check the syntax and generate the pack. Once you are familiar with this process you can create new packs very quickly.

Software Pack Utilities

Before we can begin to create a software pack, it is necessary to install a number of programs to help design the pack.

Download and install the following software programs given in Table 12.3.

Table 12.3: Utilities required to generate a CMSIS Pack

Utility	Description	Location
7Zip	Compression utility	www.7-zip.org
Notepad++	XML editor	notepad-plus-plus.org

Open the project directory of the GPS component Exercise using windows explorer.

Move to the subdirectory "pack" (Fig. 12.22).

Figure 12.22
From the GPS project move into the Pack directory.

The directory containing all the files necessary to create a software pack is given in Table 12.4.

Table 12.4: CMSIS Pack build files

File	Description
Gen_pack.bat	Pack generation batch file
PackChk.exe	Pack syntax checker
Pack.xsd	Pack XML schema
Vendor.pack.pdsc	Pack description template

Now we need to follow a few simple steps to create a pack. There are four steps required. First, you need to decide what the content of the pack is going to be. In our case, we want to make a software pack for the GPS component. So the first step will be to remove the code from a project and make it into a standalone component. This will typically be a "C" code module or library and an include header file. Our GPS component consists of files listed in Table 12.5.

Table 12.5: Software component source files

File	Description
gpsThread.c	Component source code
gps.h	API header file
Gps_config.h	Library configuration file
gpsUserThread.c	Component User template

Next, we need to organize our software component and any associated documentation, examples, and template files into a sensible directory structure. Once we have all the software component files arranged, we need to create an additional XML file that contains a description of the software component contents. Then the final stage is to create the pack by running the "gen_pack.bat" batch file, this checks the XML pack description against the component files. If there are no errors, the software pack will be generated (Fig. 12.23).

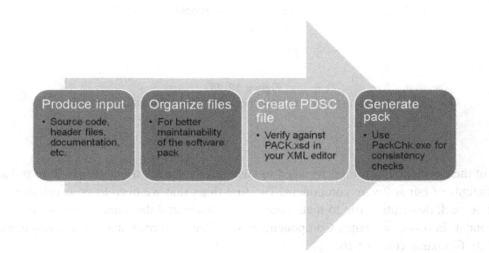

Figure 12.23
A software pack can be created in four steps: (1) Isolate the code you want to reuse; (2) Create a file structure for the software component; (3) Create an XML pack description file; (4) Run the pack generation utility.

The "files" directory contains all of the files that will be included in our pack. We can create any directory structure we want to arrange the pack files (Fig. 12.24).

Figure 12.24
Example Component directory structure.

This directory structure is completely free form, so here, we have decided on a layout that consists of top-level directories for the key component elements such as documentation, license information, and a directory for the GPS source files themselves (Fig. 12.25).

Figure 12.25
The pack contents are described by the .pdsc file in the main pack directory.

Now in the development folder, we need to customize the pack description file to reflect the contents of our software component. The first thing that we need to do is rename the template pack description file to match our vendor name and the name of software component, in this case "Hitex.Components.pdsc". The final pack name is created using the template file name (Fig. 12.26).

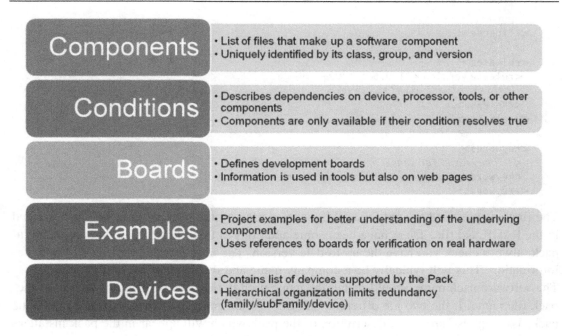

Components
- List of files that make up a software component
- Uniquely identified by its class, group, and version

Conditions
- Describes dependencies on device, processor, tools, or other components
- Components are only available if their condition resolves true

Boards
- Defines development boards
- Information is used in tools but also on web pages

Examples
- Project examples for better understanding of the underlying component
- Uses references to boards for verification on real hardware

Devices
- Contains list of devices supported by the Pack
- Hierarchical organization limits redundancy (family/subFamily/device)

Figure 12.26
The software pack description file is an XML file. It has a number of containers which can be used to describe the contents of a software pack.

The pack description file consists of a number of XML containers which describe both the software component contents and how it should be added to the RTE. You can also specify conditions for its use within a project. These conditions can be used as rules that are applied within the RTE that govern how each component is added to a project.

Now open the pack description file, with notepad++ or another XML editor and we will examine how the description file is structured.

The first section of the file provides an initial description of the file.

```xml
<?xml version = "1.0" encoding = "utf-8"?>

<package schemaVersion = "1.2" xmlns:xs = "http://www.w3.org/2001/XMLSchema-instance"
xs:noNamespaceSchemaLocation = "PACK.xsd">
  <vendor>Hitex</vendor>
  <name>Components</name>
  <description>A collecton of device driver libraries</description>
  <url>www.hitex.co.uk/repository</url>
  <supportContact>tmartin@hitex.co.uk</supportContact>
  <!-- optional license file -->
  <license>
```

```
License\License.txt
</license>

<releases>
 <release version = "1.0.0">
  Initial Release with GPS driver
 </release>
</releases>

<keywords>
 <!-- keywords for indexing -->
 <keyword>insert_keyword_for_search_engines</keyword>
</keywords>
```

The first entry is used to apply the XML schema used by the pack system. This is described in the PACK.xsd file. There have been some additions to this file so if you are producing a pack, always check you have the up-to-date version. The next section contains the vendor information. This will typically be a company name and the name of the software pack. The software pack name should match the name used in the pack description section of the pack file name. If the two are different, an error will be generated when we try to create the pack. We can also provide a description of the pack which will appear in the pack installer. The URL entry describes the location where the pack can be located. This can be a company file server or an internet webserver. Once the pack is installed, this location will be checked for an upgrade pack each time μVision IDE or the pack installer is started. This allows you to release a new version of your pack by simply updating the pack file on your webserver, now the new version will be available to any user who has installed the original pack. There is also a section for the release number which is structured as shown in Table 12.6.

Table 12.6: Naming convention for the pack version number

Driver Release Number	Format <MAJOR>.<MINOR>.<Patch>
MAJOR	Major release, may not be backward compatible
MINOR	Minor release, backward compatible
PATCH	Incremental release for bug fixes

If you want to distribute the pack over the internet, you can also include license information and search keywords. Here, we have the path to our license directory and the license document. This will be installed alongside the component and will also be displayed when the component is installed. In the keywords section, you can provide a range of search words to make your pack more visible to search engines.

```
<taxonomy>

  <description Cclass = "Components"> Driver Libraries for common hardware components</
description>

</taxonomy>
```

Now, we can start to customize the pack description file to describe our software components. The RTE has a number of standard categories that we can add our components too. However, our GPS component does not really fit into any existing category. Within the pack description language there is a taxonomy section that allows us to extend the range of component categories. Here, we are creating a "Components" section within the RTE (Fig. 12.27).

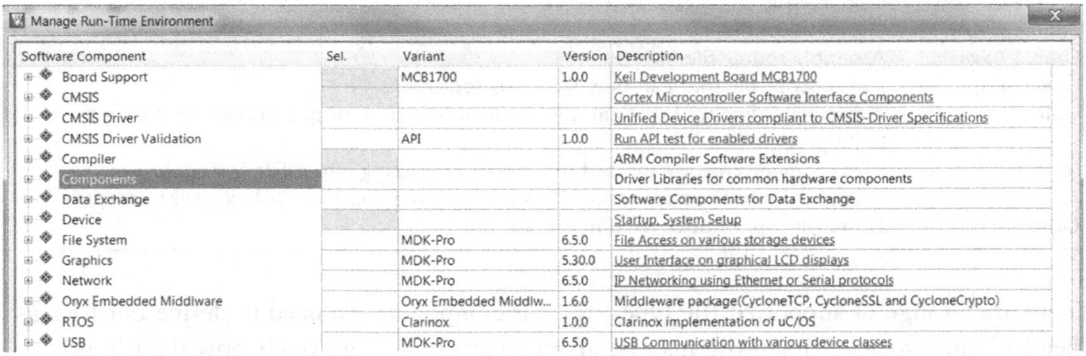

Software Component	Sel.	Variant	Version	Description
Board Support		MCB1700	1.0.0	Keil Development Board MCB1700
CMSIS				Cortex Microcontroller Software Interface Components
CMSIS Driver				Unified Device Drivers compliant to CMSIS-Driver Specifications
CMSIS Driver Validation		API	1.0.0	Run API test for enabled drivers
Compiler				ARM Compiler Software Extensions
Components				Driver Libraries for common hardware components
Data Exchange				Software Components for Data Exchange
Device				Startup, System Setup
File System		MDK-Pro	6.5.0	File Access on various storage devices
Graphics		MDK-Pro	5.30.0	User Interface on graphical LCD displays
Network		MDK-Pro	6.5.0	IP Networking using Ethernet or Serial protocols
Oryx Embedded Middleware		Oryx Embedded Middlw...	1.6.0	Middleware package(CycloneTCP, CycloneSSL and CycloneCrypto)
RTOS		Clarinox	1.0.0	Clarinox implementation of uC/OS
USB		MDK-Pro	6.5.0	USB Communication with various device classes

Figure 12.27
The taxonomy tag creates a new category in the Run-Time Environment Manager.

Once the pack is installed, the new category will be visible when the RTE manager is opened.

Now, we can start to describe the software components. Each component has its own section within the component container.

```
<components>
  <component Cclass = "Components" Cgroup = "GPS Module" Csub = "Ublox NEO-6M"
Cversion = "1.0.0" >
    <description>GPS Library for Ublox NEO-6M receiver</description>
    <files>
      <file category = "source" name = "Component\GPS\gpsThread.c" />
        <file category = "header" name = "Component\GPS\gps.h" />
        <file category = "header" name = "Component\GPS\gps_config.h" attr = "config" />
    </files>
  </component>
</components>
```

First, we define its location in the RTE. In this example, the GPS component will be added to Components section and a subgroup called GPS, we can then provide a title and description for the component. Next, we can add the files that make up the component. This means defining their type and the path and name of the file (Table 12.7).

Table 12.7: Supported CMSIS Pack file types

Category	Description
Doc	Documentation
Header	Header file used in the component. Sets an include file path
Include	Sets an include file path
Library	Library file
Object	Object file that can be added to the application
Source	Startup-, system-, and other C/C++, assembler, etc. source files
Source C	C source file
Source Cpp	C++ source file
Source ASM	Assembly source file
Linker script	Linker script file that can be selected by toolchains
Utility	A command line tool that can be configured for pre- or postprocessing during the build process
Image	Files of image type are marked for special processing into a File System Image embedded into the application. This category requires the *attr* being set to *template*
Other	Other file types not covered in the list above

There are a range of supported file types, for our component we need to define our files as "header" and "source." Each file may be given an attribute that affects how the file is added to a project (Table 12.8).

Table 12.8: CMSIS Pack source file attributes

Attribute	Description
config	The file is a configuration file of the component. It is expected that only configuration options are modified. The file is managed as part of the component, as a project-specific file typically copied into the component section of the project
template	The file is used as a source code template file. It is expected to be edited and extended by the software developer. The file can be copied into a user section of the project

In each of our components, the header file has been given the attribute config. This means that when we use the RTE Manager to add the component to our project a copy of the header file will be copied to the project directory so we can edit it. The source file has not been given and attribute so it will be held as a read only file in the pack repository within the toolchain. When the component is selected, a path will be set to include the source file. This means that if you are using a version control system you will need to include all the packs you are using to be able to fully recreate a project. This is the minimum amount of information we need to add to the pack description file to allow a pack to be generated.

An initial pack can be generated by opening a command line window in the development directory and then running the "gen_pack" batch file. This will parse the pack description file and check against the contents of the software component directories. If there are no errors, then a pack file will be created. This pack file is a compression file of the software component directories and files plus the pack description file (Fig. 12.28).

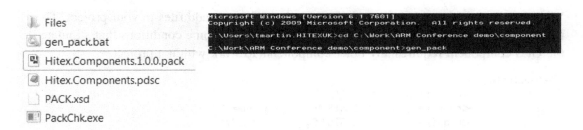

Figure 12.28
The gen_pack batch file checks the pack files against the pack description file then generates the final pack file.

Once it has been generated, we can install our GPS component by double clicking on the pack file (Fig. 12.29).

Figure 12.29
The pack file can be installed locally by double clicking on the pack file or by selecting file/import in the pack installer.

Now, we can open the RTE manager and see the new modules section with the GPS component. If we select the GPS component its files will be added to the project (Fig. 12.30).

Figure 12.30
Once the pack is installed, the GPS component is available to be added to a new project.

While this is useful at this level, it is just a "pretty" way to add files to your project. To make our component more intelligent, we can start to add some conditions that stipulate if the GPS component requires any other components to work within a project.

```
<conditions>
            <condition id = "USART">
            <require Cclass = "CMSIS Driver" Cgroup = "USART"/>
            <require Cclass = "CMSIS" Cgroup = "CORE"/>
            <require Cclass = "CMSIS" Cgroup = "RTOS"/>
            </condition>
</conditions>
```

Within the conditions container, we can create a rule that defines what additional components must be selected. Here, we have created a condition called "USART" that requires a CMSIS RTOS and the CMSIS-Core specification be selected along with a CMSIS USART driver.

```
<component Cclass = "Components" Cgroup = "GPS Module" Csub = "Ublox NEO-6M"
Cversion = "1.0.0" condition = "USART">
```

Now in the components description of the GPS component we can add the conditions. If the pack is new, regenerated, and reinstalled, we can test the rule by selecting the GPS component (Fig. 12.31).

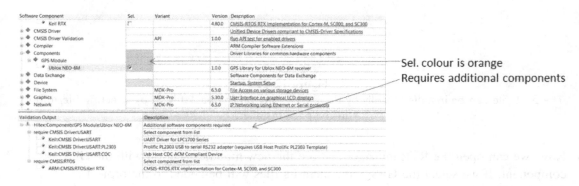

Figure 12.31
Now when the component is selected, it will require a CMSIS USART driver.

When the GPS selection (Sel.) box is ticked it will turn orange meaning it requires a subcomponent and the verification window tells us that we must include an RTOS and CMSIS Driver::USART in our project. Now press the Resolve button and the RTOS will be selected. There are several options for the CMSIS USART driver, so we must manually select a suitable driver (Fig. 12.32).

Figure 12.32
Select an available CMSIS USART Driver.

Once we have selected all the required components, all of the active elements in the Selection (Sel.) column will be colored green. Now, we can click OK and the finished component selection will be added to our project.

```
<releases>
  <release version = "1.0.0">
   Initial Release with GPS driver
  </release>
</releases>
```

As we create new versions of the software pack, we can increment the version number. As we install new versions of the component, the version history can be seen in the pack installer. Once installed, each version is held within the pack repository so it is possible to maintain projects built with different versions of a software component.

```
<files>
        <file category = "source" name = "Component\GPS\gpsThread.c" />
        <file category = "header" name = "Component\GPS\gps.h" />
        <file category = "header" name = "Component\GPS\gps_config.h" attr = "config" />
        <file category = "doc" name = "Component\Documentation\GPS\nmea.htm" />
</files>
```

We can expand the component description by adding a documentation link. This uses the doc file category and provides a link to the description of the NMEA protocol.

```
<files>
        <file category = "source" name = "Component\GPS\gpsThread.c" />
        <file category = "header" name = "Component\GPS\gps.h" />
        <file category = "header" name = "Component\GPS\gps_config.h" attr = "config" />
        <file category = "source" name = "Component\GPS\gpsUserThread.c" attr = "template"
select = "GPS User Thread" /
</files>
```

It is also possible to add a template file. This is a fragment of code that provides a typical pattern that shows how the software component is used. The file is described as a source file but has a template attribute. We can also provide the selection criteria to be shown within the file manager (Fig. 12.33).

Figure 12.33
The description entry is now a hyperlink to the component documentation.

Now when the pack is regenerated and reinstalled, the documentation will be available through a hyperlink (Fig. 12.34).

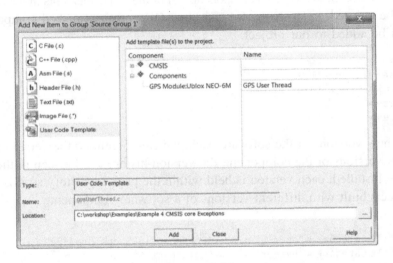

Figure 12.34
Template code is available in the "Add New Item" dialog.

If we add the GPS component to a project its template will be available in the "Add new Item to Group" dialog when we add a new source file to the project.

CMSIS Pack also makes it possible to include example projects to demonstrate the software component features. The example project must be placed in a suitable folder within the software pack file structure. Then we can add an entry to the pack description file to make it available through the pack installer.

```
<example name="GPS component configuration Example" folder="Examples\GPS
\Configuration" doc="Abstract.txt" version="1.0">
    <description>This example demonstrates how to configure the GPS component</
description>
    <board vendor="Keil" name="MCB1700"/>
    <project>
     <environment name="uv" load="configuration.uvprojx"/>
    </project>
    <attributes>
     <component Cclass="CMSIS" Cgroup="CORE"/>
     <component Cclass="Device" Cgroup="Startup"/>
    </attributes>
</example>
```

The pack template contains an example container which allows you to add any number of example projects. Each example is described as shown above. The example name and location of the example folder is defined along with a documentation file. Next, a description of the example is added; this will appear in the pack installer example description column. To complete the example description, we provide information about the board vendor the project development IDE being used and the project name to load. In the attributes section, we can provide details of any additional components that need to be loaded to allow the project to be rebuilt (Fig. 12.35).

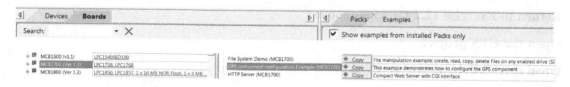

Figure 12.35
Component examples are available through the pack installer.

Now, when the pack is installed our examples will be visible in the pack installer examples tab.

Configuration Wizard

Once the component files have been added to a project, we will first need to set up any configuration options. In a typical component, the configuration options will be grouped together as a set of #defines in a header file. To make the component more intuitive to use it is possible to add some annotations within comments that allow the header file to be viewed as a configuration wizard. Then selections made in the configuration wizard will modify the associated #define. The Table 12.9 provides an overview of the available tags.

Table 12.9: XML tags are used to build a configuration wizard

Template Tag	Description
<h>	Create header section
<e>	Create header section with enable
<e.i>	Create header section with enable and modify specific bit
<i>	Tooltip text
<q>	Bit values set by a text box
<o>	Text box for numeric value
<o.i>	Modify a single bit in the following #define
<o.x..y>	Text box for numeric value with range
<s>	ASCII string
<s.i>	ASCII string with character limit
<qi>, <oi>, <oi.x>, <si>, <si.x>	Skip i #defines and modify the next #define

Once you are familiar with the configuration wizard annotations, creating new template files is a relatively quick process.

Exercise 12.4 Configuration Wizard

In this exercise, we will add the configuration wizard annotations to the GPS_Config.h header file.

Return to Exercise 12.2 GPS component.

Open the GPS project.

Open the header file gps_config.h.

The configuration options for our software component are held in the header file gps_config.h. This file contains a set of #defines that are used by the main application code. This file also contains the necessary annotations to be viewed as a configuration wizard. In this example, we will read through this file to see how they work.

First, we need to enable this file as a configuration wizard file. This is done by adding the following comment in the first 100 lines of the header file.

```
// <<< Use Configuration Wizard in Context Menu >>>
```

At the end of the file, add the following comment to close the configuration wizard:

```
// <<< end of configuration section >>>
```

The closing comment is optional but it is good practice to add it.

Now, we can create some logical sections to group our configuration settings. A section header can be created with the following comment:

```
// <h> section header
```

The end of the section must be closed by;

```
//</h>
```

It is also possible to create a section header with an enable option.

```
//<e> section with enable
#define ENABLE 0
```

This tag must be followed by a #define. This define will be modified to a 1 if the selection is ticked. This section must be closed by:

```
</e>
```

By adding section headers, we can start to structure the include file template (Fig. 12.36).

```
//<<< Use Configuration Wizard in Context Menu >>>
//<h>NEO-6M GPS configuration
//<h>Select NMEA Sentences
//</h>
//<h>Select Serial Interface
//</h>
//</h>
```

Option
- NEO-6M GPS configuration
  - Select NMEA Sentences
  - Select Serial Interface

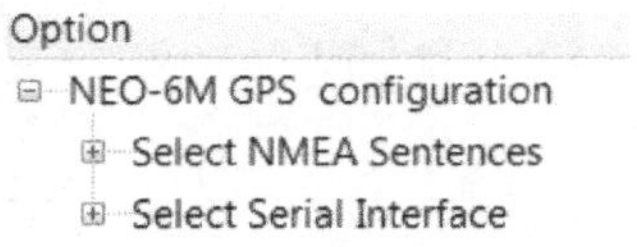

Figure 12.36
Configuration wizard tags are placed in comments. The editor can then view the source code as a configuration wizard.

Now, we can start to populate the sections with configuration options. For the NMEA sentence section, we can specify the GPS messages we want to receive. Using the following tag, it is possible to create tick boxes which modify the #define that immediately follows them.

```
//<q> select option
#define OPTION 0
```

Using this tag, we can expand the sentences section with configuration options to select the NMEA messages that will be processed.

```
//<h>Select NMEA Sentences
//<q> GPGGA
#define GPS_GPGGA 1
//<q> GSA
#define GPS_GSA 0
```

```
//<q> GSV
#define GPS_GSV 0
//<q> PRMC
#define GPS_PRMC 0
//</h>
```

Next, we can extend the serial interfaces section. Here, we can create a drop down selection box by using the following tags:

```
// <o>Selection box
//        <1 = > selection 1
//        <2 = > selection 2
//        <3 = > selection 3
//        <4 = > selection 4
#define SELECTION 1
```

Using the selection box tag, we can now create selection options for the USART serial interfaces.

```
//<o>Select USART Peripheral
//        <1 = > USART 1
//        <2 = > USART 2
//        <3 = > USART 3
//        <4 = > USART 4
#define GPS_USART_PERIF 3
//<o>Select USART baud rate
//        <2400 = > 2400
//        <4300 = > 4300
//        <9600 = > 9600
//        <34000 = > 34000
#define GPS_USART_BAUD 9600
```

Once we have a functioning configuration wizard, we can add some tooltips to provide additional assistance to the user. After any of the configuration, we can add further help information as shown in Fig. 12.37.

```
//<q> GBS
//<i> Time Position and Fix data
```

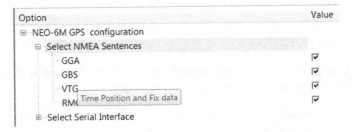

Figure 12.37
Configuration wizard with tooltip.

While the configuration wizard is an easy way to display complex configuration options, it does rely on the user having a basic understanding of the software component. If, for example, the GPS unit had two types of serial interfaces, SPI and USART which could not be used at the same time. The configuration wizard would show you the options for both but would do nothing to stop the user from selecting both. To give the user some warning, it is possible to add error messages that will be emitted during the build process. So, for example, in the case of two serial interfaces:

```
#if (GPS_SPI_EN == 1 && GPS_USART_EN == 1)
   #error "::GPS Driver:Too many serial interfaces enabled"
#endif
```

If an end user of the software component selects both serial interfaces, the project will fail to build and a warning will be issued in the build output window.

Deploying Software Components

Once a software pack has been created, you will need a way to distribute it to your developers. As we have seen, you can copy the pack file to a new PC and then simply double click on the file to install it locally. You can also publish the pack to a team by installing it on a fileserver or intranet. In this case, the location that the pack is installed at should match the URL defined in the pack. Once a user has downloaded and installed your pack, the URL location will be checked when the pack installer is started. If a new version is found, the pack installer will show that a new version is available. This allows you to maintain and update the pack and any updates are pushed out to your users. If you want the pack to be available to download by anyone, you can simply place it on a webserver to make it publically available. This is a great way to share software components and it allows users to add useful code to their projects from a variety of sources.

Currently, the pack installer will automatically check the Keil website for the current list of public and third party pack. You can view the current listing here:

http://www.keil.com/dd2/pack/

Any new packs added to this listing will automatically be displayed within the pack installer. You can submit your pack to be validated and then it can be added to this list making it visible to any tool using the pack system. The pack can be hosted in two ways. Once the pack is validated, it can be listed and stored on the Keil website. This is the easiest way but is harder to maintain. The second method is to list the pack on the Keil website, so it can be discovered but the pack can be hosted on your own server. This way you can make your pack visible to the entire user base but still be able to maintain and update the pack on your own local server.

Conclusion

In this chapter, we have looked at an important step forward for microcontroller-based embedded systems. Together, CMSIS Driver and CMSIS Pack for the first time provide a standardized API for common peripherals and a means of making and distributing software components. At the time of writing, ARM has just published a pack plugin for the Eclipse IDE. This will also make the pack system available to many other toolchains and open source tools. As the pack system becomes available in multiple development tools, it will spur the adoption of the CMSIS Pack standard as a way to share and distribute Cortex-M software components.

CHAPTER 13

ARMv8-M

Introduction

As we saw in the introduction, the current Cortex-M processors are based on the ARMv6-M (Cortex-M0/M0+) and ARMv7-M (Cortex-M3/M4/M7) architectures. Toward the end of 2015, ARM announced the next generation of Cortex-M processors which are based on a new architectural standard ARMv8-M. The ARMv8-M architecture is a 32-bit architecture based on the existing ARMv6-M and ARMv7-M architectures and inherits most features of the existing Cortex-M programmer's model. This allows most application code to be easily ported onto the new architecture with minimal changes. While ARMv8-M is the latest processor design from ARM it does not make ARMv7-M or ARMv6-M devices obsolete. They will be around for a long time yet but ARMv8-M paves the way for highly scalable, cost effective families of microcontroller and introduces a hardware security model that is a foundation for secure connected devices (Fig. 13.1).

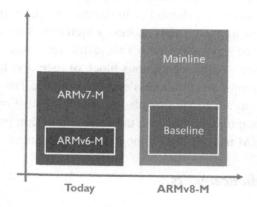

Figure 13.1

The ARMv8-M architecture consists of two profiles Mainline and Baseline which are analogous to the existing ARMv7-M and ARMv6-M architectures.

The ARMv8-M is designed to be a highly scalable architecture and consists of a Mainline profile and a Baseline subprofile. While these two profiles are analogous to the original ARMv7-M and ARMv6-M split, the ARMv8-M creates a unified architecture for microcontroller processors with significant new features. The ARMv8-M architecture is also designed to be highly modular for silicon designers; this allows more flexibility in

The Designer's Guide to the Cortex-M Processor Family.
DOI: http://dx.doi.org/10.1016/B978-0-08-100629-0.00023-2
© 2016 Elsevier Ltd. All rights reserved.
445

processor feature selection within the two profiles. The ARMv8-M architecture also introduces an important new hardware security extension called TrustZone which is available in both the Mainline and Baseline profiles.

Common Architectural Enhancements

The two ARMv8-M profiles have a common set of architectural enhancements (Table 13.1).

Table 13.1: Common architectural enhancements for the mainline and baseline profiles

Load acquire store release instructions
New instructions for TrustZone support
New style memory protection unit programmers' model
Better debug capability

ARMv8-M introduces a new group of "load acquire store and release" instructions which are required to provide hardware support for the latest "C" language standards. These instructions provide atomic variable handling that is required by the latest "C" and "C++" standards. The ARMv8 architecture includes the memory protection unit (MPU) as an optional unit in both the Mainline and Baseline profiles. In ARMv8-M, the MPU has a new programmer's model called the "Protected Memory System Architecture" (PMSAv8). The new PMSAv8 model allows greater flexibility in defining memory protection regions. In the ARMv7 MPU, a region has to have a start address which is a multiple of the region size and the region size has to be a power of two. This restriction could often mean using two or more MPU regions to cover a single contiguous block of memory. In the PMSAv8 model, an MPU region is defined with a start address and end address. The only limitation is that the start and end address has a 32-byte granularity. There are also enhancements to the data watch and breakpoint debug units along with the inclusion of comprehensive trace features. At the time of writing ARM has not fully announced these features.

ARMv8 Baseline Enhancements

In addition, the Baseline profile has the following enhancements (Table 13.2).

Table 13.2: Architectural enhancements for the baseline profile

Hardware divide instruction
Compare and Branch and 32-bit branch instructions
Exclusive access instruction
16-bit Immediate data handling instructions
Support for more interrupts

The ARMv6 architecture uses the original 16-bit Thumb instruction set. In ARMv8-M Baseline, the Thumb instruction set is replaced with a subset of the Thumb-2 32-bit instruction set. This will mean something like a 40% increase in performance over the current Cortex-M0 and M0+. The Baseline Thumb-2 instruction set now adds a number of useful instructions that ARMv6-M was missing. Some of these instructions such as hardware divide and the improved branch instruction will give an obvious boost to the baseline performance. However, the inclusion on the 16-bit data handling instructions (MOVT and MOVW) is more subtle. These two instructions are used to load two 16-bit immediate values into the upper and lower half of a register to form a 32-bit word. This allows you to build large immediate values very efficiently but more importantly since the immediate value is part of the encoded instruction it can be run from eXecute Only Memory (XOM). This allows both the Mainline and Baseline profiles to build application code that can run in the XOM as protected firmware that cannot be reverse engineered through data or debug accesses. The ARMv8-M architecture also adds the exclusive access instructions to the Baseline profile. This is to increase support for multicore devices. There have already been a few multicore Cortex-M-based microcontrollers notably the NXP LPC43xx and LPC15000 families. These are asymmetric multiprocessor (AMP) designs which consist of a Cortex-M4 and a Cortex-M0 (LPC43xx) or a Cortex-M4 and a Cortex-M0+. In both cases, the developers at NXP had to implement their own interprocessor messaging system because the current Cortex-M0 and M0+ do not have the exclusive access instructions. ARMv8-M-based devices will be much easier to integrate into a multicore system be it a standard microcontroller or a custom System-On-Chip. Finally, the ARMv8-M baseline will also extend the number of interrupt channels available on the NVIC. The ARMv6-M-based processors (Cortex-M0 and Cortex-M0+) are limited to 32 interrupt channels and on some devices this is becoming a limitation.

ARMv8-M Mainline Enhancements

The Mainline profile has the following enhancements (Table 13.3).

Table 13.3: Architectural enhancements for the mainline profile

Floating Point Extension Architecture v5
Optional addition of the DSP instructions

While the Mainline profile includes all of the common architectural improvements listed above, it also gets a new floating point unit and the option to include the DSP instructions with or without the floating point unit.

TrustZone

With an ARMv6-M or ARMv7-M-based device it is perfectly possible to design a secure software architecture. However, ARMv8-M provides hardware support for a protected zone in the form of TrustZone. TrustZone is an established technology on Cortex-A processors and with ARMv8-M it will be available to Cortex-M microcontrollers for the first time. TrustZone is a set of hardware extensions that allows you to create a boundary between Secure and Non-Secure resource with a minimum of application code. TrustZone has negligible impact on the performance of an ARMv8-M processor, this all means simple efficient code that is easier to develop, debug, and validate (Fig. 13.2).

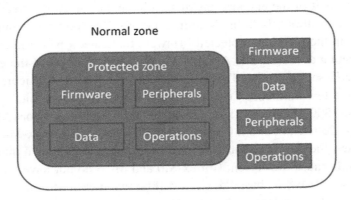

Figure 13.2
TrustZone creates a protected zone within the microcontroller system.

TrustZone extends the programmers model to include Secure and Non-Secure states each with their own Handler and Thread mode. This means that each state has its own set of stack pointers, MPU, and SysTick timer. The processor memory map can also be partitioned into Secure and Non-Secure regions. When the processor is running in Secure state, it can access Secure and Non-Secure memory areas. When it is running in the Non-Secure state, it can only access Non-Secure memory areas. Both the Secure and Non-Secure states have their own MPU which can be configured to police access to different memory region depending on the processor state (Fig. 13.3).

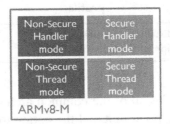

Figure 13.3
TrustZone creates two new states, Secure state and Non-Secure state.

We can build code to run in the Non-Secure state. Call functions access Non-Secure peripherals and service Non-Secure interrupts as normal. Non-Secure code can also make function calls to access code in the Secure state. When Non-Secure code makes a call to Secure code it must make an entry to Secure code through a valid entry point. These valid entry points are a new instruction called the Secure Gateway (SG) instruction. If Non-Secure code tries to enter the protected zone by accessing anything other than an SG instruction, a SecureFault exception will be raised. The SG instruction also clears the least significant bit of the Link register (normally set to one). Two new instructions are provided to return from the Secure state to the Non-Secure state. The Secure code can use a Branch Exchange to Non-Secure code (BXNS) or Branch Link Exchange to Non-Secure code (BLXNS). These instructions use the least significant bit (LSB) of the return address to indicate a transition from Secure to Non-Secure state (Fig. 13.4).

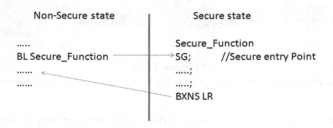

Figure 13.4
Non-Secure code can call Secure code but it must enter Secure state via an SG instruction.

It is also possible for Secure code to call Non-Secure code. When this happens, the Secure state return address is stored on the Secure stack and a special code FNC_RETURN is stored in the Link register. When the Non-Secure function ends, it will branch on the Link register. This branch instruction will see the FNC_RETURN code this triggers the unstacking of the true return address from the Secure stack and a return to the Secure function (Fig. 13.5).

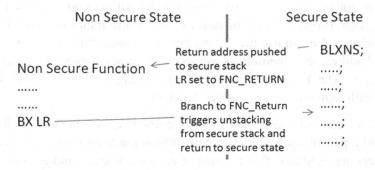

Figure 13.5
Secure code can call Non-Secure code. Its return address is hidden from Non-Secure code.

At the heart of the TrustZone technology is the Security Attribution Unit (SAU) and an optional Implementation Definition Attribution Unit (IDAU). An ARMv8-M processor with TrustZone will always boot in Secure mode, and the Secure software can then define the Secure and Non-Secure partitions by programming the SAU in a similar fashion to the MPU. The SAU can define memory region with three levels of security. We can define a memory region as Secure or Non-Secure but there is also an additional level of "Non-Secure Callable" (NSC) (Fig. 13.6).

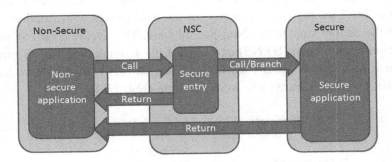

Figure 13.6

The security model implements three zones. Secure, Non-Secure, and Non-Secure callable.

In a real application, it is possible that constant data may inadvertently have the same binary pattern as an SG instruction. This would create a false entry point into the protected zone which could be exploited. The SAU allows us to define an NSC region which will contain all the SG instructions and can be more easily policed for false entry points. Any SG instruction pattern outside of the NSC region will raise a security exception if accessed. In practice, the NSC region is populated by a set of branch tables to the Secure code. These locations are fixed in the NSC memory and allow us to publish a fixed Secure API for the Non-Secure code to access. This also means that we can modify and update the Secure code without having to update the call addresses in the Non-Secure code. In practice, the NSC region is likely to occupy a small region of memory which can be easily analyzed for spurious SG patterns.

TrustZone also includes an additional instruction which allows you to check the bounds of a data object. The Test Target (TT) instruction can be used to return the SAU region number for a selected address. We can use the TT instruction to request the SAU region number for the start and end address of a memory object to ensure it does not span across Secure and Non-Secure regions. The TT Instruction allows us to create efficient bounds checking code for any objects with dynamic ranges such as pointers.

The Secure state adds an additional pair of stack pointers (MSP_S and PSP_S) for the secure stacks and a pair of stack limit registers in both the Secure and Non-Secure states. The limit registers are used to define the size of each stack space and a security fault will be generated if any stack space is exceeded (Fig. 13.7).

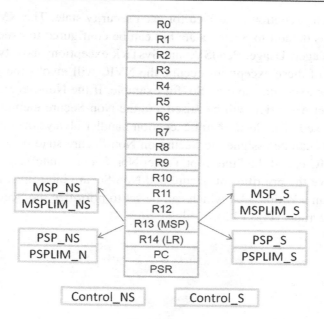

Figure 13.7

The CPU register file is extended with Secure and Non-Secure stack pointers and control registers. Stack limit registers are provided for each of the four stacks.

The ARMv8-M processor peripherals are also extended with Secure and Non-Secure MPU, SysTick and System Control Blocks. When the processor is in the Secure state, it can access the Non-Secure peripherals through alias registers.

Interrupts and Exceptions

The ARMv8-M architecture retains the NVIC as the interrupt processing unit and provides some extensions to support the TrustZone Secure and Non-Secure states. The processor exceptions are the same as for ARMv7-M with the addition of a SecurityFault exception (Table 13.4).

Table 13.4: ARMv8-M processor exception table. A new securefault exception is added

Exception	Processing State
NMI	Default: Secure state can be routed to Non-Secure state
Hard Fault	Default: Secure state can be routed to Non-Secure state
MemManager	Banked
Bus Fault	Default: Secure state can be routed to Non-Secure state
Usage Fault	Banked
SecureFault	Secure State
DebugMonitor	Can target Secure or Non-Secure state (Defined by hardware)
PendSV	Banked
SysTick	Banked
Peripheral interrupts	Programmable through new NVIC "Interrupt Target Non-Secure State" registers

The behavior of each exception is defined for each security state. The NMI, Hard Fault, and Bus Fault exceptions default to Secure state but can be configured to execute in Non-Secure state. The MemManager, Usage, PendSV, and SysTick exceptions have two service routines, when one of these exceptions occurs, the NVIC will invoke the service routine corresponding to the exception source. So, for example, if the Non-Secure MPU generates a fault a MemManager exception will be raised and the Non-Secure memory manager service routine will be invoked. The SecureFault exception handler always operates in Secure state. Peripheral interrupts can be assigned to Secure on Non-Secure state by a new group of registers in the NVIC called the "Interrupt Target Non-Secure State" registers. It is also possible to interleave the priorities of Secure and Non-Secure interrupts. However, the AICR register has an additional bit which can be set to prioritize the Secure interrupts over the Non-Secure interrupts (Fig. 13.8).

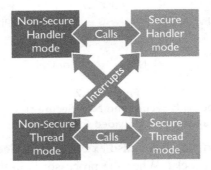

Figure 13.8
Interrupts can be configured as Secure or Non-Secure and may be served irrespective of the current processor state.

It is possible for Non-Secure code to call Secure functions and for Secure code to call Non-Secure code. Similarly any interrupt can run in Secure or Non-Secure state irrespective of the current Thread operating state. The TrustZone security model does not introduce any additional hardware latency except when the processor is running in Secure state and has to service a Non-Secure interrupt. In this case, the processor pushes all the CPU registers to the Secure stack and then writes zero to the CPU registers to hide the Secure state data from the Non-Secure code. This gives the processor more work to do and introduces a "slightly longer" interrupt latency.

If we have a mixture of Secure and Non-Secure interrupts and exceptions, it follows that we must isolate the Secure interrupt vectors so they are not visible from the Non-Secure code. In order to do this TrustZone creates two interrupt vector tables a Secure vector table and a Non-Secure vector table. The location of each vector table in the processor memory map is controlled by a pair of vector table offset registers, one for Non-Secure state (VTOR_NS) and one for Secure state (VTOR_S) (Fig. 13.9).

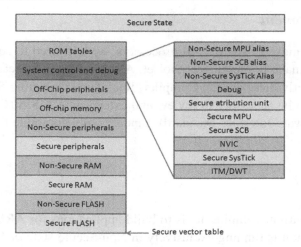

Figure 13.9

In Secure state all the processor resources are available. Aliased addresses are provided for the Non-Secure processor peripherals.

The Secure vector table will be located at the start of the memory map, so the processor can boot into Secure mode. The Secure startup code can then program the SAU to define the Secure and Non-Secure regions and then we can locate the Non-Secure vector table at the start of the Non-Secure code space (Fig. 13.10).

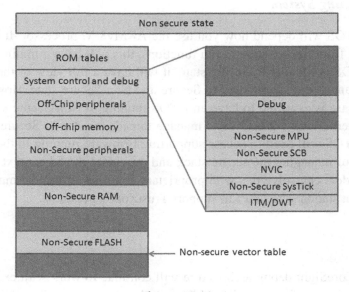

Figure 13.10

In Non-Secure state, only the Non-Secure resources are visible.

Software Development

In order to make the move to ARMv8-M processors, we will need to consider the changes that we will need to make to our current toolset. ARMv8-M will affect our toolchain in a number of areas, chiefly these are the Compiler, Real-Time Operating System (RTOS), and Debugger. It will also be necessary to review and extend the CMSIS (Cortex Microcontroller Software Interface Standard) standards to incorporate the new ARMv9-M instructions.

Compiler

It is possible to use current compiler tools to build applications for ARMv8-M processors provided the application is running exclusively in Non-Secure state or Secure state. Any code running on Baseline devices will benefit from being rebuilt by a compiler that supports the full Baseline instruction set. If you are planning to use the TrustZone security extension to mix Secure and Non-Secure code, then you will need a compiler that supports the full ARMv8-M instruction set. At the time of writing, the ARM C compiler (version 6) is the reference compiler for ARMv8-M. This version of the compiler also supports the C++11, C++14, and C11 standards. By the time real devices have been released, it is to be expected that all the mainstream compilers will support the new processors.

Real-Time Operating System

Changes to an RTOS will depend how you use the ARMv8-M processor. It is possible to make calls to and from Secure state to run functions, this will have a minimal impact on an RTOS. If the RTOS is running in Secure state, it will have a full view of the device and will be able to manage threads running in Secure and Non-Secure state. However, the most common design case is expected to have an RTOS running in Non-Secure state. In this case, the RTOS kernel will not be able to manage threads running in Secure state. This means that the RTOS will need to access supporting firmware running in the Secure state which is capable of managing the secure stack and providing thread context switch support. While RTOS vendors can solve this with proprietary solutions, it will be much more effective to have a standardized API to support TrustZone operations.

Debugger

The ARMv8-M CoreSight debug architecture will continue to work with existing debug adapters. The debugger software will need to be updated to support the additional CPU registers and additions to the processor system control block.

Cortex Microcontroller Software Interface Standard

As noted above, the CMSIS specification will need to be updated to support the new features introduced by the ARMv8-M architecture. As discussed above, the CMSIS-RTOS API will require a standard API that allows designers to incorporate Secure and Non-Secure threads within an application. This will also allow software vendors to develop RTOS-based software components that will run in Secure state. The CMSIS-Core will also need to be updated to support the additional registers of the ATMv8-M processor. The CMSIS-DSP library will also need to be optimized to take advantage of the ARMv8-M processor.

Conclusion

The ARMv8-M architecture is more evolution than revolution. It provides an easy migration to the next generation of Cortex-M microcontrollers. The introduction of TrustZone into the Cortex-M profile provides an easy to use security model which is vitally important as the devices become more connected and the Internet of Things starts to become a reality.

Appendix

Contact Details

I hope you enjoyed working through this book and have found it useful. If you have any questions or comments please contact me on the email address below:

tmartin@hitex.co.uk

Further information and updates are published on my blog at:

www.while-1.co.uk

Appendices

This appendix lists some further resources that are worth investigating once you have worked through this book.

Debug Tools and Software

There are lots of development tools available for ARM Cortex-M microcontrollers. The following is by no means an exclusive list. I have just tried to list some of the key software resources available.

Commercial GNU-Based Toolchains

Table A.1: Commercial GNU toolchains

Atollic	www.atollic.com
Rowley	www.rowley.co.uk

Commercial Toolchains

Table A.2: Commercial toolchains

IAR	www.iar.com
Keil	www.keil.com
Tasking	www.tasking.com

Rapid Prototyping and Evaluation Modules

Table A.3: Modules

Mbed	mbed.org
Embedded artists	www.embeddedartists.com/

MBED is a range of low-cost Cortex-M-based modules with a web-based development environment. It has a thriving community with lots of free resources.

MBED OS is an open source embedded operating system designed for the "Things" in the Internet of Things.

Real-Time Operating Systems

Like the compiler toolchains, there are many Real-Time Operating System (RTOS) for Cortex-M processor. Here, I have listed some of the most widely used RTOS vendors.

Table A.4: Real-time operating system (RTOS)

Free RTOS	www.freertos.org
Keil	www.keil.com
Micrium	www.micrium.com
Segger	www.segger.com
CMX	www.cmx.com

Digital Signal Processing

Table A.5: Digital signal processing (DSP)

DSP resources	www.dspguru.com
Filter design	www.iowegian.com
DSP algorithm development	www.mathworks.co.uk
	www.scilab.org/

Books

Programming

Table A.6: C programming books

The C Programming Language Kernighan and Ritchie	ISBN-10: 0131103628
Test Driven Development for Embedded C James W. Grenning	ISBN-10: 193435662X
Clean Code: A Handbook of Agile Software Craftsmanship Robert C Martin	ISBN-10: 0132350882

Cortex-M Processor

Table A.7: Cortex-M processor books

Cortex-M Technical Reference Manuals	www.arm.com/products/processors/cortex-m/index.php
The Definitive guide to the Cortex-M3 and M4 Joseph Yui	ISBN-10: 0124080820
The Definitive guide to the Cortex-M0 and M0+ Joseph Yui	ISBN-10: 0128032774
Insiders Guide to the STM32 Trevor Martin	www.hitex.co.uk/index.php?id=download-insiders-guides00

Standards

Table A.8: Coding standards

CMSIS specification	http://www.arm.com/products/processors/cortex-m/cortex-microcontroller- software-interface-standard.php
MISRA C	www.misra.org.uk
Barr coding standard	www.barrgroup.com/

Digital Signal Processing

Table A.9: Digital signal processing (DSP) books

Digital Signal Processing: A Practical Guide for Engineers and Scientists Stephen Smith	ISBN-10: 075067444X
Understanding Digital Signal Processing Richard G Lyons	ISBN-10: 0137027419

Real-Time Operating System

Table A.10: Real-time operating system (RTOS) books

Real-Time Concepts for Embedded Systems	ISBN-10: 1578201241
Qing Li and Caroline Yao	
Little Book of Semaphores	www.greenteapress.com/semaphores/
Downey	

Silicon Vendors

There are now over 3500 Cortex-M-based microcontrollers and the list is still growing. The link below is a database of devices supported by the MDK-ARM. This is an up-to-date list of all the mainstream devices and is a good place to start when selecting a device. Cortex-M-device database.

Table A.11: Silicon Vendors

Keil device database	www.keil.com/dd/

Training

The following companies provide hands on embedded systems training courses:

Table A.12: Training

Hitex	www.hitex.co.uk/index.php?id=3431
Feabhas	www.Feabhas.com
Doulos	www.Doulos.com

Index

Note: Page numbers followed by "*f*" and "*t*" refer to figures and tables, respectively.

Printed in the United States
By Bookmasters